2/1/91

D0877384

Polymer Characterization

PHYSICAL TECHNIQUES

Polymer Characterization

PHYSICAL TECHNIQUES

D. CAMPBELL PhD, MEng, FPRI
*Department of Mechanical Engineering and
Manufacturing Systems, Newcastle upon Tyne
Polytechnic, UK*

J.R. WHITE ARCS, DIC, PhD, FPRI
*Department of Metallurgy and Engineering Materials,
University of Newcastle upon Tyne, UK*

London   New York
CHAPMAN AND HALL

First published in 1989 by
Chapman and Hall Ltd
11 New Fetter Lane, London EC4P 4EE
Published in the USA by
Chapman and Hall
29 West 35th Street, New York NY 10001
© 1989 D. Campbell and J.R. White

Phototypeset in 10/12 Photina by
Thomson Press (India) Limited, New Delhi.
Printed in Great Britain at The University Press, Cambridge.

ISBN 0 412 27160 5 (hardback)
ISBN 0 412 27170 2 (paperback)

British Library Cataloguing in Publication Data

Campbell, D.
  Polymer characterization.
  1. Polymers. Characterisation
  I. Title II. White, J.R.
  547.7'046

  ISBN 0-412-27160-5

Contents

Preface

The spectacular growth in the number of polymers, collectively covering a vast range of properties, has created the need for a wide range of methods of characterization. Many of the important structural and microstructural features of polymers have no counterpart in other classes of materials and require special characterization techniques. For example, molecular weight has no significance with condensed phase metals and ceramics whereas it has an important influence over the properties of a polymer. Furthermore, when a standard method of characterization, developed initially for more traditional materials, is applied to polymers, considerable modification may be needed. An example of this is the study of crystals by transmission electron microscopy, where the electron beam sensitivity of polymers has necessitated the development of procedures that are unnecessary with metals and ceramics. This book concentrates on those techniques peculiar to polymers and on those aspects of more general techniques that are of particular value in polymer studies. The aim is to introduce the reader to the major physical techniques for polymer characterization. The book is intended for use as the basis of a course offered to second or third year undergraduates in Materials Science, Polymer Technology and Chemistry, and it is hoped that it will be of value to many scientists and technologists working in the polymer industry.

Although the book is meant to be self contained, a certain knowledge of physics and of polymeric materials is assumed. The emphasis is on polymer applications and the basic theory behind the various techniques is introduced only sparingly. A bibliography is provided at the end of each chapter to help the interested reader to find out more about both the background theory and the practical aspects of the techniques described, and to indicate sources of modern and advanced polymer applications. Finally, it should be noted that the scope of this book has been enlarged to include polymer composites whenever there are good examples of applications with these materials.

D. Campbell
J.R. White

1
Preliminary survey

1.1 The purpose of characterization

There are many varied motives for characterizing a polymeric material. These include identification, material design at the molecular level, microstructural design and processing control, and failure analysis. Let us consider each of these briefly in turn.

1.1.1 Identification

It may sometimes be of interest to identify the polymer from which an article is made. Many of the techniques described in later chapters can be used to assist in the process, though little emphasis is placed on this aspect of characterization. Simple identification of a polymer is often of little value in solving a problem. For example, if a component is found to have a particularly poor outdoor performance, knowing that it is made from poly(vinyl chloride) is of little use because this polymer can have excellent weatherability when compounded with a small addition of a suitable ultraviolet stabilizer. Thus it might be more pertinent to attempt to analyse the additive content. Some consideration is given to the analysis of additives but it is not attempted to present a comprehensive discussion of this topic.

1.1.2 Material design at the molecular level

When a research chemist prepares a polymer it is essential that the means to characterize the material produced are available. The detailed molecular structure determines the properties of an assemblage of molecules, whether in the form of a solid, a liquid (or melt) or a solution. Accurate characterization of a range of polymers permits correlations to be made between features of the molecular structures and (macroscopic) properties. Having established such correlations the research chemist may then 'design' a molecule with a structure favouring a particular property or combination of properties; after devising and executing a method or preparation, the product must then be characterized to confirm that the target structure has indeed been achieved. Thus it is desirable to have methods capable of identifying the functional

1

groups present within a polymer (and their distribution) and determining the steric arrangements of the bonds within the molecule. Molecular weight and molecular weight distribution influence property and need to be measured.

1.1.3 Microstructure, and the influence of processing

The arrangement of molecules in a solid polymer can be modified considerably by the presence of additives and also by the processing conditions used during shaping operations such as moulding, fibre drawing etc. In the case of spherulitic crystallizing polymers, the spherulite size can be influenced considerably by small amounts of nucleating agent, often added deliberately for this purpose, and by the cooling rate. Thus it is partly controlled by the formulation and partly under the control of the fabricator. Crystallinity may vary widely, and can be influenced by moulding conditions and by post-moulding conditioning. Molecular flow during moulding creates anisotropy and for crystallizing polymers may influence the crystal morphology. Fibre orientation in short fibre reinforced thermoplastics is similarly sensitive to moulding conditions. Molecular orientation may also be present in mouldings made from glassy polymers, causing properties to be anisotropic. Thus methods of characterization are needed in order to be able to determine how formulation and processing conditions control the microstructure. The microstructure has an important influence on properties, especially mechanical properties, and thus the link is established between processing and properties.

1.1.4 Failure analysis

When a component fails it is frequently required to investigate why. Apart from the rather negative requirement of determining blame, it is important to know how failure can be avoided in the future. Was the failure caused by the wrong choice of material? Was there an inadequacy in the design? Can failure be traced to a fabrication flaw? Was it the result of an accidental event, such as an overload, and if so, is a similar problem likely to recur? Did the material degrade in service due to weathering or some other environmental factor? Some of the methods of characterization described in the following chapters can help to answer these questions. For example, one of the most common ways in which a component fails is by fracture, and inspection of the fracture surface by one or more of the microscopical techniques can lead to identification of the mechanisms of fracture (often by comparison with fracture surfaces produced under strictly controlled conditions in the laboratory) and hence determine whether failure was a fault of the material, the fabrication process (e.g. if a stress concentrating void or agglomerate of filler is present and has nucleated the crack) or unreasonable service conditions.

1.2 Molecular architecture

The properties of a polymer are determined not only by the atomic constituents of each monomer unit but also by the stereochemical arrangement of the monomer units within the macromolecule. Thus *cis-trans* isomers can have quite different properties, and polymers that can be produced with different tacticities may also display a range of properties.*

The constituents of the monomer and the shape that they take within the polymer molecule determine to a large extent the packing characteristics of the molecules. Thus in an atactic polymer the disruption of periodicity along the polymer molecule inhibits crystallization. Another example is the effect of a bulky side group which prevents the close approach of parallel molecule segments and so prevents the formation of stable crystals. In the amorphous phase, movement of the side group is required to allow conformation changes to take place and the more difficult this is the higher is the glass transition temperature. In fact all relaxation mechanisms depend on the structure of the transforming group and on the environment provided by the neighbouring molecule segments.The packing density of the molecules (including any crystallinity) will determine such properties as the stiffness, strength and permeability (to gases and to liquids).

Optical properties depend on molecular structure. For example, strong birefringence is obtained if highly polarizable bonds (multiple bonds) are present and possess preferred orientation.

1.3 Crystallizing polymers

1.3.1 Crystallization conditions

Many polymers crystallize and many others do not. Since crystals have a well-ordered periodic structure (by definition) we would not expect to be able to crystallize those polymers which do not possess a periodic structure (e.g. random copolymers). On the other hand it can be anticipated that linear polymer molecules with regular periodic structures might arrange themselves side by side to form crystals. Crystals formed in this way will have only secondary bonding holding molecules together. Thus the internal energy is not very great and melting temperatures tend to be quite small, sometimes below room temperature. Side groups have to be accommodated and may force

---

*It is assumed that the reader is familiar with the concepts of *cis-trans* isomerization and tacticity. These topics and other relevant discussions are found in standard texts on organic chemistry and polymer science (e.g. references 1 and 2). In later chapters some specific definitions are recalled when special attention is drawn to particular features, e.g. in chapter 6, in connection with nuclear magnetic resonance structure analysis.

the molecule to take on a helical conformation in order to achieve efficient packing. If the side group is too large to permit close packing of the molecules the internal energy can never be sufficient to produce a stable structure and the polymer does not crystallize. Sometimes a polymer that shows little or no tendency to crystallize at a particular temperature may be made to crystallize by orienting it (e.g. by shearing it in solution or by extending it in the rubbery state). This causes the molecules to take up conformations that are more nearly straight and parallel and from this state they can readily form crystals.

Thus for crystallization to occur the polymer molecule must have a regular and relatively simple structure and there must be sufficient freedom for molecular motion to permit the molecule, or segments of it, to achieve the conformation required for it to attach to a growing crystal. The temperature must be below that at which spontaneous melting occurs, but not so far below it to inhibit molecular motion. Nucleating agents may act as templates to encourage molecule segments to arrange in the correct manner to form a crystal nucleus. These may be included specifically for this purpose, but nucleation may often take place on a reinforcing fibre or other additive included for some other purpose.

Cooling rate also influences crystallization. In many melt processing operations a large temperature gradient is established across the material and the cooling rate differs from one position to another. For example, in injection moulding the hot melt enters a cold mould cavity and the material near the cavity wall cools and solidifies rapidly to form a skin that contains molecules that have very little time to recover from the extended conformations caused by flow into the mould. Both the cooling rate and the flow-induced orientation may influence the crystal morphology. The material in the interior of the moulding cools much more slowly as a result of the low thermal conductivity of the polymer and usually produces a quite different crystal morphology.

The pressure exerted during moulding operations (up to 250 MPa in injection moulding) has a secondary influence on crystallization. Some experiments at much higher pressures have produced quite different morphologies [1]. These can be examined by the techniques described in this book but have not been exploited commercially and will not be discussed further here.

1.3.2 Crystal structure

It is relatively straightforward to determine the crystal structure but much less so to determine the types of defects present and their abundance. Dislocations can occur in polymer crystals just as with crystals of other materials, but they are less easy to observe and characterize (see Chapter 9). There are types of defect that do not have counterparts in other materials, such as chain ends where the molecule end is located within the crystal, and buried loops where the molecule folds within the crystal instead of at the surface. Another type of defect occurs if there is an interruption in the molecule repeat sequence, for example if a propylene unit is included in a polyethylene molecule. In this case the methyl group may be forced into the polyethylene crystal lattice. No satisfactory method exists for the observation of such defects, though general levels of disorder can be deduced from X-ray diffraction measurements (Chapter 8).

1.3.3 Crystal morphologies

When polymer crystals are grown from solution they normally form as chain-folded lamellae [1]. There are many questions arising out of the concept of chain-folded crystallization. Does the molecule double back on itself in the minimum possible number of chain bonds? Does it form a loose loop at the crystal surface? Does it re-enter the crystal at the nearest ('adjacent') site, or somewhere else ('switchboard')? What determines the fold length? These questions are not easily answered, but many careful laboratory investigations involving several of the characterization techniques described in this book have been mounted to investigate the structure.

When polymers crystallize from the melt they again usually form lamellae, in this case frequently arranged into spherulites [1]. The internal structure of the spherulite and the spherulite size have an important influence over properties and analyses of these characteristics are of great value. Some molecules or parts of molecules are always rejected from the crystal phase and the fractional crystallinity (i.e. the fraction of material that is actually present as crystals) is another important characteristic that has a significant influence over property. This is because the crystalline and non-crystalline regions have very different properties and the overall bulk properties will therefore depend on the fraction of each one present. Hence a number of methods have been devised to measure fractional crystallinity, most of them unique to polymer studies.

When crystallization takes place in a flow system the molecular orientation caused by flow can influence the crystal morphology. In extreme examples chain extended crystallization occurs in which very little molecular folding is present. Sometimes chain folded crystals grow out from the fibre-like extended chain crystals to form a 'shish-kebab' structure. This can be promoted in solution by simply rotating a rod about its axis in the solution contained in a co-axial cylindrical beaker. The fibrous growth attaches to the rod. Similar structures may well form from the melt during moulding operations that involve flow.

Crystal morphology can be altered by deformation in the solid state. For example fibre drawing causes pronounced crystal orientation with the crystal axis that coincides with the molecule axis tending to lie parallel to the draw direction. Crystal orientation produces strong property anisotropy and occurs in many fabrication processes. Sometimes significant benefits accrue. In fibre drawing great enhancement of stiffness and strength are produced parallel to the fibre axis, with little reduction in property in the transverse direction. Thus methods of measuring orientation are of importance.

1.4 Survey of characterization techniques

One of the most important characteristics of a polymer is its molecular weight. Measurement techniques have been developed almost exclusively for polymers, though colloidal suspensions can contain inorganic particles that are not conventionally described as macromolecules yet which lie within the same mass range as polymers. The techniques for molecular weight determination are not particularly versatile and are normally used only for that purpose. The exception is gel permeation chromatography in which different molecular weight fractions are separated as part of the procedure to determine molecular weight distribution, and this therefore enables the preparation of (tiny) samples of selected size with a very narrow molecular weight distribution ('monodisperse'). Molecular weight determination is dealt with in Chapter 2.

Spectroscopic techniques are used in many branches of chemistry for structure analysis. Many different structural features can be examined and each technique has a range of applications. Further explanation is given in Chapter 3, then Chapters 4 to 7 are devoted to individual techniques. Ultraviolet–visible spectroscopy is placed first (Chapter 4) because it uses electromagnetic radiation of the lowest wavelength among those techniques discussed. It has wide application in organic chemistry for identifying molecules and molecular structures. With polymers it cannot provide a complete analysis, but may be used in conjunction with other methods. It has special value in determining the conjugation length within unsaturated molecules. Another use is the measurement of the carbonyl group concentration which can be used to indicate the extent of degradation by oxidation/weathering. Chapter 5 deals with vibrational spectroscopy, both infra-red (IR) and Raman. These are often described as 'fingerprinting' techniques and are the most popular methods for identifying a polymer. They can provide information about the three dimensional arrangement of atoms within the molecule. It is possible to determine information about orientation and crystallinity by appropriate experimental arrangement and analysis.

Nuclear magnetic resonance (NMR: Chapter 6) is also used for molecular structure analysis. Again details concerning the precise three-dimensional arrangement of atoms in the molecules (tacticity, *cis-trans* isomer discrimination etc.) can be deduced from the measurements. The response is very different in the liquid than in the solid, and a significant change is observed on passing through the glass transition temperature, T_g. Thus if the spectrum is monitored continually while the temperature is changed, T_g may be determined. The electron spin resonance signal comes from paramagnetism associated with unpaired electron spins, and can be used to measure the concentration of free radicals (and obtain a clue to their species). It therefore finds application in studies of molecular fracture since mechanical scission leaves behind radical molecule ends (Chapter 7).

Whereas the techniques outlined so far are concerned primarily with the size of molecules and the arrangement of the atoms within them, Chapters 8 to 11 deal with the arrangements of molecules and structural features on a larger scale. Many polymers form crystals, and methods of determining crystal structure are of great value. X-ray diffraction (Chapter 8) is foremost in this field, and electron diffraction (Chapter 9) is also important. Orientation and fractional crystallinity can also be measured using X-ray diffraction. Details of crystal morphology can often be inferred from continuum analytical measurements, but the most positive morphological determinations are made by microscopy. Transmission electron microscopy (Chapter 9) and scanning electron microscopy (Chapter 10) can provide detailed and unambiguous morphological information, though great care must be taken to avoid artefacts that might lead to serious errors. These techniques can sometimes be used to measure lamellar thickness and other details of crystal morphology directly, though the accuracy of the method must always be carefully assessed. X-ray line broadening can also be used to estimate lamellar thickness though in this case an average value is obtained whereas a microscopic measurement is based on a feature (or features) selected by the observer. Raman spectroscopy provides yet another technique for lamellar thickness measurement. The average spherulite size can be obtained by small angle X-ray scattering (Chapter 8) or small angle light scattering (Chapter 11), or can be obtained directly by microscopy, most commonly light microscopy (Chapter 11) or scanning electron microscopy.

X-ray diffraction can be used to measure crystallinity whereas transmission electron microscopy can be used to show details of the arrangement of the crystal lamellae and the non-crystalline regions. Crystal orientation in the bulk is usually determined using X-ray diffraction, though crystal habit analysis is most elegantly performed in the transmission electron microscope using a combination of microscopy and selected area electron diffraction. Molecular orientation in the amorphous state is often measured using birefringence (Chapter 11).

Fracture mechanisms can be determined with the help of microscopic examination of fracture surfaces using either light or scanning electron microscopy. These techniques are also suitable for studying the distribution of fillers, including reinforcing fibres.

Methods of thermal analysis (Chapter 12) provide continuum (average) information on the thermal properties of polymers. Information can be obtained about important temperatures, such as the crystal melting temperature and T_g, but in addition calorimetric measurements such as determination of the heat of melting can be made using the differential scanning calorimeter (DSC). The thermal expansion behaviour is also of interest; it will often be anisotropic if there is structural anisotropy (orientation) present. Thermogravimetric analysis is useful in determining the presence of adsorbed volatiles and for identifying the temperature at which degradation begins. Other thermal techniques involve making a measurement as the temperature is scanned through a wide range (e.g. as in dynamic mechanical thermal analysis) and are often directed at understanding relaxation processes.

When choosing topics for a book of this kind it is difficult to decide exactly where to stop and several valuable characterization techniques are not covered in Chapters 2 to 12. We have selected some of them for brief treatment in Chapter 13. Of these selected techniques the one that looks most likely to merit a separate chapter in any future publication of this kind is ESCA (electron spectroscopy for chemical analysis/applications). This method can produce unique information about the surface of a material (namely the composition and the bonding state) and for some applications is indispensable. It is, however, rather specialized and not as widely available as the other methods described here. Neutron diffraction is even less accessible, but can provide unique information about the conformational state of polymer molecules. Conversely, density measurement is quite common (and the equipment is relatively cheap) yet is sufficiently straightforward to deal with in a few paragraphs and does not merit expanding into a chapter. Measurements of density can be related to crystallinity or, in the case of glassy polymers, to the thermomechanical history.

It is frequently found that several different techniques are required to provide information to enable a particular characterization to be executed and examples are given in Chapter 14. In some cases it is possible to conduct characterizations with two or more techniques simultaneously. It is often possible to examine a specimen in a light microscope during the performance of an experiment. For example, a microscope heating/cooling stage in which the specimen is located in a DSC cell with optical windows can be purchased permitting calorimetric data to be gathered while observing morphological changes, e.g. when passing through a melting temperature. Similarly, commercial instruments are available that can perform differential thermal analysis and thermogravimetric analysis simultaneously.

Finally we note that although our emphasis is firmly on characterization of polymers, many of the techniques described here are of value in detecting and identifying the additives present in commercial plastics, and we take the opportunity of pointing out examples whenever they arise.

Reference

[1] Bassett, D.C. (1981) *Principles of Polymer Morphology*, Cambridge University Press.

Further reading

[1] Young, R.J. (1981) *Introduction to Polymers*, Chapman and Hall, London.
[2] Billmeyer, F.W. Jr (1984) *Textbook of Polymer Science*, 3rd edn, Wiley, New York.

2
Molecular weight determination

2.1 Introduction

The essential distinguishing characteristic of polymeric materials is their molecular size. The properties which have enabled polymers to be used in a diversity of applications derive almost entirely from their macro-molecular nature. Non-network polymers such as thermoplastics and uncured elastomers and thermosetting resins consist of an assembly of molecules having a distribution of molecular sizes, i.e. they are polydisperse. In order to characterize fully these materials it is essential to have some means of defining and determining their molecular weights and molecular weight distributions. It is more correct to use the term relative molecular mass rather than molecular weight but as the latter is used more generally in polymer technology, we will follow established practice. For crosslinked network structures the molecular size is considered to be essentially infinite so that the concept of molecular weight is less useful and their properties are determined largely by the density of the crosslinks.

It is not always practicable to determine molecular weight distributions, although as we shall see in section 2.3.2, this is becoming more common practice using chromatographic techniques. Rather, recourse is made to expressing molecular size in terms of molecular weight averages.

2.1.1 Molecular weight averages

If we consider a simple molecular weight distribution as shown in Fig. 2.1 which represents the weight fraction (w_i) of molecules having relative molecular masses (M_i), it is possible to define several useful average values. Averaging carried out on the basis of the number of molecules (N_i) of a particular size (M_i) gives the Number Average Molecular Weight

$$\bar{M}_n = \frac{\sum N_i M_i}{\sum N_i} \tag{2.1}$$

An important consequence of this definition is that the Number Average Molecular Weight in grams contains Avogadro's Number of molecules. This

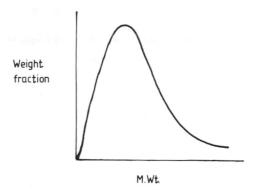

Weight fraction

M.Wt.

FIGURE 2.1 Typical molecular weight distribution curve.

definition of molecular weight is consistent with that of monodisperse molecular species, i.e. molecules having the same molecular weight. Of more significance is the recognition that if the number of molecules in a given mass of a polydisperse polymer can be determined in some way then $\bar{M}_n$ can be calculated readily. This is the basis of colligative property measurements (section 2.2.1(b)).

Averaging on the basis of the weight fractions (w_i) of molecules of a given mass (M_i) leads to the definition of Weight Average Molecular Weights

$$\bar{M}_w = \frac{\sum w_i N_i}{\sum w_1} = \frac{\sum N_i M_i^2}{\sum N_i M_i} \tag{2.2}$$

$\bar{M}_w$ is probably a more useful means for expressing polymer molecular weights than $\bar{M}_n$ since it reflects more accurately such properties as melt viscosity and mechanical properties of polymers. It should be emphasized, however, that knowledge of the molecular weight distribution is essential to explain physical properties fully.

There are two other molecular weight averages which find uses in describing polymer properties. The first is the z-average

$$\bar{M}_z = \frac{\sum N_i M_i^3}{\sum N_i M_i^2} \tag{2.3}$$

which can be determined experimentally using sedimentation/diffusion measurements, but will not be discussed further in this text. The second is the Viscosity Average

$$\bar{M}_v = \left[\frac{\sum N_i M_i^{1+\alpha}}{\sum N_i M_i} \right]^{1/\alpha} \tag{2.4}$$

where α is an empirical constant for a given polymer, solvent and temperature. $\bar{M}_v$ is not a widely quoted molecular weight average but solution viscosity measurements are extensively used for determining molecular weights (section 2.3.1).

The ratios of molecular weight averages are useful indicators of molecular weight distributions. For monodisperse samples the molecular weight averages are identical so that the ratios are all unity. For those step condensation polymerizations in which there is statistically the most probable distribution of molecular sizes then $\bar{M}_z : \bar{M}_w : \bar{M}_n = 3:2:1$. With increasingly broader molecular weight distributions, such as those produced by chain addition polymerization, particularly at high levels of conversion, the ratio $\bar{M}_w : \bar{M}_n$, for example, can increase to > 25.

When considering instrumental methods for determining polymer molecular weights, we should distinguish those primary or absolute methods which are capable of determining molecular weights from first principles (although in practice comparative methods may be employed) and secondary methods which require calibration with known molecular weight standards. Secondary methods usually offer additional advantages including experimental convenience.

2.2 Primary methods

Primary methods include colligative property measurements to determine $\bar{M}_n$, and light scattering techniques to determine $\bar{M}_w$. Secondary methods include solution viscosity measurements and gel permeation chromatography. The latter method is particularly useful in providing a direct measure of molecular weight distributions.

2.2.1 Number average molecular weights

When we considered the definition of $\bar{M}_n$, it was pointed out that where it is possible effectively to count the number of molecules in a known weight of a polymer, then $\bar{M}_n$ could be readily determined. This relationship provides the basis of the primary methods for determining $\bar{M}_n$. The experimental methods employed differ in the ways by which the numbers of molecules are determined. The simplest approach may be applied to those polymers possessing characteristic chain end groups which can be analysed quantitatively. The method, termed end group analysis, will only be considered briefly and is included in this section for completeness. Other primary methods are based on the colligative properties of solutions.

(a) END GROUP ANALYSIS

In simple step condensation polymerizations of, for example, polyesters

$$n\text{HO--R--OH} + n\text{HOOC--R'COOH} \rightarrow$$
$$\rightarrow \text{HO--R--O--[OC--R'--COO--R--O]OC--R'COOH} + (n-1)\text{H}_2\text{O}$$

it can be presumed that on average each polymer chain possesses one –COOH and one –OH group. Analysis of the polymer using spectroscopic techniques such as infra-red or nuclear magnetic resonance enables the concentration of these characteristic groups to be determined relative to the in-chain ester links. The same general procedures may be applied to other condensation polymers such as polyamides. The essential requirement in every case is that the structure of the chain end groups be known.

For chain addition polymerizations which proceed with no measurable chain transfer the chain end initiator fragment may be sufficiently distinct to enable identification and quantitative analysis to be carried out. Provided that the mode of termination is known, i.e. combination or disproportionation, $\bar{M}_n$ can be determined. For example polymerization of styrene initiated with AZBN (azobisisobutyronitrile) would yield the following chain structure when the mode of termination is by combination

It is possible to distinguish the in-chain phenyl protons and the aliphatic protons including the chain end *t*-butyl groups by means of proton NMR (Chapter 6). Absorption peaks at 7.2 ppm can be assigned to phenyl protons (intensity $= A$) and absorptions in the range 0.6 to 2.8 ppm can be assigned to the aliphatic protons (total integrated intensity $= B$). Assuming that

$$\frac{A}{B} = \frac{5(n+m)}{12 + 3(n+m)}$$

then $(n+m)$ may be calculated, and hence, from the molecular weights of the respective groups, $\bar{M}_n = 104(n+m) + 132$.

The major problem encountered in end group analysis is that of sensitivity which decreases markedly with increasing molecular weight. The method relies on measurement of the concentration of only one or two groups per molecule and the sensitivities needed to determine polymer molecular weights of 10^4 and 10^5 with an accuracy of 10% are respectively 10^{-5} and 10^{-6} moles. Spectroscopic techniques such as IR and NMR or the use of tracer techniques by labelling with, for example, ^{14}C can give high accuracy but nevertheless end group analysis is restricted to molecular weight averages $< 20\,000$. Since useful thermoplastics normally have molecular weights greater than this range, other methods for determining molecular weights are preferred for routine measurements.

(b) COLLIGATIVE PROPERTIES

Solution properties such as lowering of vapour pressure, elevation of boiling point, depression of freezing point and osmotic pressure are termed colligative properties. The name derives from the fact that at infinite dilution, the properties depend on the number of independent kinetic units in the solution and hence on the number of solute molecules. The colligative properties may all be expressed in terms of the free energy of dilution of the solution $\Delta \bar{G}_1$ which is the total Gibb's Free Energy when 1 mole of a solvent is added to a very large volume of solution.

$$\Delta \bar{G}_1 = \bar{G}_1 - \bar{G}_1^0 \qquad (2.5)$$

where $\bar{G}_1$ and $\bar{G}_1^0$ are the partial molar free energies of a solution and in the pure solvent respectively.

For ideal solutions at low concentration the colligative property relationships may be expressed by

$$\Delta \bar{G}_1 = RT \ln \frac{p_1}{p_1^0} \qquad (2.6)$$

where p_1 and p_1^0 are the vapour pressure of solution and solvent respectively

$$\Delta \bar{G}_1 = -\Pi V \qquad (2.7)$$

where Π is the osmotic pressure and V the molar volume of the solvent.

$$\Delta \bar{G}_1 = \frac{L_f}{T_f} \Delta T_f \qquad (2.8)$$

$$\Delta \bar{G}_1 = \frac{L_e}{T_e} \Delta T_e \qquad (2.9)$$

where L_f and L_e are the latent heats of fusion and evaporation, T_f and T_e are the freezing point and boiling point of the solvent, and ΔT_f and ΔT_e are the depression of the freezing point and elevation of the boiling point respectively.

Since for ideal solutions the free energy of dilution can be expressed by

$$\Delta \bar{G}_1 = -RT \frac{V_1}{M_2} c \qquad (2.10)$$

where M_2 is the molecular weight of solute and c its concentration in g dl^{-1}, it follows that

$$\ln \frac{p_1}{p_1^0} = -\frac{V_1}{M_2} c \qquad (2.11)$$

$$\Pi = \frac{RT}{M_2} c \qquad (2.12)$$

$$\Delta T_f = \frac{RT_f^2}{L_f} \frac{V_1}{M_2} c \qquad (2.13)$$

and

$$\Delta T_e = \frac{RT_e^2}{L_e}\frac{V_1 c}{M_2} \qquad (2.14)$$

so that measurement of colligative properties permits the solute molecular weight to be determined. As with end group analysis, the accuracy with which the various measurements can be made imposes limitations on their use. In addition, it is not expected that polymer solutions would behave ideally so that the above expressions require modification to take into account specific polymer–solvent interactions. This is done by expressing the colligative property such as osmotic pressure in the form of a virial equation, i.e.

$$property = A_1 c + A_2 c^2 + A_3 c^2 + \cdots \qquad (2.15)$$

where the higher power terms express the departure from ideality and may be used to provide additional information on polymer–solvent interactions.

Applications of colligative property methods will be considered by specific reference to ebulliometry and osomotic pressure measurements since these have been used most widely and are more likely to be encountered in the laboratory. Some reference will also be made to vapour pressure osmometry which is a practical means of determining vapour pressure lowering.

(i) Ebulliometry

As we have already seen, the elevation in boiling point (ΔT) of a solution of known concentration (c) of solute in a particular solvent is related to the number average molecular weight $\bar{M}_n$. For a dilute, ideal solution the molecular weight is given by

$$\bar{M}_n = \frac{K_E c}{\Delta T} \qquad (2.16)$$

where K_E is the ebullioscopic constant $= \dfrac{RT_e^2 M_1}{L_e}$

R is the ideal gas constant, and M_1 the relative molecular mass of the solvent.

For non-ideal solutions, such as for polymers, additional terms need to be introduced to account for the departure from ideality. An expression of the form

$$\Delta T = A_1 c + A_2 c^2 + A_3 c^3 + \cdots \qquad (2.17)$$

where A_1, A_2 and A_3 are the first, second and third virial coefficients respectively and $A_1 = K_E/\bar{M}_n$, is found to express the behaviour adequately. For ideal solutions the second and third terms in the expression, which takes into account heat and entropy of mixing, vanish and the expression is then identical to equation (2.16). $\bar{M}_n$ for polymers can then be determined by measuring ΔT as a function of c at low concentrations. A plot of $\Delta T/c$ against c

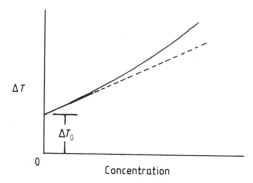

FIGURE 2.2 Ebulliometry; zero error plot.

will be approximately linear as $c \to 0$ provided that A_2 and A_3 are not too great. The intercept when $c = 0$ is then equal to $K_E/\bar{M}_n$ so that knowing K_E, $\bar{M}_n$ can be obtained.

In practice, errors will arise in applying this procedure if there is 'zero error'. Zero error can be expressed by considering the equation to be modified by the inclusion of an additional term ΔT_0

$$\Delta T = \Delta T_0 + A_1 c + A_2 c^2 + A_3 c^3 + \cdots \qquad (2.18)$$

A plot of ΔT against c should be linear with intercept equal to ΔT_0 (Fig. 2.2). The reasons for the zero error are not fully understood but arise from differences in the manner of the boiling process for dilute solutions possibly due to surface tension effects.

When zero error effects are noticeable, treatment of the data needs to be modified in order to yield $\bar{M}_n$. The recommended procedure is the divided difference method in which the ratio of differences in boiling points, $(T_{n+1} - T_n)$ due to successive small changes in concentration $(c_{n+1} - c_n)$, are related in terms of the virial equation in the following manner:

$$\frac{T_{n+1} - T_n}{c_{n+1} - c_n} = A_1 + A_2(c_{n+1} + c_n) \qquad (2.19)$$

A plot of the left hand side of equation (2.19) versus $(c_{n+1} + c_n)$ should be linear with intercept equal to A_1 which ought now to be independent of any zero error. Extrapolation of the data to zero concentration difference should be carried out using linear regression analysis to ensure accuracy.

Although it is possible in principle to determine K_E from known solvent properties, i.e. T_e, L_e, M_1 (equation (2.16)) it is more normal practice to employ comparative methods using polymers of known molecular weights as reference standards. In this way problems which may arise from instrumental variables can be minimized. These may include factors such as solvent held up in the reflux system which effectively changes the solution concentration, and superheating due to the effects of hydrostatic pressure. In addition, there is some doubt about the constancy of K_E with changes in molecular weight. Determination of the boiling point elevations produced by a series of dilute solutions of a solute of known molecular weight enables a calibration curve to be constructed by plotting the data in the form of equation (2.19). From the calibration curve an effective value of K_E may be obtained which takes into account some of the instrumental variables that may exist. Measurements of boiling point elevations for polymer solutions using the same experimental procedure enable $\bar{M}_n$ to be determined using the previously determined K_E value.

Inspection of equation (2.16) shows that ΔT is directly proportional to solute concentration and inversely proportional to polymer molecular weight. Since it is necessary to work with dilute solutions, the magnitude of ΔT is small and the practical difficulties in measuring small temperature differences places limits in the range of applicability of the technique. For example, polymer molecular weights of 10^4 and 5×10^4 give ΔT values of approximately 2×10^{-3} and 5×10^{-4}°C respectively in 1% solutions. Measurement of such small temperature differences is best achieved using bridge circuits employing thermocouples or thermistors. Some highly sensitive temperature sensing devices have been reported in the literature with accuracies down to 2×10^{-6}°C and have enabled polymer molecular weights up to 10^5 to be determined. For most systems, however, the practical limit for molecular weight determination by ebullioscopic techniques is considered to be approximately $3–4 \times 10^4$ and even then, great care is needed to ensure accuracy. Thus the method is most suitable for low molecular weight materials such as novolak resins and other intermediates.

Measurement of boiling point elevation is best carried out by means of a differential ebullioscope. That is to say that the boiling point of the solution is directly compared with the boiling point of the refluxing solvent. There are two particular experimental arrangements currently used. The first utilizes a single refluxing column in which the temperature of the boiling solution is compared to the temperature of the refluxing vapour. The second arrangement employs two identical columns placed side by side with one containing pure solvent and the other the solution of known concentration. For very high sensitivity, the former arrangement is considered to be more accurate since instrumental variables are reduced but for many routine applications the twin column system is adequate and is likely to be simpler in construction and operation.

A typical arrangement is shown in Fig. 2.3. A particular feature is the Cottrell pump which causes the boiling solution to flow over the thermistor or thermocouple pocket so reducing effects of possible superheating of the boiling vapour. The experimental procedure involves placing pure solvent in both columns and balancing the Wheatstone bridge network to give a null reading on the galvanometer. Accurately weighed samples of solute are added to one column and the system allowed to reach equilibrium when the bridge is again

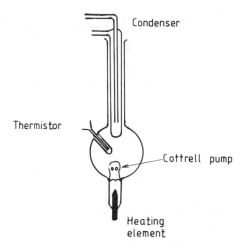

FIGURE 2.3 Ebulliometry column showing Cottrell pump.

balanced by means of a calibrated variable resistor or potentiometer. The magnitude of the change in the variable resistor is directly proportional to the boiling point elevation. The results may be plotted directly as potentiometer dial reading against sample weight or as the divided difference discussed earlier and then compared with results obtained from the reference standard.

Errors in temperature measurement can arise due to excessive foaming of the solution. For high accuracy, steps need to be taken to reduce the effects of foaming including the use of mechanical devices to disrupt the foam.

(ii) Osmometry

The chemical potential of a solvent in solution is lower than that in the pure solvent in proportion to the concentration of the solute species. If it is so arranged that the solution is separated from pure solvent by means of a semipermeable membrane, i.e. one which only permits passage of solvent and not solute molecules, then solvent will diffuse through the membrane in order to equalize the chemical potentials and so establish equilibrium. The result of the flow of solvent is to generate a pressure on the solution side of the membrane which at equilibrium is just sufficient to prevent any further net flow of solvent. The pressure difference is termed the osmotic pressure Π. This is illustrated schematically in Fig. 2.4, where the osmotic pressure is indicated by the difference in heights of the liquids in the side arms. The Pinner–Stabin osmometer operates with this principle (Fig. 2.5). Although the Pinner–Stabin and similar osmometers are useful in molecular weight determination and in explaining the principles involved, they suffer from the major drawback that lengths of time up to several hours are necessary to achieve equilibrium. It is also difficult to establish that equilibrium has actually been attained not least

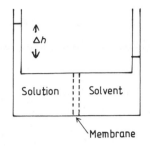

FIGURE 2.4 Schematic representation of osmotic pressure.

because there is the possibility of diffusion of low molecular weight polymer molecules through the membrane.

These limitations are overcome to a great extent in modern instruments which operate such that either solvent flow is prevented or where the osmotic pressure is detected directly using electromechanical devices. The Knauer membrane osmometer will serve to show the general procedures employed. The essential part of the instrument is the steel measuring cell which is divided into two halves by the semipermeable membrane. In operation the lower half of the cell is filled with degassed solvent and sealed by the semipermeable membrane which forms a seal when the upper half of the cell is placed in position (Fig. 2.6).

The membrane materials are typically regenerated cellulosics such as cellophane, which are required to be conditioned in the solvent prior to use.

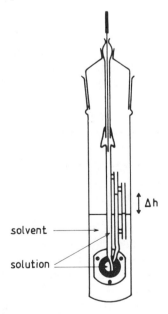

FIGURE 2.5 Pinner–Stabin osmometer.

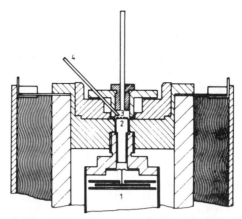

FIGURE 2.6 Measuring cell of Knauer membrane osmometer (courtesy of Owens Polyscience Ltd). 1. Capacitive pressure measuring system. 2. Mushroom over which the membrane is placed. 3. Semi-permeable two layer membrane. 4. Extractor device for rinsing.

The membranes are supplied in a swollen, wet state and should not be allowed to dry out. Conditioning of the membranes involves successive immersions in a series of intermediate solutions containing increasing amounts of the required solvent until 100% solvent is reached. Recommended immersion times can vary significantly depending on the membrane and solvent so suppliers' instructions should be followed.

Sample solutions of known concentration in the range 0.2 to 2.0 g/100 ml are injected successively into the upper half of the cell using a syringe. Diffusion of solvent through the membrane into the solution results in a negative pressure being developed on the solvent side of the membrane. A steel membrane in the lower half of the cell is caused to deflect under the action of the negative pressure and this in turn is detected as a change in capacitance of an electronic circuit which is therefore related directly to Π.

The volume of the solvent side is kept as small as possible (about 120 μl) so as to reduce effects due to temperature variations. Likewise only small volumes of sample solution are necessary (about 1.5 ml) for each reading. Membrane osmometers are capable of operating up to 130°C which enables polymer of low solubility to be examined.

Reference to the virial equation

$$\Pi = \frac{RTc}{\bar{M}_n} + A_2c^2 + A_3c^3 + \cdots \tag{2.21}$$

shows that a plot of Π/c versus c should be linear for dilute solutions, when the second and third terms vanish, with the intercept being equal to RT/M_n from which M_n is readily determined.

As with all colligative property measurements, the accuracy of the technique is dependent on the linearity of the curve at low concentration in order to permit extrapolation to zero concentration. This in turn depends on the magnitude of the deviations from ideal solute behaviour as indicated by the magnitude of the second and higher order virial coefficients. For most systems the third and higher virial coefficients are negligibly small and may usually be ignored. The second virial coefficient which takes into account specific polymer–solvent interactions, is more important in determining how good particular solvents are. Thermodynamically good solvents correspond to high values of A_2 (typically of the order of $10^{-4} \, \mathrm{m^3 \, mol \, kg^{-2}}$), and as A_2 decreases in value, the solvent becomes increasingly poorer. Theta solvent conditions arise when $A_2 = 0$ corresponding to the situation when an infinitely high molecular weight polymer just swells but does not dissolve in the solvent. In this case a plot of Π/c versus c has zero slope. When A_3 becomes significantly large, a plot of Π/c versus c is no longer linear and shows some curvature with increasing concentration. Extrapolation to zero concentration is then more uncertain. However, if the data are plotted in the form of $(\Pi/c)^{1/2}$ versus c, the graph should be linear permitting extrapolation to give A_2 as the intercept.

Membrane osmometry is particularly useful in that absolute molecular weight averages can be determined over a much wider molecular weight range than is possible for the other colligative property measurements. Molecular weights up to one million can be determined, although accuracy diminishes with increasing molecular weights. Unfortunately the technique has limitations for polymers having low molecular weight averages due to the possibility of diffusion of low molecular weight molecules through the membrane. When this occurs, the measured osmotic pressure is not the true equilibrium value and so constitutes a major source of error. In addition, it is necessary to completely replace the contaminated solvent which may involve dismantling and reassembling the osmometer and replacing the membrane which, to say the least, is very inconvenient.

Osmotic pressure measurements are also very temperature dependent so that careful thermostatting of the equipment is essential. This is, of course, standard practice and standards specify the temperature at which measurements must be made.

(iii) Vapour phase osmometry

The limitations placed on osmotic pressure measurements due to the presence of low molecular weight fractions are not encountered in methods based on lowering of vapour pressure. Consequently vapour pressure measurements are more accurate for low molecular weights. Practical methods for determination of molecular weights based on lowering of vapour pressure involve measurement of small temperature changes caused by the preferential condensation of solvent from the vapour on to a dilute polymer solution. Transport of solvent to solution then proceeds via the vapour phase, thus the method is termed Vapour Phase Osmometry.

A typical vapour phase osmometer (Fig. 2.7) consists of a thermostatted sample chamber containing two matched thermistors connected to a bridge

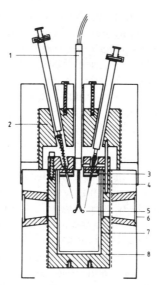

FIGURE 2.7 Schematic of vapour phase osmometer (courtesy Owens Polyscience Ltd). 1. Measuring probe. 2. Aluminium block for thermostatic control of syringes. 3. Seal. 4. Wick. 5. Thermistors. 6. Windows. 7. Glass container for solvent. 8. Measuring cell (aluminium).

circuit capable of measuring small temperature differences with high accuracy. The chamber is initially saturated with solvent vapour at the given temperature. A small amount of solvent is placed by means of a syringe on to one of the thermistors and a small amount of a dilute polymer solution on to the other. Differences in vapour pressure between the solvent and solution cause solvent to condense on to the solution with a resultant rise in temperature of the solvent due to the evolution of the latent heat of vaporization. The measured temperature difference ΔT is directly proportional to the lowering of the vapour pressure and hence is related to the polymer molecular weight.

In practice, the equipment is calibrated using known molecular weight standards and recording differences in electrical resistance ΔR rather than ΔT directly so that the molecular weight is obtained from

$$\frac{\Delta R}{K'c} = \frac{1}{\overline{M}_n} + A_2c \qquad (2.22)$$

where K' is a calibration constant.

As with other methods R values are obtained for a series of dilute solutions of concentration c and $\Delta R/K'c$ is plotted against c with extrapolation to zero concentration to give $\overline{M}_n$.

2.2.2 Determination of weight average molecular weights by light scattering

When a parallel beam of light passes through a transparent system, a small part of the light is scattered elastically (Rayleigh Scattering). The scattering arises due to optical discontinuities in the medium which in gases and liquids can be due simply to the molecules themselves. For solutions of polymer molecules, additional scattering arises from the presence of the solute molecules and this may be shown to be a function of the concentration of the polymer molecules, as well as their size and shape. When measurements are made of the differences in intensity of the scattered light between the solvent and a series of dilute solutions, then it is possible to determine the average size of the solute scattering centres, and hence their molecular weights. As we shall see, it is the weight average value that is determined from light scattering measurements.

The scattering of light by polymer solutions may be expressed in terms of the Debye equation

$$H\frac{c}{\tau} = K\frac{c}{R_\theta} = \frac{1}{\bar{M}_w P(\theta)} + 2A_2 c \tag{2.23}$$

where

$$H = \frac{32\pi^3 n^2 (dn/dc)^2}{3N_0 \lambda^4} \tag{2.24}$$

and τ is the turbidity (from $I = I_0 e^{-\tau l}$); n is the refractive index; dn/dc is the concentration dependence of the refractive index of the polymer solution (a constant); N_0 is Avogadro's number; λ is the wavelength of light in air; K is given by the equation

$$K = \frac{2\pi^2 n^2 (dn/dc)^2}{N_0 \lambda^4} \tag{2.25}$$

and $\bar{R}_\theta = R_\theta$ (solution) $- R_\theta$ (solvent) where R_θ, the Rayleigh Factor, is the ratio of the scattered light intensity per unit volume of the scattering solution, per unit solid angle of the detector, to the incident light intensity (I_0), i.e.

$$R_\theta = \frac{i_\theta r^2}{I_0} \tag{2.26}$$

where r is the distance from the scattering centre to the detector and i_θ is the amount of light scattered per unit volume at angle θ to the incident beam by one scattering centre.

First dn/dc is determined by measuring the refractive indices of a range of solutions of known concentrations. Secondly the scattered light intensity is measured at different angles to the incident beam for a similar range of solution concentrations. In principle, for small molecules a plot of Kc/R_θ versus c is linear with intercept equal to $1/\bar{M}_w$ so that determination of molecular weight is relatively straightforward. However, as the molecular size increases destructive interference effects arise from light scattered from different parts of the molecule. These interference effects are taken into account by inclusion of the function $P(\theta)$ in equation (2.23). $P(\theta)$ describes the angular dependence of the scattered light and is also related to the dimensions of the scattering entities.

For small angles:

$$\frac{1}{P(\theta)} = 1 + \frac{16\pi^2}{3\lambda_s^2} \overline{s_z^2} \sin^2(\theta/2) \tag{2.27}$$

where $\overline{s_z^2}$ is the z-average mean square radius of gyration and $\lambda_s = \lambda/n$ is the wavelength of light in the solution. Note that when $\theta = 0$ then $P(\theta) = 1$ and use is made of this in determining $\bar{M}_w$ from the experimental data.

The experimental procedure followed for determining $\bar{M}_w$ for polymers involves measuring the intensity of light scattered over a range of angles relative to the incident beam (typically $35°$ to $145°$). Calculation of $\bar{M}_w$ makes use of the Zimm plot which is a method of plotting the data in such a way that the extent of light scattering at $\theta = 0$ may be obtained by extrapolation to $P(\theta) = 1$. From the equation

$$\frac{Kc}{\bar{R}_\theta} = \frac{1}{\bar{M}_w} + \frac{1}{\bar{M}_w}\left(\frac{16\pi^2}{3\lambda_s^2}\right)\sin^2\left(\frac{\theta}{2}\right)\overline{s_z^2} + A_2 c \tag{2.28}$$

double plots are constructed of $Kc/\bar{R}_\theta$ against $\sin^2(\theta/2)$. These are a series of plots obtained by first keeping c constant and varying θ, then keeping θ constant whilst varying c, as shown in Fig. 2.8. Extrapolation for $c = 0$ and $\theta = 0$ gives $1/\bar{M}_w$ as the intercept on the $Kc/\bar{R}_\theta$ axis.

In addition to $\bar{M}_w$ the graphs yield values of A_2 which is given by the slope of the $\theta = 0$ curve, and the radius of gyration $\overline{s_z^2}$ which is given by the slope of the $c = 0$ curve. Both of these quantities are useful in describing the behaviour of polymer molecules in solution.

There are a number of experimental arrangements available for determining light scattering of polymer solutions and these have been described in some detail in the literature so that only an outline description will be given here. Light scattering photometers (Fig. 2.9) consist essentially of a high intensity monochromatic light source (S) (usually a mercury arc lamp) and suitable filters (F), a plane polarizer (P), a sample cell (SC) so constructed to permit passage of the incident beam and to enable the scattered light to be detected, and a rotatable photomultiplier tube (PM) to permit measurement of the scattered light intensity over a range of scattering angles. Photometers differ in the detailed construction of the sample cell in attempts to reduce or eliminate entirely scattering from the container walls which causes spurious results.

The most serious problem that arises in determining molecular weights by light scattering techniques is that due to the presence of impurities such as dust particles which contribute to the overall scattering intensity. Great care is required in preparing the solutions to remove any extraneous matter. The recommended procedures require filtration of the solvent and solutions through ultrafine sintered glass filters and ensuring that there is no further contamination before transferring samples to the light scattering cell. The concentration of the solutions is best determined gravimetrically by evaporat-

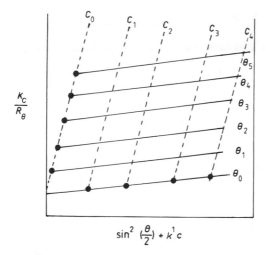

$$\frac{K_c}{R_\theta} =$$

$$\sin^2\left(\frac{\theta}{2}\right) + k^1 c$$

FIGURE 2.8 Zimm plot showing double extrapolation technique; extrapolated points ●. K^1 is a scaling factor.

ing a known volume of the filtered solution and weighing the residues to give the weight of dissolved polymer.

Recent developments in light scattering photometers have involved laser light sources. Because of the much higher light intensity, it is possible to detect light scattered at small angles ($< 10°$) even for dilute polymer solutions. Since $P(\theta) \rightarrow 1$ as $\theta \rightarrow 0$ we can rewrite equation (2.23)

$$\frac{Kc}{\bar{R}_\theta} = \frac{1}{\bar{M}_w} + 2A_2c \tag{2.29}$$

Measurement of $\bar{R}_\theta$ for a single angle then permits $\bar{M}_w$ to be determined from the intercept of the graph of Kc/R_θ versus c. This markedly reduces the number of measurements required so greatly simplifying the procedure. With the increased availability of laser light sources, such instruments are becoming the norm and, when allied to on-line data processing, lead to more rapid molecular weight determination.

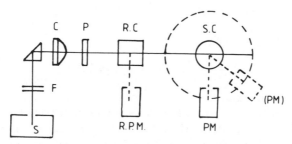

FIGURE 2.9 Schematic of typical light scattering photometer. S = source, F = filter, C = condenser, P = polarizer, R.C. = Reference chamber, R.P.M. = Reference photomultiplier, S.C. = sample chamber, PM = Rotatable photomultiplier.

2.3 Secondary methods

2.3.1 Solution viscometry

A secondary method widely used for the determination of polymer molecular weights is that based on solution viscosity measurements. It is reasonable to presume that the viscosity of a polymer solution would be dependent both on concentration and on the average molecular size of the sample and hence the molecular weight. Although there is no complete theoretical treatment of the dependence useful empirical relations have been established which have found general applicability. For instance, we have already seen that the viscosity average molecular weight $(\bar{M}_v)$ may be expressed by equation (2.4).

The procedure employed requires determination of the intrinsic viscosity for the polymer/solvent system $[\eta]$ which is related to $\bar{M}_v$ through the Mark–Houwink equation

$$[\eta] = K\bar{M}_v^{\alpha} \tag{2.30}$$

where K and α are empirical constants for the particular system. We have noted that solution viscosities are dependent in part on the size of the polymer molecules but this statement requires some amplification. In the present context the effective size of the molecules depends on the extent of interactions with the solvent. In good solvents the molecules assume expanded conformations but as the solvent becomes poorer, intramolecular interactions become more significant until for theta solvents the chains assume their unperturbed dimensions. Consequently the solution viscosity is expected to show quite different dependence on molecular weight in different solvents. In addition even for a particular polymer/solvent combination effects such as chain branching will influence the variation of $[\eta]$ with $\bar{M}_v$ so that the empirical parameters K and α may not be constant for the same nominal polymer type.

The intrinsic viscosity is determined experimentally by measurements of flow times of solvent (t_0) and a series of dilute polymer solutions of known concentration (t_c) in a standard capillary viscosity. The specific viscosity is calculated from the equation

$$\eta_{sp} = \frac{\eta_c - \eta_0}{\eta_0} = \frac{t_c - t_0}{t_0} \tag{2.31}$$

and from the Huggins equation

$$\frac{\eta_{sp}}{c} = [\eta] + k^1[\eta]^2 c \tag{2.32}$$

where k^1 is a positive constant, the intrinsic viscosity is obtained by plotting η_{sp}/c versus c and extrapolating to zero concentration (Fig. 2.10).

The Ubbelohde viscometer (Fig. 2.11) offers a convenient method for determining the flow times under essentially constant conditions. The experimental procedure involves introducing filtered solvent in the lower bulb

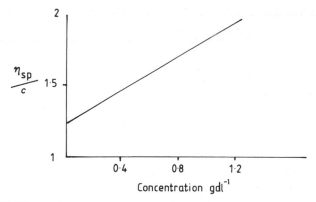

FIGURE 2.10 Dilute solution viscometry; Huggins plot of η_{sp}/c against concentration.

of the viscometer via the filling tube. The viscometer is then mounted vertically in a water bath thermostatted to $25° \pm 0.02°C$ and left to equilibrate at that temperature. It is advisable to immerse the solvent container in the water bath itself to reduce the time required to reach temperature equilibrium. The solvent is then drawn up the central capillary tube by suction using a hand aspirator until it is above the upper etched mark. The time for a constant volume of the solvent to flow between the two etched marks is determined to high accuracy using a stopwatch. This value is t_0.

The viscometer is emptied, cleaned, dried and a known volume of a filtered solution just sufficient to be drawn up the capillary without introducing air bubbles, and containing approximately 1% w/v polymer in the same solvent, is

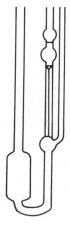

FIGURE 2.11 Ubbelohde viscometer.

introduced. The flow time for the solution is determined in the same manner as for the solvent. The stock solution is then progressively diluted by the addition of known amounts of the solvent to give at least 5 different concentrations, care being taken at each stage to mix thoroughly. This can be achieved by repeated lowering and raising the solution up the capillary tube using the hand aspirator. The flow times for each dilution are determined (t_c) and then $[\eta]$ and $\bar{M}_w$ can be calculated with known values of K and α.

Values of the empirical constants K and α have been determined experimentally by measurements of $[\eta]$ for polymer samples of known molecular weights which themselves have been determined using one of the absolute methods. For most polymer/solvent systems reference can be made to tabulated values. It should be emphasized that K and α values vary with polymer, solvent and temperature so that care must be exercised in interpreting values of molecular weight from solution viscosity measurements. In general, $\bar{M}_v$ will be close to, but somewhat lower, than $\bar{M}_w$.

For most practical situations manually operated viscometers such as the Ubbelohde capillary viscometer are adequate and are relatively simple to operate. It is, of course, quite feasible to measure the flow times automatically using photoelectric techniques, and to employ automatic filling and diluting of the solution and practical systems have been reported.

Automatic on-line viscometers have also been reported in which the viscosity η is determined by accurately monitoring the pressure drop ΔP across a capillary of known length, L and radius, R, and applying Poiseuille's formula

$$\left[\frac{dV}{dt} = \frac{\pi R^4 \Delta P}{8 \eta L} \right]$$

where dV/dt is the output flow rate.

Instruments based on this principle find use particularly in gel permeation chromatography for continuously monitoring the effluent enabling the molecular weight distribution to be determined.

2.3.2 Gel permeation chromatography

Although not a primary method for determining molecular weights, gel permeation chromatography (GPC) or size exclusion chromatography (SEC) has developed into one of the most useful methods for routine determination of average molecular weights and molecular weight distributions of polymers. It is the latter facility which makes the technique so attractive to polymer scientists and technologists allied to the development of on-line microprocessor based data handling which facilitates the calculation of the molecular weight averages from the size distribution chromatograms. In addition, only very small sample sizes are required permitting, for example, studies of the effects of weathering on the surface of polymers.

GPC is a form of liquid chromatography in which the molecules are separated according to their molecular size. The procedure involves injecting a dilute solution of a polydisperse polymer into a continuous flow of solvent passing through a column containing tightly packed microporous gel particles. The gel has particle sizes in the range $5-10\,\mu m$ in order to give efficient packing and typically possesses a range of pore sizes from 0.5 to $10^5\,nm$, which correspond to the effective size range of polymer molecules. Separation of the molecules occurs by preferential penetration of the different sized molecules into the pores; small molecules are able to permeate more easily through the pores compared to the larger sized molecules so that their rate of passage through the column is correspondingly slower. The continuous flow of solvent leads to separation of the molecules according to size with the larger molecules being eluted first and the smaller molecules, which have penetrated more deeply into the pores, requiring longer elution times. It follows that the time or, more usually volume of elution (V_e), is inversely proportional to the molecular size. If the pore size is too small to permit penetration by any of the molecules, or if the pore size is so large that all of the molecules can penetrate with the same relative ease, there would be little or no separation of the molecules. Consequently, selection of the column packing material to have the appropriate pore size distribution is crucial and different columns are usually required for polymers having widely different molecular weight distributions. The recent availability of gels of mixed pore sizes which can operate over four decades of molecular weight has made this a less demanding requirement.

Different molecular weight fractions are characterized by the elution volume or peak retention volume (V_R) which is the volume of solution eluted from the time of injection of the polydisperse sample into the GPC column to peak of the chromatogram for the particular fraction (Fig. 2.12).

V_R has been related empirically to V_0, the interstitial volume and V_i, the volume of liquid within the pores by

$$V_R = V_0 + kV_i \qquad (2.33)$$

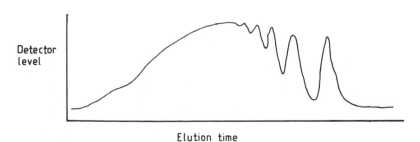

FIGURE 2.12 Example of GPC chromatogram for an epoxy cresol resin (Perkin Elmer).

where k is a distribution coefficient which indicates the relative ease of penetration of the solute molecules into the pore structure. When $k = 0$ there is no penetration and when $k = 1$ there is total unrestricted penetration. In principle, k values could be associated directly with molecular size but it is common practice to relate V_R to molecular weight empirically. The calibration procedures which are followed will be discussed in due course. Before doing so we will need to outline the general procedures followed to produce GPC chromatograms.

The essential requirements for a GPC chromatograph (Fig. 2.13) are:

1. Solvent delivery system: capable of maintaining a constant linear velocity flow;
2. Column(s) containing suitable microporous gel particles to produce the necessary size separation;
3. Injection system: capable of delivering accurately small volumes of sample solutions without disturbing the solvent flow;
4. Detection system to monitor output from the columns and to provide continuous quantitative and possibly qualitative data on the fractions being eluted;
5. Recorder to give continuous output traces.

In addition, most modern GPC systems include an automatic data handling facility to convert output data into useful average values.

The retention times show a logarithmic relationship with molecular weight so it is essential that the flow rate remains constant, even with the inevitable

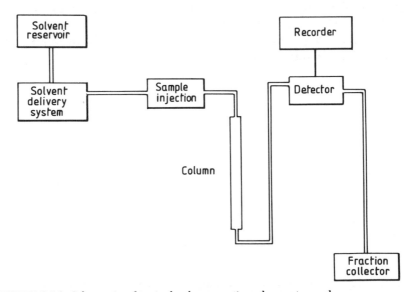

FIGURE 2.13 Schematic of typical gel permeation chromatograph.

changes in solution viscosity between samples. Typical flow rates are in the range 0.01 to 10.0 ml min^{-1} and commercial instruments are reproducible to about 0.3%.

The columns are packed with microporous crosslinked styrene-divinylbenzene gels or porous glasses or silica. The inorganic packing materials are preferred for aqueous solution work. Styrene-divinylbenzene gels are usually employed for non-polar organic solvents such as toluene or tetrahydrofuran, although improvement in the stability of crosslinked gels permits the use of more polar solvents such as dimethylformamide, *o*-chlorophenol and *N*-methylpyrrolidone. This means that some polymers such as nylon and poly(ethylene terephthalate) (PET) which previously required elevated temperatures may now be analysed at ambient temperatures. However, the ability to function effectively at elevated temperatures is usually an important requirement of the packing material. Rates of molecular diffusion increase with increasing temperature giving enhanced GPC separations in shorter times, and in addition, semi-crystalline polymers such as polyethylene require high temperatures for dissolution. The efficiency of modern packing materials enables complete separation to be achieved in the order of minutes. This facility explains in part the great interest in the technique. GPC columns are available with a range of pore sizes from 0.5 to 10^5 nm, enabling molecular weights from < 100 to approximately 4×10^7 to be discriminated.

The columns are about 30 or 60 cm long and 7.5 mm in diameter, constructed of stainless steel and threaded to permit ready assembly. Fig. 2.14 shows calibration curves for column packings having a range of pore sizes. The use of mixed pore sized packing is shown to give a linear calibration from 500 to 5×10^6 and this is particularly useful when determining molecular weight distributions.

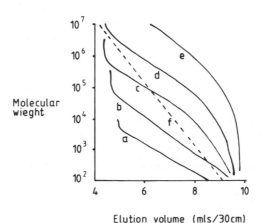

FIGURE 2.14 Calibration curves for different pore size gels; a, 10 nm; b, 10^2 nm; c, 10^3 nm; d, 10^4 nm; e, 10^5 nm; and f, mixed pore size (polystyrene standards).

The detector is required to monitor the concentration of the solute molecules in the solvent continuously and ideally to permit qualitative analysis of the various constituents in the compound. The most common types of detector are differential refractometers, which may be used for all types of molecules, and ultraviolet/visible absorption instruments having fixed or variable wavelengths which find particular application for detecting additives which absorb in the ultraviolet or visible spectral regions. Infra-red absorption detection systems are also available. Fourier transform infra-red spectroscopy is particularly useful since it makes possible the determination of complete spectra in very short times.

The spectroscopic detectors may be used in such a way that the entire spectrum of collected fractions can be determined for identification purposes or, by monitoring absorption of a single wavelength, as a means of following changes in concentration continuously. Other detector techniques that may be employed include the use of radioactive tracer elements, e.g. ^{14}C, evaporation and direct weighing of solute fractions and density measurements but these are much less important and are not used in commercial instruments. Possibly of more interest are methods for continuous measurement of the solute molecular weight using viscosity or light scattering techniques which are available commercially although not as yet in general use.

Although GPC separates molecules according to their molecular size, with the result being presented as a size distribution curve, the technique does not give absolute values of molecular weight and there is a need to calibrate with polymer standards of known molecular weight. This is one of the major limitations of the technique since only a limited number of polymer standards are available. Polystyrene standards are most commonly used. Monodisperse samples of the polymer having molecular weights in the range 500 to 15×10^6 with $\bar{M}_w/\bar{M}_n \approx 1.05$ are available from suppliers. Other polymer standards including poly(methylmethacrylate), poly(α-methylstyrene), *cis*-polyisoprene, poly(ethylene oxide), poly(ethylene glycol) and polyethylene may also be supplied for calibration purposes.

Various methods have been employed to construct GPC calibration curves. Most are based on direct calibration using polymer standards having known narrow (monodisperse) or broad (polydisperse) molecular weight distributions. The molecular weights of the standards are determined using absolute methods such as light scattering and osmotic pressure measurements, or by solution viscosity measurements.

For narrow molecular distribution standards the elution volume for each of the standards is determined and a calibration curve of $\log M$ versus V_e is constituted. Curve fitting procedures permit the determination of the molecular weight at any elution volume within the range covered. Figure 2.14 shows the use of a mixed pore sized gel to give an essentially linear calibration curve such that

$$\log M = A - BV_e \qquad (2.34)$$

However, the use of a calibration curve based upon linear polystyrene standards is not likely to be strictly applicable to other polymer types and errors will undoubtedly arise in some cases. For instance, since GPC separates molecules according to their effective size, it may not adequately distinguish between branched and linear chain polymers of the same molecular weight. Polymer–solvent interactions also influence the effective size or hydrodynamic volume of polymer molecules, and since these interactions are also concentration dependent, they constitute possible sources of error.

Calibration procedures using polydisperse standards of the same polymer type as those to be analysed require first of all that $\bar{M}_n$, $\bar{M}_w$ and/or $\bar{M}_v$ be determined for the standards using established techniques. It is also necessary to construct an initial calibration curve using monodisperse standards as previously described. Since the aim is to relate the peak values of GPC chromatograms to molecular weight averages the required relationship needs to be established. The procedure is as follows. Apparent molecular weight averages ($\bar{M}_n$, $\bar{M}_w$, $\bar{M}_v$) are calculated from the GPC curves of the polydisperse standards using the initial calibration curve (see end of this section). The calculated values are then compared with the previously determined molecular weight averages (absolute values) and are likely to show discrepancies. The calibration curve is then adjusted by an iterative method until the estimated and true molecular weight averages agree most closely over the optimum calibration range.

The most widely used calibration method is termed the universal calibration procedure. This relies on the dependence of the effective molecular size or more correctly the hydrodynamic volume on the product of the intrinsic solution viscosity [η] and the molar volume of the solute M which is proportional to the molecular weight. It is presumed also that the separation of the molecular fractions is governed solely by the hydrodynamic volume of the molecule. A calibration curve produced for one polymer, usually polystyrene, can then be used to construct the calibration curves for other polymers.

A universal calibration curve of $\log([\eta]M)$ versus V_e is constructed from solution viscosity and GPC measurements for standard monodisperse polystyrene samples. Typically the curve is linear over most of the range so that

$$\log([\eta]M) = C - DV_e \qquad (2.35)$$

where C and D are empirical constants.

The linear relationship appears to hold, with some notable exceptions, for a wide range of linear polymer types and molecular weight ranges (Fig. 2.15). By making use of the Mark–Houwink equation (equation (2.28))

$$[\eta] = KM^\alpha$$

so that
$$[\eta]M = KM^{\alpha+1} \qquad (2.36)$$

where K and α are empirical constants, so that knowledge of α and K for the particular polymer and solvent allows the molecular weights to be readily determined. Care is still required when trying to apply the universal calibration

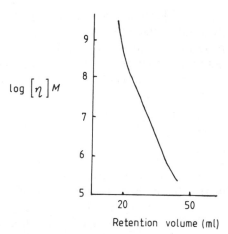

FIGURE 2.15 GPC Universal calibration curve.

method to polymer/solvent combinations other than those used in constructing the calibration curve.

For polydisperse samples, the GPC data may be used directly to determine the various molecular weight averages. This may be shown by the following. The GPC chromatogram is a record of the concentration of the solute at a particular elution time or volume. This data can be summed to give the integral or cumulative weight distribution curve which is conveniently normalized to give a total weight of unity. The differential of this curve gives the differential weight distribution $W(M)$, which may be expressed by

$$W(M) = \frac{\mathrm{d}W_V}{\mathrm{d}M} = \frac{\mathrm{d}W_V}{\mathrm{d}V} \frac{\mathrm{d}V}{\mathrm{d}(\log M)} \frac{\mathrm{d}(\log M)}{\mathrm{d}M} \tag{2.37}$$

$$= \frac{\mathrm{d}W_V}{\mathrm{d}V} \frac{\mathrm{d}V}{\mathrm{d}(\log M)} \frac{1}{M} \tag{2.38}$$

where W_V = weight fraction eluted up to volume V, i.e. with molecular weights $< M$

$\dfrac{\mathrm{d}W_V}{\mathrm{d}V}$ = height of the chromatogram

$\dfrac{\mathrm{d}(\log M)}{\mathrm{d}V}$ = gradient of calibration curve at volume V (if the calibration curve is linear then this is a constant).

The various molecular weight averages are then related to $W(M)$ in the following manner

$$\bar{M}_n = \frac{1}{\displaystyle\int_{M_1}^{M_\alpha} (1/M) W(M)\, \mathrm{d}M} \tag{2.39}$$

$$\bar{M}_w = \int_{M_1}^{M_\alpha} MW(M)\,\mathrm{d}M \tag{2.40}$$

$$\bar{M}_z = \frac{\displaystyle\int_{M_1}^{M_\alpha} M^2 W(M)\,\mathrm{d}M}{\displaystyle\int_{M_1}^{M_\alpha} W(M)\,\mathrm{d}M} \tag{2.41}$$

$$\bar{M}_v = \left[\int_{M_1}^{M_\alpha} M^\alpha W(M)\,\mathrm{d}M \right]^{1/2} \tag{2.42}$$

Computer software is available to perform the necessary calculations on the chromatographic data to yield the various molecular weight averages. Commercial equipment is now supplied with the automatic data processing capability built-in.

The ability of GPC to produce molecular weight distribution curves directly and to enable calculation of the average molecular weights, makes this an invaluable technique for polymer characterization. The ability to separate and identify low molecular weight fractions such as monomers, oligomers and additives such as stabilizers plasticizers etc. finds applications in a number of areas of polymer science and technology. The technique is finding increased uses for quality control purposes as well as detailed analysis in polymer syntheses and polymer processing. With some commercial instruments analysis of additives is facilitated by the use of gradient reverse phase chromatography using mixed solvents and gradient solvent facilities. In principle, the use of mixed solvents whose composition is varied during the course of the run permits better separation of otherwise difficult-to-separate compounds.

2.4 Conclusions

In this Chapter an attempt has been made to outline in some detail the various instrumental methods which are routinely available for the determination of polymer molecular weights. The need to determine molecular weights of polymers cannot be overstated since the essential characteristics of polymers derive from their macromolecular nature. The treatment has not been comprehensive but it is hoped that sufficient detail has been provided to enable the interested student to appreciate the essential principles and practice of molecular weight determination. Inevitably, some details have been omitted and the reader is referred to the texts listed at the end of this Chapter for more comprehensive treatments.

Further reading

1. Billmeyer, F.W. (1984) *Textbook of Polymer Science*, 3rd edn, Wiley, New York.
2. Billingham, N.C. (1977) *Molar Mass Measurements in Polymer Science*, Wiley, New York.
3. Collins, E.A., Bares, J., Billmeyer, F.W. (1973) *Experiments in Polymer Science*, Wiley, New York.
4. Johnson, J.F. and Porter, R.F. (1970) Gel Permeation Chromatography in *Progress in Polymer Science*, Vol. 2 (ed. A.D. Jenkins), Pergamon Press, Oxford.
5. Davison, G. (1982) Ebullioscopic Methods for Molecular Weights in *Analysis of Polymer Systems* (eds L.S. Bark and N.S. Allen), Applied Science, London.
6. Cooper, A.R. (1982) Analysis of Polymers by Gel Permeation Chromatography in *Analysis of Polymer Systems* (eds L.S. Bark and N.S. Allen), Applied Science, London.

Exercises

2.1 A sample of styrene is polymerized using, as initiator, ^{14}C labelled AZBN having an activity of 10^7 counts per minute per mole in a scintillation counter. Polystyrene (3.50 g) is found to have an activity of 105 counts per minute under the same conditions. Given that the mode of termination is by two-thirds combination and one-third disproportionation calculate the number average molecular weight of the polymer.

2.2 The following data were obtained for the osmotic pressure of dilute solutions of polystyrene in toluene at 378 K:

concentration (g l^{-1})   1.35 2.00 2.70 3.71 4.52 5.94
osmotic pressure (cm toluene) 1.46 2.24 3.16 4.52 5.74 8.10

Given that $R = 8.3143 \, \text{JK}^{-1} \, \text{mol}^{-1}$, and the density of toluene $= 785 \, \text{kgm}^{-3}$, calculate $\bar{M}_n$ for the polystyrene.

2.3 Solution viscosity measurements were carried out on a series of dilute solutions of polystyrene in toluene at 30°C

concentration
  (g per dl)   0   0.402   0.505   0.595   0.804   1.207
flow time (s)   67.94 107.70 121.05   132.77   161.39   227.84

Given that for polystyrene in toluene the Mark–Houwink constants $K = 12 \times 10^{-5}$ cm^3 g^{-1} and $\alpha = 0.71$, calculate the average molecular weight for the polymer.

2.4 The Rayleigh ratio for a series of dilute solutions of polymethylmethacrylate in ethylene dichloride at 25°C, was determined in a light scattering photometer at various angles θ.

The table shows values of $c/\Delta R_\theta$ for the various concentrations and scattering angles.

θ	c			
	0.0096	0.0048	0.0024	0.0012
30	56.3	35.9	26.4	21.4
45	57.1	36.4	26.7	21.5
60	57.5	36.8	26.8	21.8
75	58.3	37.5	27.6	22.6
90	59.1	38.4	28.3	23.6

Given $n = 1.5$, $dn/dc = 0.11 \, \text{cm}^3 \text{g}^{-1}$, $\lambda = 436 \, \text{nm}$ and Avogadro's Number $= 6.03 \times 10^{23}$. Calculate $\bar{M}_w$.

What other information may be derived from the measurements?

3

Introduction to spectroscopic techniques

3.1 General remarks

Many characterization techniques contain the word 'spectroscopy' in their title. In this book we concentrate on those in which measurements are made of the intensity of absorption or emission of electromagnetic (EM) radiation by the polymer samples under examination. For example, in the next chapter we consider the absorption of ultraviolet (UV) and visible light as it passes through a polymer: different parts of the EM spectrum are absorbed to different extents depending on the molecular structure of the sample. Measurement of the wavelengths at which absorptions take place can be interpreted in terms of the molecular structure.

Absorption or emission of photons occurs when an atomic or molecular system undergoes a transition between energy states. When a photon is absorbed its energy is used to promote the molecular system to a higher energy state; if the molecular system is already at an elevated energy it can drop to a lower state with the emission of a photon. Electronic, vibrational and magnetic states are all relevant in the spectroscopic analysis of polymers.* The electronic states have large energy separations and the photons involved in transitions between them lie in the UV–visible part of the EM spectrum. In round terms this corresponds to a range of wavelengths (λ) of approximately 100–1000 nm (Chapter 4). Vibrational states have lower energy separations and correspond to the infra-red (IR) range ($\lambda \sim 1\,\mu m$ to 1 mm; Chapter 5). Some rotational motions are located here too, but some have lower energy and appear in the microwave region. Transitions between different magnetic states (of a spinning nucleus or an unpaired electron) also require less energy, and can be promoted by photons in the microwave regions ($\lambda \sim 1$ mm to 1 cm; Chapters 6 and 7).

---

*The total energy E_{tot} of the system equals the sum of the several contributions,

i.e. $E_{tot} = E_{el} + E_{vib} + E_{rot} + E_{mag}$

where the subscripts stand for electronic, vibrational, rotational and magnetic respectively. The spectroscopy techniques described here concern transitions involving a change in only one of these.

3.2 Strategy in spectroscopy studies

Spectroscopic techniques may be used to identify a completely unknown polymer, to determine configurational states (e.g. in *cis-trans* isomers), to confirm the presence of a particular functional group and to detect free radicals or additives. Which spectroscopic technique is most appropriate to use depends on the nature of the problem. Most common examples are indicated in the following chapters and the reader is left to develop 'experience' with this. We do not attempt to present a set of rules, for this cannot be easily and succinctly achieved to cover every possible case.

Having decided which types of energy transitions are likely to reveal the structural features of interest (and hence which technique is to be used) the next step may be to do some theoretical modelling and make rough estimates of the energy levels of interest. This is possible with atoms and with some of the smaller molecules by means of simple models of the structure. Next must be investigated whether transitions between particular pairs of energy levels are allowed. For this 'selection rules' that are derived from quantum mechanics are required. When dealing with polymers complete theoretical analyses cannot be contemplated, but it is often possible to consider parts of the molecule (e.g. a functional group) as if in isolation, showing energy states that are not significantly perturbed by the existence of, and their attachment to, the rest of the molecule. These ideas will be expanded in the following chapters. Next it is necessary to conduct the measurement. The equipment used must contain a source of radiation in the appropriate range in the EM spectrum. A source having a range of wavelengths from which can be selected a narrow (monochromatic) sample, together with a suitable detection system, is often required.

Finally, the analysis of the results may be conducted entirely with reference to the theoretical energy level predictions mentioned above, (possibly with some modification derived from complementary experimental studies) but more often it is made with reference to data obtained on standard substances. The reference data can be presented in the form of tables indicating the prominent absorption (or emission) regions for particular atomic arrangements, or as graphs each showing a continuous spectrum for a particular compound. Recently it has become common to store the reference data in the memory of a microcomputer, often linked directly to the spectroscopic measurement equipment, facilitating rapid comparison of the subject material with a wide range of standard substances.

3.3 Energy level calculations

It is beyond the scope of this book to deal with this topic in any detail and we shall restrict coverage to very simple introductory models. No attempt is made to present the relevant quantum mechanics, and selection rules are introduced when needed without further explanation. There is a wealth of material published on these topics, but the reader with no background in this subject is referred to Atkins [1] for a basic introduction.

3.4 Properties of electromagnetic radiation

For a full description of the properties of EM radiation the reader should consult books on light optics. Here we shall simply summarize those relationships that are of importance in spectroscopy.

The energy (E) of a photon which has frequency v is

$$E = hv \qquad (3.1)$$

where h is Planck's constant, $= 6.626 \times 10^{-34}$ Js. The product of the frequency and wavelength (λ) is equal to the velocity of the radiation, (u), i.e.

$$u(n) = v\lambda(n) \qquad (3.2)$$

The velocity and the wavelength are both functions of the refractive index (n) of the medium in which the wave is travelling. In a vacuum all EM waves have the same speed (c) and if the wavelength refers to the value in vacuum we shall simply use the symbol λ so that $c = v\lambda$. It is popular in many branches of spectroscopy to present results in terms of the wavenumber ($\bar{v}$) which is the spatial frequency (i.e. the number of waves in a given distance)

$$\bar{v} = \frac{v}{c} = \frac{1}{\lambda} \qquad (3.3)$$

3.4.1 Quantification of absorption

When EM radiation passes through a medium, absorption causes the intensity to fall. Consider the passage of a beam of EM radiation through a parallel-sided block of material of thickness l, (Fig. 3.1). Let the intensity at depth x be $I(x)$. The intensity absorbed on passage through a further thin slice, length δx, will be proportional both to $I(x)$ and to δx, and can be expressed as $(-\delta I) = AI(x)\delta x$, where A is a proportionality constant. The intensity absorbed will also be proportional to the concentration of the absorbing species (c) so A can be replaced by αc where α is a measure of the effectiveness of the absorbing species, giving $(-\delta I) = \alpha c I(x)\delta x$.

Rearranging and taking δx and δI to limitingly small values we have $-dI/I = \alpha c dx$. Integrating and setting $I = I_0$ at the entrance ($x = 0$) and $I = I$ at the exit ($x = l$) we find

$$I = I_0 e^{-\alpha c l} \qquad (3.4)$$

This is known as the Beer–Lambert law, and is often written in the alternative

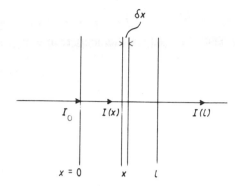

FIGURE 3.1 Passage of EM radiation, incident intensity I_0, through a parallel-sided medium of thickness l.

form

$$I = I_0 \, 10^{-\varepsilon c l}$$

where $\varepsilon = \alpha/\ln 10$ and is called the molar absorption coefficient, or the extinction coefficient of the absorbing species at the given frequency. (Note here that c is chosen to represent concentration despite its use earlier in this chapter to represent the velocity of light. This should not cause confusion, and is consistent with the nomenclature used in many other publications.)

3.5 Double beam optics

Most aspects of instrumentation are specialized and will be discussed in the chapters dealing with particular techniques. Double beam optics follows a general principle and is used in UV–visible and in IR spectrometers as well as in other related instruments such as microdensitometers. In a spectrometer that measures the intensity absorbed as a function of the wavelength of the EM radiation there are two problems that can be resolved by using double beam optics. First radiation sources do not have uniform intensity across the whole of the wavelength range, and this must be taken into account when assessing the spectral absorption characteristics of the sample. Secondly, in most spectrometers the wavelength range is scanned sequentially so that if the (overall) source intensity changes in time, the signal recorded for different wavelengths will contain an error corresponding to different values of incident intensity.

To overcome both of these problems the incident beam is split into two. One part passes through the sample and the other (the 'reference beam') passes through an optical path identical to the first one except that the cell through which the reference beam passes is filled with pure solvent. The difference in intensity between the sample beam and the reference beam is measured and provides information about the absorption of the sample free from errors arising from source beam intensity. In some instruments it is arranged that both sample and reference beam are measured in rapid sequence using the same detector to avoid problems with drift and variations in spectral response in the detector.

Reference

1. Atkins, P.W. (1986) *Physical Chemistry*, 3rd edn, Oxford University Press.

4

Ultraviolet–visible spectroscopy

4.1 Introduction

The UV–visible part of the EM spectrum corresponds to electronic excitation and the energy levels depend on the chemical bonds within the specimen. σ-electrons, involved in covalent bonds, absorb high energy photons in the UV region, whereas π electrons absorb photons at longer wavelengths, often in the visible region. The interaction between polymer molecules and the UV component in sunlight can lead to dissociation of σ-bonds and is the cause of photodegradation of polymers.

Although UV–visible spectroscopy is a common routine analytical technique in organic chemistry it is not capable alone of completely identifying an unknown compound, requiring other information, or measurements using other techniques. Its use in polymer analysis is still more restricted, though it performs certain tasks with great efficiency. It has the advantage of using very small samples, though the preferred state is in dilute solution, and with many polymers this cannot be conveniently arranged. Modern instrumentation permits very fast analyses to be performed with good accuracy and minimal training as long as a few simple rules relating to clean handling of the sample and all components in the optical path are carefully observed. The rapid analysis capability of the technique means that it can be adapted to make measurements on-line on flowing systems, one example of relevance to polymer analysis being eluent in GPC columns (see Chapter 2). It should be noted that solvent–solute interaction can shift the energy of the absorption bands for π-bonding and this must be taken into account when studying polymers in solution.

In addition to providing information about polymer molecules, the UV–visible technique may also be of value in studying additives in polymers. Pigments and UV stabilizers are two classes of additive which depend on their response to EM radiation within this range to provide the very property for which they are included. In such applications the choice of additive may be guided by reference to UV-visible studies, though the technique can be used equally in a diagnostic manner to indicate the concentration of such an additive and, possibly, to identify it. It is important to realize that such compounds may be present in any commercial plastic presented for analysis and that the spectrum obtained may contain large contributions from relatively small amounts of additive, a hazard unlikely to be met when analysing the product of a carefully controlled laboratory polymer preparation.

4.2 Instrumentation

The range of wavelengths with which we are concerned here is approximately 200–800 nm, of which 400–800 nm corresponds approximately to the visible region. Below about 185 nm wavelength the absorption by gas is strong, requiring the optical path to be kept under a vacuum, making instrument design more difficult and increasing the price considerably. Furthermore, few solvents are transparent in this region so that specimen form is often restricted to thin films. Thus studies in the 'vacuum ultraviolet' (approximately 100–185 nm wavelength) are less popular and used only when the information cannot be obtained in any other way; this is especially true with polymers, for which interpretation of the spectra is often not straightforward.

The general layout of a UV–visible spectrometer is shown in Fig. 4.1. Sources covering the full wavelength range 200–800 nm cannot be obtained so it has to be divided between two separate sources. Most instruments are arranged to scan smoothly through the full spectral range and switching from one source to the other can be achieved smoothly using a moving mirror, though in some instruments the light sources themselves are moved. The EM radiation next passes through a monochromator which can consist of a prism or a diffraction grating. This spreads out the spectral components, permitting a very narrow range to be sampled by a slit. The slit passes a finite bandwidth

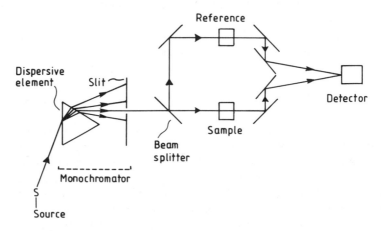

FIGURE 4.1 Layout of double-beam UV-visible spectrometer.

(the 'effective spectral width') which must be less than 10% of the natural bandwidth of the spectral line to be measured to avoid significant error in the peak intensity. This selected radiation then passes into the sample which, in the case of a solution, is held in a standard cell with flat windows. The intensity of radiation passing through the specimen (and the cell windows) is measured by a detector, commonly a photomultiplier. The sample compartment may have thermostatic control for the most accurate work. With polymers a thin parallel-sided solid specimen may be mounted in place of the solution cell if a suitable solvent cannot be found.

Double beam optics is usually employed, sometimes using a rotating sector mirror beam chopper that permits both sample and reference beams to be fed into the same detector in rapid succession. The optical path is controlled within the instrument using mirrors rather than lenses to avoid differential absorption by the instrument. When passage of EM radiation through a component is unavoidable, as with the sample cell windows, the material chosen must be transparent in the UV as well as in the visible range.

Photomultiplier performance has improved steadily over the years and has contributed greatly to the performance of UV–visible spectrometers. A fairly recent development in some advanced instruments has been the replacement of the photomultiplier detector by a diode array containing many tiny detectors that can sample simultaneously a large spectral range spread out across it, permitting very fast analysis. (With this modification, the 'monochromator' should contain only the dispersive element, but no selector slit at the exit.)

Further details of the design and operation of UV–visible spectrometers can be found in [1].

4.3 Theoretical estimation of electronic energy levels

One of the most useful applications of UV–visible spectroscopy to polymers is the measurement of the length of conjugation in unsaturated molecules. A modern application is in the analysis of rigid rod molecules with liquid crystal properties. The stiffness of the molecule derives from the presence of cyclic structures along the backbone. The uninterrupted conjugation lengths can be obtained using UV–visible spectroscopy.

The π-electrons in a conjugated system are delocalized,which means that they are not restricted to an atomic orbit, but can move freely within the conjugated region. The theory starts with the proposal that the behaviour of a π-electron can be modelled by a particle in a square potential well with infinite sides and whose length is chosen to coincide with the conjugation length within the molecule. Such a particle can exist anywhere within the well, but it cannot escape. This represents a good approximation because a π-electron has a very low probability of escape. To proceed further the reader must accept the

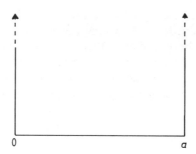

FIGURE 4.2 Square well potential function with infinite sides at $x = 0$ and $x = a$.

validity of the principle of particle–wave duality and that the state of the particle in the box must be capable of representation by a solution to Schrödinger's equation

$$\frac{-h^2}{8\pi^2 m} \frac{d^2\psi}{dx^2} + U(x)\psi = E\psi \qquad (4.1)$$

where h is Planck's constant; m is the mass of the particle (electron); ψ is the wave function describing the probability of its existence as a function of position and x is the spatial variable. $U(x)$ is the potential function (for a square well it is a constant within the region of interest and can be set equal to zero for convenience); and E is the total energy of the particle.

Let the conjugation length be a and choose the potential well to be located at $0 \leqslant x \leqslant a$ (Fig. 4.2). Thus ψ must disappear when $x < 0$ and $x > a$ because a particle cannot exist in regions of infinite potential. If only physically sensible solutions are to be accepted, the function describing ψ must be continuous and the solution within $0 \leqslant x \leqslant a$ must fall to zero at the boundaries $x = 0$, $x = a$. A solution to equation (4.1) which satisfies this condition is

$$\psi = A \sin \frac{n\pi x}{a} \qquad 0 \leqslant x \leqslant a \qquad (4.2)$$

where n is a whole number. No truly different solutions can be found. On substitution of equation (4.2) into equation (4.1) we find that

$$E = \frac{n^2 h^2}{8ma^2} \qquad (4.3)$$

This describes the possible energy states for the particle, and n is called the principal quantum number. Thus the energy levels are in the sequence shown in Fig. 4.3, which also shows how the functions ψ fit into the interval $0 \leqslant x \leqslant a$ at each level.

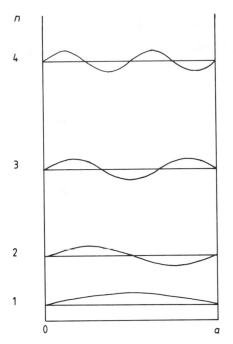

FIGURE 4.3 Wave functions, ψ, for principal quantum numbers $n = 1, 2, 3, 4$.

4.3.1 Application to molecular structures

In the majority of cases of interest there will be more than one π-electron present in the conjugated region. Although it might be expected on classical grounds that the lowest energy state for such a system would be obtained if each electron occupied the lowest energy level ($n = 1$), quantum mechanics predicts that this cannot happen. The quantum mechanical rule that describes the behaviour is the Pauli exclusion principle which states that no two electrons may occupy the same state. Electrons have spin and can exist in two possible spin states ('up' and 'down') so that each of the energy levels (orbitals) shown in Fig. 4.3 can be occupied by two electrons without violating the Pauli exclusion principle. Thus for 1, 3, 5-hexatriene which has 6 π-electrons, the lowest energy state that can be chosen is that in which two electrons are in each of the levels corresponding to $n = 1, 2, 3$.

The next task is to consider how a transition to another state might be achieved, with the absorption of a photon. In the case of 1, 3, 5-hexatriene, an electron from the $n = 3$ level could be promoted to the $n = 4$ level (Fig. 4.4). The energy difference between these two levels can be estimated using equation (4.3) as

$$E_4 - E_3 = \frac{(16 - 9)h^2}{8ma^2} = h\nu \qquad (4.4)$$

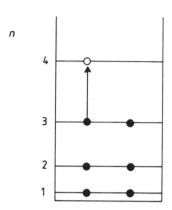

FIGURE 4.4 Energy level diagram for π electrons in 1, 3, 5-hexatriene.

where v is the frequency of the photon required to promote this transition. Using $a = 0.73$ nm to represent the length of the conjugated region in this molecule, and substituting for h and m ($= 9.1095 \times 10^{-31}$ kg) the frequency is estimated to be 1.195×10^{15} s^{-1}. This has wavelength 251 nm in a vacuum, or can alternatively be expressed as a wavenumber: 39 850 cm^{-1}.

4.4 Application to polymers

The measurement of the conjugation length is an important application of UV–visible spectroscopy that follows directly from the theoretical analysis presented in section 4.3. This clearly relates to the backbone of the molecule, and in the case of rigid rod molecules may indicate the length of the rigid part (between flexible units, if present).

Another application of the measurement of the conjugation length (a) is in the study of *cis-trans* isomers. In general, the *cis*-form is more likely to show steric hindrance, restricting the three-dimensional arrangement of the atoms, whereas the *trans*-form can exist as an extended coplanar structure more easily, allowing an extensive π-electron system with a larger value of a than is observed with the *cis*-form, giving a correspondingly high absorption wavelength ($\lambda \propto a^2 \propto 1/v$, see equation (4.4)). The presence of substituents may alter the absorption wavelength, and may also alter the difference in the spectra obtained with *cis*- and *trans*-isomers.

One of the commonest conjugated systems is the six-membered carbon ring, as found in benzene or as a phenyl group (a constituent in many polymer molecules). This gives a characteristic family of absorptions, though overlap with other absorptions produced by other structures on the polymer molecule may sometimes obscure it or cause unambiguous identification to be impossible. Other groups of atoms may give characteristic absorptions that may be used to establish their presence, and, if the spectrum is not cluttered with absorptions from other sources, may be used to measure the abundance of the groups in question. A group giving rise to such a characteristic absorption is known as a 'chromophore' and an example is the carbonyl group, C=O. Here the electron transition is of the type $n \rightarrow \pi^*$. In addition to the characterization of polymers that contain carbonyl groups, this may be utilized in studying degradation in other polymers. Carbonyl groups form in many polymers when they degrade in air, often simply during outdoor weathering. Poly(vinyl chloride) (PVC) is an example, though the degradation process also leads to main chain unsaturation which itself leads to UV absorption. Thus with PVC an absorption band in the wavelength range 270–285 nm may be used to follow the progress of degradation.

Another application of UV–visible spectroscopy is in the analysis of copolymers. If the two (or more) monomers possess characteristic absorptions that do not overlap then the relative intensities of the absorptions measured on the copolymer formed from them can be used to estimate composition. It is not important whether the absorption bands correspond to conjugated backbone structures or to pendant chromophores, merely that they are distinguishable in the copolymer absorption spectrum. It is, of course, necessary that there are no major shifts in intensity or spectral position of one unit due to the presence of the other. Suppose the copolymer contains two monomer species, '1' and '2'. If the molar absorption coefficients are measured for the two monomers (or the corresponding homopolymers) at the characteristic wavelengths as ε_1 and ε_2, and for the copolymer as ε_c, then

$$\varepsilon_c = x\varepsilon_1 + (1 - x)\varepsilon_2$$

where x is the mole fraction of monomer '1' present in the copolymer. Rearrangement gives

$$x = (\varepsilon_c - \varepsilon_2)/(\varepsilon_1 - \varepsilon_2).$$

Application of this method has been found of value when different techniques for determining copolymer composition have been in disagreement.

Another similar application is in the determination of the fraction of unpolymerized styrene in a polystyrene sample. In this case x represents the weight fraction of unreacted styrene, ε_c is replaced by the molar absorption coefficient of the monomer polymer mixture, and ε_1 and ε_2 represent molar absorption coefficients for unreacted styrene and polystyrene respectively.

Reference

1. Knowles A. and Burgess C. (eds) (1984) *Practical absorption spectrometry.* Chapman and Hall, London.

5

Vibrational spectroscopy: infra-red and Raman spectroscopy

5.1 Fundamentals of vibrational spectroscopy

In attempting to identify and characterize polymeric materials almost invariably recourse will be made to some form of vibrational spectroscopy. When used either alone or in conjunction with other physicochemical techniques, vibrational spectroscopy is capable of providing detailed information on polymer structure [1, 2].

Vibrational spectroscopy is concerned with the detection of transitions between energy levels in molecules which result from vibrations of the interatomic bonds. The vibrational frequencies are shown to be characteristic of particular functional groups in molecules. They are sensitive to the molecular environment, chain conformations and morphology and so afford a useful method for polymer analysis.

At room temperature most molecules exist in their ground vibrational states and in order to excite them to higher vibrational states energy must be absorbed. The two experimental techniques which are used to detect changes in vibrational energy states are infra-red spectroscopy (IR) and Raman spectroscopy. Although both spectroscopic methods are concerned with vibrations in molecules, they differ in the manner by which interaction with the exciting radiation occurs. The following sections are intended to give a general understanding of the principles and practice of vibrational spectroscopy. It is not the intention to present a comprehensive treatment of the basic theory of vibrational spectroscopy so the interested reader is advised to consult more detailed texts [3, 4].

5.1.1 Infra-red spectroscopy

When molecular vibrations result in a change in the bond dipole moment, as a consequence of change in the electron distribution in the bond, it is possible to stimulate transitions between energy levels by interaction with EM radiation of the appropriate frequency. In effect, when the vibrating dipole is in phase with the electric vector of the incident radiation the vibrations are enhanced and there is transfer of energy from the incident radiation to the molecule. It is the detection of this energy absorption which constitutes IR spectroscopy. In practice, the spectral transitions are detected by scanning through the frequency whilst continuously monitoring the transmitted light intensity. The energies of molecular vibrations of interest for analytical work correspond to EM wavelengths in the range 2.5–25 μm or, when expressed conventionally in terms of wavenumber, 4000–400 cm^{-1} (frequency in wavenumber (cm^{-1}) = 10^4/wavelength in μm). Some spectrometers operate in the near infra-red region (NIR) 0.7 to 2.5 μm (4000 to 1400 cm^{-1}) and others in the far infra-red (FIR) 50 to 800 μm (200 to 12 cm^{-1}).

5.1.2 Raman spectroscopy

Raman spectroscopy is concerned with detection of light scattered inelastically due to interaction of molecules with incident monochromatic radiation. When EM radiation interacts with matter a certain fraction of the incident radiation is scattered elastically (Rayleigh scattering) such that its frequency remains unchanged, but a very small fraction of the radiation will interact inelastically and is scattered with a different frequency (Raman scattering). The differences between the wavelengths of the scattered and incident radiation arise due to induced transitions of the vibrational states of the molecules. The transitions which occur are illustrated in Fig. 5.1, where $v = 0$ represents the ground state and $v = 1$ the excited vibrational state. Interactions with the incident monochromatic radiation of frequency v_0 results in light being scattered with frequencies v_0 (Rayleigh scattering), $v_0 - v_{vib}$ (Stokes lines) when the transitions from $v = 0$ to $v = 1$ occur and $v_0 + v_{vib}$ (anti-Stokes lines) when

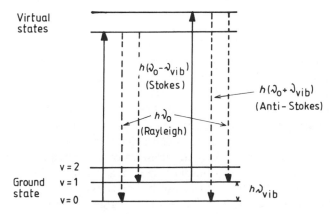

FIGURE 5.1 Energy transitions for elastic (Rayleigh) and inelastic (Stokes) light scattering.

transitions are induced from the excited to the ground state. As a consequence of the interactions the scattered radiation contains information on the vibrational states in the molecules. The essential prerequisite for Raman scattering is a change in the polarizability of the bond when vibrations occur. Polarizability may be thought of as a measure of the ease with which the electron distribution in the bond can be changed under the action of an external stimulus such as an electric field.

In practice the sample under examination is illuminated with monochromatic radiation in the visible region and light scattered inelastically with wavelengths close to that of the incident radiation is detected. In most cases the scattered radiation is observed at 90° to the incident beam but for some sample types back scattering is preferred. The scattered light is about 9 orders of magnitude less intense than the incident light, so that high intensity sources (lasers) and very sensitive detection techniques are required. The Stokes lines are more intense than the anti-Stokes lines so that measurements are normally made on the low frequency side.

5.1.3 Near infra-red spectroscopy (NIR)

Although not a widely used technique, NIR has been applied to the characterization of polymers. NIR spectra arise from overtones of the fundamental vibrations and hence the technique is complementary to IR and Raman spectroscopy and has found some limited applications for polymer analysis. NIR has the advantage of relatively simple sample preparation. Glass has a low absorption in the near infra-red region so glass sample holders may be used. Much thicker samples, up to about 10 mm, may also be used. In addition, since most of the spectral lines detectable in the NIR arise from vibrations of hydrogen-containing bonds, hydrogen-free solvents such as CCl_4 may be used with little difficulty.

5.1.4 Vibrations in molecules

If we consider the number of degrees of freedom that a molecule, treated as a rigid body, possesses we can see that there are normally three translational and three rotational degrees of freedom. For a molecule consisting of n atoms, the number of vibrational degrees of freedom is then $3n - 6$ (for linear molecules, where two rotational modes are equivalent, this is increased to $3n - 5$). For large molecules such as polymers it is fortunate that many of the vibrations are equivalent so that the number of distinguishable vibrational states is reduced and the problem of assignments is greatly simplified. Put in another way, the vibrational spectrum becomes essentially independent of polymer molecular weight for relatively low degrees of polymerization.

The vibrations of interest in macromolecules may be considered by reference to the methylene group which occurs in many polymers, notably polyethylene. These are of the form:

(a) stretching   (i) symmetrical   (ν_s)

                              (ii) asymmetrical   (ν_{as})

(b) in-plane bending   (i) scissor   (δ)

                              (ii) rocking   (δ_r)

(c) out-of-plane bending   (i) twisting   (γ_t)

                              (ii) wagging   (γ_w)

Similarly, for other commonly encountered functional groups such as –OH the following vibrational modes are important.

(a) in-plane bending   —C—O   (δ)

(e) out-of-plane bending —C—O   (γ_w)

For polymers such as polystyrene aromatic ring expansion or breathing give useful characteristic absorptions.

It is possible to predict from first principles the expected vibration spectra for a molecule whose basic structure is known or presumed. The approach which is taken for relatively simple molecules involves application of group theory which permits the expected normal vibrations to be determined from the symmetry of the molecule and also predicts which vibrations are IR and which are Raman active. A normal coordinate analysis then permits the force constants to be estimated for each vibration and by comparison with observed spectra detailed assignments of the absorption peaks can be made. For polymers the lower degree of symmetry makes the theoretical analysis more difficult although a number of detailed investigations have been carried out which have facilitated the interpretation of vibrational spectra. It is not considered appropriate to develop this aspect in the present text and the reader is referred to other comprehensive treatments [5].

It has already been stated that activity in the IR requires changes in dipole moments whereas Raman activity involves changes in polarizability of the bonds so that not all vibrations are detectable by both techniques. In general it may be shown that where a molecule possesses a centre of symmetry, for example O=C=O, no vibration gives rise to both IR and Raman absorptions. For less symmetrical molecules some vibrations are found to be active in IR, some are active in Raman whereas others are inactive. It is also shown that for vibrations which are active in both Raman and IR, asymmetric vibrations are strong in IR but give only weak absorptions in Raman. For symmetric vibrations, the reverse is found to be the case. It follows that detailed analysis of the vibrational modes in polymers requires the complementary use of both IR and Raman spectroscopy.

5.1.5 *Group frequencies*

Although it is possible to predict the expected vibrational spectra of molecules, most spectroscopic investigations rely on a much more empirical approach. The procedure followed makes use of the concept of group frequencies and is based on the fact that the vibrational frequencies of particular chemical groups in molecules, e.g. C=O, $-CH_3$, tend to behave largely independently of the rest of the molecule of which they are a part. In other words the degree of mechanical or electrical coupling between vibrations of atoms in the group in question to those in neighbouring groups is minimal. Consequently the absorption frequencies for a particular chemical group are essentially constant and the absorption bands appear in roughly the same region of the spectra. This means that it is possible in most cases to assign particular absorption bands to vibrations in the groups by reference to standard correlation tables [6]. However it is very important to recognize that where coupling between neighbouring groups occurs to a significant extent, the spectral shifts which result provide valuable information on the molecular environment of the particular chemical group. The effect of coupling will be discussed in slightly more detail with reference to chain conformations in section 5.3.1(c). The magnitude of the coupling effect is smallest when the atoms of neighbouring groups have significantly different masses or when the bond vibrations possess relatively large force constants. Since the force constants are greater for bond stretching vibrations than for bending deformations it follows that the absorption frequencies for stretching modes are essentially invariant whereas those for bending modes show more significant shifts due to coupling. Some understanding of these effects is necessary when assigning bands to particular groups in molecules.

5.1.6 *Quantitative analysis*

The variation of light intensity on passing through an absorbing medium may be expressed in terms of the Beer–Lambert Law:

$$A = \log \frac{I_0}{I} = \varepsilon c l$$

where A is the absorbance; I_0 is the incident light intensity; I the intensity at a depth l in the absorbing medium; c the concentration of the absorbing species; and ε is a constant for the material termed the absorptivity or extinction coefficient.

Values of absorbance range from zero when there is no absorption ($I = I_0$) to infinity when there is complete absorption of the incident radiation ($I = 0$). Most spectrometers display the spectrum as percentage transmittance against wavelength or wavenumber.

Transmittance is defined as $T = I/I_0$ so that percentage transmittance $= 100I/I_0$ and absorbance $A = \log 1/T$. In principle determination of the concentration of the absorbing species can be made by direct application of the Beer–Lambert Law provided that the extinction coefficient for the species at the particular wavelength is known. It is more usual, however, to use comparative methods for quantitative analysis of polymers as is the case even for low molecular weight materials. Nevertheless, there are some situations where extinction coefficients may need to be determined to facilitate quantitative analysis. A problem arises with the physical form in which the polymeric materials are available. Quantitative analysis is best carried out in solution but although solutions of some polymers may be prepared at appropriate concentrations (section 5.2.3(a)), this is not generally done. More usually there is need to examine polymers in the solid state, e.g. as thin films or in the form of KBr discs or Nujol mulls. Difficulties encountered in ensuring uniform film thickness or homogeneous distribution of absorbing species (particularly additives) and optical effects such as extraneous light scattering means that there is a need for great care when attempting absolute measurements.

More reliable results are usually obtained when it is possible to compare the sample under investigation with a standard reference material located in the reference beam of dual beam spectrometers. In some cases it is possible to use internal reference standards. A particular example of the latter approach is in the analysis of copolymers where absorption peaks characteristic of the different comonomers may be compared directly. Thus the vinylacetate content of EVA copolymers can be determined by measurement of the relative intensities of the acetate absorption at 9.8 μm and the methylene absorption at 13.8 μm and comparing with calibration curves for known compositions determined using other absolute methods.

5.2 Experimental techniques

5.2.1 Instrumentation

(a) INFRA-RED SPECTROMETERS

Almost all commercial IR spectrometers operate in the double beam mode and only these will be considered here. The essential components of double beam spectrometers have been described in Chapter 3. The source emits radiation over the whole IR region and is usually a mercury lamp (globar).

It is necessary to use more than one monochromator for the following reasons: (i) the photon energy decreases with increasing wavelength; (ii) as the spectral range is extended higher order reflections interfere; (iii) occurrence of extraneous absorption arising from lens and window materials and by water vapour in the atmosphere. Typical detectors include thermopiles, bolometers or Golay cells. Any absorption of radiation by the sample results in a difference in intensity between the two beams. The absorption is usually measured using an optical null method. The pulsating signal is amplified to give an AC signal which is used in turn to drive an attenuator into the path of the reference beam in order to equalize the sample and reference beam intensities. The movement of the attenuator is arranged to be directly proportional to the sample absorption and by suitable electronic means can be used to display the percentage transmission as a function of wavelength or wavenumber. In some spectrometers two separate choppers are used for the sample and reference beams in order to compensate for any extraneous IR emissions from spectrometer components.

(b) FOURIER TRANSFORM SPECTROMETERS

Interferometry was first used to investigate the otherwise inaccessible far infra-red (FIR) region of the spectrum. Developments in Fourier transform spectroscopy (FTIR) have extended applications of interferometry to the whole IR region particularly for rapid sampling in such applications as gel permeation chromatography (Chapter 2).

The basic construction of interferometer instruments is illustrated in Fig. 5.2. The light source is a high intensity mercury lamp or heated wire which emits an approximately continuous spectrum. The light from the source passes into the Michelson interferometer which is at the heart of the spectrometer (Fig. 5.2(b)). The essential components of the interferometer are the two mirrors, one fixed and the other moveable and the beam splitter which is set at an angle of 45° to the path of the collimated beam. The beam is partially transmitted and partially reflected at the thin dielectric beam splitter. The transmitted and reflected beams are incident normal to the two mirrors and following reflection, are recombined at the beam splitter where they produce interference effects. The beam is then passed through the sample and on to the detector. If the mirrors are located equidistant from the beam splitter or when the optical path difference is an integral number of wavelengths, the reflected beams are in phase and so produce constructive interference. Conversely when the optical path difference is an odd number of half wavelengths, destructive interference will occur. When the moveable mirror is moved axially, relative phase displacement occurs resulting in an oscillatory

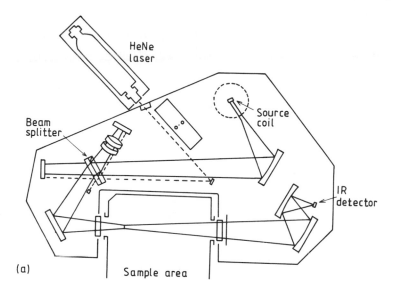

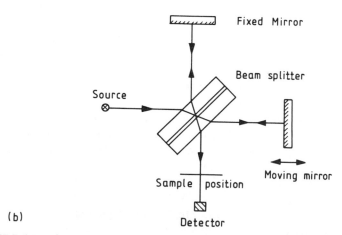

FIGURE 5.2 Schematic of IR interferometer (a) general arrangement of optics (b) Michelson Interferometer (courtesy Perkin Elmer).

pattern or interferogram, which is a representation of the spectral distribution of the absorption signal, reaching the detector.

The interferogram is an intensity function of the path difference x. If the moveable mirror is moved at a constant velocity, the output signal is modulated in a manner dependent on the velocity of the mirror and the spectral distribution of the absorption signal. The form of an interferogram is

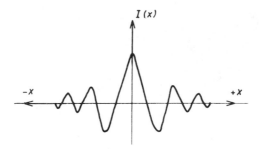

FIGURE 5.3 Typical form of IR interferogram.

shown in Fig. 5.3. The true absorption spectrum is derived from the inter-ferogram by a Fourier transform of the form

$$G(v) = \int_{-x_1}^{+x_1} I(x) \cos 2\pi v x \, \mathrm{d}v$$

where $G(v)$ is the spectral distribution at a given frequency (v). The transformation is carried out on a range of intensity values $I(x)$ at different path lengths x to give the spectral intensity at the given frequency. The signal must be sampled at precise path differences. This is achieved by using a He/Ne laser whose beam traverses the same optical path and the interference produced by the laser is used to trigger the sampling of the IR signal. Typical detectors include Golay cells or the more recent lithium tantalate or deuterated triglycine sulphate detectors. The Fourier transformation is usually carried out by means of a built-in dedicated microprocessor, enabling the absorption spectrum to be produced in a relatively short time.

The movement of the mirror determines the resolution of the spectra in the different IR frequency ranges. Rapid scan speeds ($3 \, \mathrm{mm \, s^{-1}}$) are used for the middle infra-red (MIR) region (4000 to $400 \, \mathrm{cm^{-1}}$), whereas much lower scan speeds ($0.005 \, \mathrm{mm \, s^{-1}}$) are necessary in the FIR region. Interferometers used in the MIR range also differ from those used for FIR spectroscopy in using a different source (Nernst filament), using semiconductor dielectrics supported on KBr as the beam splitter, and employing a different detector system (pyroelectric device).

(c) RAMAN SPECTROMETERS

The essential components of a Raman spectrometer are shown in Fig. 5.4. We have seen that when light passes through a transparent medium a fraction (1 in 10^4) is scattered elastically (Rayleigh scattering) and only a very small fraction (1 in 10^8) is scattered inelastically (Raman scattering). The very low intensity of the Raman scattering necessitates the use of high intensity monochromatic light. Most spectrometers now employ laser light sources. The

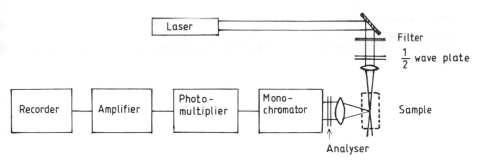

FIGURE 5.4 Schematic of typical Raman spectrometer.

three most common lasers used in Raman spectrophotometers are argon ion, krypton ion and helium–neon gas lasers. Table 5.1 lists the more intense plasma emission lines for the different lasers used in Raman spectroscopy.

Because of the low intensity of the Raman scattering, the laser beam must be focused very accurately on to the specimen; focused beams of the order of microns in diameter are not untypical. The scattered light is usually observed at 90° to the incident beam with the 180° configuration being used less frequently for transparent specimens. Measurement of the low intensity Raman frequencies against the background of the Rayleigh scattered incident light requires the use of very sensitive and high resolution monochromators. In practice double and even triple monochromators are used to give the necessary discrimination and to reject stray light. Both grating and prism monochromators find use in Raman spectrometers. Great care is essential in setting up the instrument to ensure efficient collection of the Raman scattered light. The intensity of the scattered light is measured using very sensitive photomultiplier tubes. A photomultiplier consists of a photosensitive cathode which emits electrons when photons strike it. The electrons are accelerated by an applied electric field and cause the emission of secondary electrons in the

TABLE 5.1 The more intense plasma emission lines of lasers used in Raman spectroscopy

Source	Source wavelength (nm)	Spectral range*(nm)
Argon ion	488	488–953
	514.5	514.5–1060
	568.2	568.2–1316
Krypton ion	530.8	530.8–1131
	647.1	647.1–1834
Helium–neon	632.8	632.8–1724

* To give relative spectral range 3–$10\,000\,\text{cm}^{-1}$

form of a cascade by striking an assembly of dynodes. The electrical pulses are amplified and when the spectral range is scanned the variations in pulse light constitutes the Raman spectrum. The slit settings are fixed so that there is no compensation for variations in photon energy with the result that the resolution of the instrument varies continuously with wavelength. Manufacturers' recommended slit settings should be followed to ensure maximum resolution for different laser frequencies. It should also be noted that the resolution decreases with increasing scan speed.

(d) ATTENUATED TOTAL REFLECTANCE (ATR)

When light is incident at an interface between materials having different refractive indices (n) such that the angle of incidence exceeds the critical angle, α_c, where $\sin \alpha_c = n_2/n_1$ then the light is reflected from the surface rather than refracted (Fig. 5.5(a)). At the interface the electric field component of the EM radiation penetrates to some small extent into the surface layers of the less dense material (n_2). The depth of penetration depends both on the wavelength of the light, increasing with increasing wavelength, and on the relative values of the refractive indices and is typically of the order of several microns. As a consequence, light having wavelengths corresponding to the absorption spectrum will be absorbed in proportion to the depth of penetration. This implies that there will be greater absorption of the longer wavelength light.

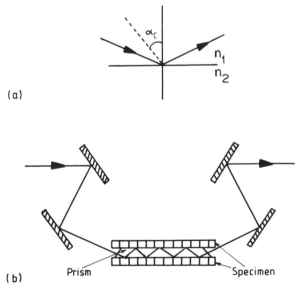

(a)

(b)   Prism   Specimen

FIGURE 5.5 (a) Total internal reflection of light when angle of incidence $> \alpha_c$, (b) optical arrangement for multiple reflection ATR.

The spectral distribution of the reflected light is then representative of the surface regions of the material. The effect can be used to examine in particular the surface layers of polymeric materials and so provide information on, for example, surface oxidation, or to obtain spectra of samples having very high absorptivities not readily examinable by transmission spectroscopy.

In order to obtain measurable absorption spectra, it is normal practice to use multiple reflection prisms (Fig. 5.5(b)) and these are available as standard spectrometer accessories. A common material for the prisms is thallium bromoiodide (KRS5). A trapezoidal prism having an angle of incidence of $45°$ giving 25 reflections is useful for polymeric materials. The sample is clamped securely in order to provide good optical contact with the prism and, where possible, the sample should be placed on both surfaces of the prism to give optimum sensitivity.

5.2.2 Infra-red dichroism and Raman polarization

Spectroscopic studies employing polarized light sources are used to provide valuable structural information particularly on chain orientation. Specifically this involves determination of the extent of IR dichroism and/or Raman depolarization.

The maximum absorption of IR radiation occurs when the electric vector of the incident EM radiation is parallel to the dipole of the vibrating group and is zero in the perpendicular direction. The actual absorbance measured is an average of all possible orientations in the sample. For a specimen in which the molecules are orientated randomly, the measured absorbance is independent of the polarization of the incident light. For orientated specimens the observed spectra will differ depending on the direction of the plane of polarization relative to the molecular orientation. The effect is conveniently expressed by the dichroic ratio $R = A_{/\!/}/A_{\perp}$, where $A_{/\!/}$ and $A_{\perp}$ are the absorbances defined with respect to reference axes in the sample, typically the orientation direction.

For IR dichroic measurements the polarizer is located in the beam just before it passes into the monochromator. Rotation of the polarizer enables the absorption spectra to be recorded with the plane of polarization successively parallel or perpendicular to the reference axis. When the degree of orientation is small such that R is approximately equal to unity, a differential method can be used to give $A_{/\!/} - A_{\perp}$ directly. The method requires the use of two polarizers, one in each of the sample and reference beams, oriented $\pm 45°$ to the monochromator slit (to reduce polarization effects of the monochromator). The sample is placed in the recombined beam so that it is illuminated in turn with light having its plane of polarization parallel then perpendicular to the reference axis. An additional measurement of, say, $A_{\perp}$ permits R to be determined from $A_{/\!/} - A_{\perp}/A_{\perp} = R - 1$.

The directional properties of Raman spectra are determined both by the symmetry of the vibrations and by the orientation of the molecules relative to the plane of polarization. Even for randomly orientated molecules such as in liquids, the intensities of Raman lines are dependent on the polarization of the incident beam and the symmetry of the vibrational modes. For orientated specimens the intensity of the scattered light is dependent also on the molecular orientation relative to the plane of polarization of the incident radiation as well as on the direction and plane of polarization of the scattered radiation. Polarization effects are expressed quantitatively by the depolarization ratio (ρ) defined as $\rho = I_{//}/I_{\perp}$ where $I_{//}$ is the intensity of the light scattered with its plane of polarization in the same direction as that of the incident light, and $I_{\perp}$ is the intensity of light having its plane of polarization at right angles to that of the incident light. A band is considered as being depolarized when $\rho = 0.75$ and polarized for $\rho < 0.75$.

It is fortuitous that the light emitted by gas lasers is highly plane polarized. For depolarization studies the polarization of the scattered light is determined relative to incident monochromatic radiation. This is achieved by placing the analyser between the sample and the entrance slit to the monochromator. The use of a half wave plate in the beam permits the plane of polarization of the light to be rotated independently of the analyser position.

5.2.3 *Specimen preparation*

(a) INFRA-RED SPECTROSCOPY

Polymers are usually examined in the solid state and the nature of the technique imposes some limitations on the form of sample which may be examined. In general solids have strong IR absorptions which means that the samples have to be relatively thin or the absorbing species only present in low concentration when dispersed in a transparent medium. The simplest physical form is thin film prepared either by solvent casting or by hot pressing. Film thicknesses are typically in the range 0.001 to 0.05 mm. Films can be cast directly from solvent on to the surface of a sodium chloride or potassium bromide disc, the salts being transparent in the IR region. Alternatively the films may be cast on to a substrate such as glass or polyethylene film from which they may be stripped for subsequent mounting in a sample holder for examination in the spectrometer. Immersion in an ultrasonic bath is useful for detaching overly adherent films.

Hot pressed films may be prepared between heated metal plates using spacers to determine the thickness. An even simpler method involves pressing the sample between glass microscope slides on a hot plate using manual pressure. It is difficult to control film thickness by this method but it is rapid and usually adequate for qualitative analysis of unknown samples.

Standard sample preparation techniques used for IR examination of non-polymeric materials including filled KBr discs are also useful when examining polymeric materials. The method requires that the KBr and the polymer sample are finely ground, dispersed to give an approximately 1% concentration of the polymer then pressed in a die using a combination of pressure and vacuum to produce a thin disc. A problem for most polymers arises due to their toughness which means that it is difficult to grind them to a sufficiently fine particle size and it is usually necessary to cool in solid CO_2 or liquid N_2. Care must be taken to desiccate the disc since KBr is hygroscopic. The use of mulls (dispersions of powder in liquids such as paraffin oil or Nujol) although feasible is not generally recommended for polymers.

Thin sections may also be removed from bulk specimens using microtomes (Chapter 11). Cooling to low temperatures is often necessary. The technique is useful when used in conjunction with dichroic ratio measurements for the investigation of orientation in mouldings.

Quantitative analysis is best carried out in solution because application of the Beer–Lambert Law is then simplified. However this presents some problems for polymer systems since they have only low solubility and it is difficult to find suitable solvents which are sufficiently transparent to IR radiation. It is possible in some cases to use several different solvents which absorb in different regions of the spectrum and carry out several measurements in order to examine the whole of the polymer spectrum. Alternatively it is possible, by placing solvent in the reference beam, to compensate for the solvent absorptions. The latter method requires high concentrations for accuracy and is not generally recommended. Additional problems arise from the need to use a container for the solution which must be transparent to the radiation. These problems are reduced considerably when using FTIR or NIR spectroscopy when it is even possible to examine aqueous solutions which absorb very strongly in the normal IR region.

(b) RAMAN SPECTROSCOPY

There are fewer limitations when preparing specimens for Raman spectroscopy. When colourless specimens are examined illumination using laser light in the visible region is fairly straightforward. Since visible light is not absorbed by glass, specimens may be contained in glass sample holders and, if necessary, liquid samples may be sealed. There are also few restrictions on specimen size provided that they can be brought into good optical contact with the sample lens. Greatest sensitivity is usually obtained by using glass capillary cells (15 cm × 1 mm ID) having rounded ends to ensure good optical contact. The scattered light is normally examined 90° to the incident monochromatic beam when using these sample holders. For polarization studies larger diameter tubes are used in order to reduce multiple reflections from the cell walls which may cause some depolarization of the light.

Solid samples may be examined in the bulk or as powders but the experimental procedure depends very much on the light scattering characteristics of the material. For solid, turbid materials the scattered light is normally

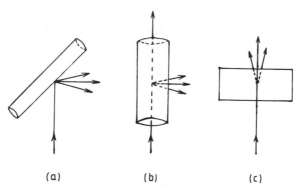

(a)   (b)   (c)

FIGURE 5.6 Schematic showing different arrangements for detecting Raman scattering: (a) front surface scattering, (b) 90° scattering, (c) forward scattering.

detected by front surface reflection whereas powders packed in glass ampoules are examined by 90° scattering (Fig. 5.6).

Problems sometimes arise due to fluorescence from impurities in the glass sample holders, or in the polymer samples. Fluorescence may be 'burned out' by prolonged exposure to the laser light prior to displaying the spectrum. The use of high purity quartz obviates much of this problem, but sample fluorescence may still be encountered.

Coloured samples may require more careful treatment since absorption of the high intensity visible light leads to localized heating and possible thermal degradation of the polymer. Heating effects can be reduced by defocusing the laser beam so reducing the overall intensity but this also results in some loss of sensitivity.

In contrast to its IR spectrum, water has only a weak Raman spectrum so that examination of aqueous solutions can be undertaken with little difficulty. This can be useful not only for water soluble polymers but also for aqueous dispersions and emulsions.

A number of accessories has been reported, some available commercially, for both IR and Raman spectroscopy, which permit measurement over a range of temperatures and pressures and for examining samples subject to mechanical stress.

5.3 Applications to polymers

Although IR and Raman spectroscopy are most widely used in qualitative and quantitative analysis of polymers and additives, the potential these techniques offer in producing information on chain conformations, stereochemical structure, crystallinity and orientation of polymers render them more generally useful in polymer characterization. Infra-red spectroscopy is the more readily available of the techniques and it is likely that Raman spectroscopy will remain a specialized technique for some time to come. Nevertheless from what we have already seen, the two techniques are complementary and taken together provide means for detailed investigation of polymer structures.

5.3.1 Infra-red analysis

(a) IDENTIFICATION OF POLYMERS AND ADDITIVES

Complete qualitative analysis of polymeric materials implies identification of all the constituents including additives such as stabilizers, fillers, plasticizers, etc. This is likely to be a fairly complex process involving a number of different analytical techniques as appropriate to the different components. For many purposes it may only be necessary to identify the polymeric components but even this will involve extraction or separation procedures in order to remove interference from overlapping spectral lines. In most cases separation procedures involve either solvent extraction of the additive or dissolution and reprecipitation of the polymer. Soxhlet extraction is a convenient means of separation when the polymer and additive(s) have different solubilities in given solvents. Use of a non-solvent for the polymer permits extraction of soluble additives whereas use of a solvent for the polymer leaves insoluble additives behind. Even if complete separation is not achieved, partial separation is often adequate to permit identification of the polymer and/or selective subtraction of the spectra.

For example, Fig. 5.7 shows IR spectra from hot pressed films of plasticized PVC before and after extraction with acetone. Comparison with the spectrum of the plasticizer reveals that the differences arise due to extraction of the plasticizer which may itself be separated from the acetone solution and its spectrum determined independently.

An alternative approach is to use a spectral subtraction technique. Knowing, for example, the total spectrum of the polymer and a particular band due to the polymer alone the contribution the polymer makes to the complex spectrum can be subtracted leaving the spectrum of the additive(s). The subtraction process is best carried out by use of a computer. In principle the procedure can be extended to enable the contributions from a number of different components to be determined.

Identification of an unknown polymer almost inevitably involves detection of characteristic absorption bands due to particular chemical groups. Reference is then made to group frequency correlation tables and the spectra compared with reference standards. Thus in effect the spectra act as a 'fingerprint' for the polymer. The almost unique nature of IR spectra may be judged by comparing spectra from a series of polyolefins (Fig. 5.8), where substitution of one of the hydrogen atoms in polyethylene by alkyl groups leads

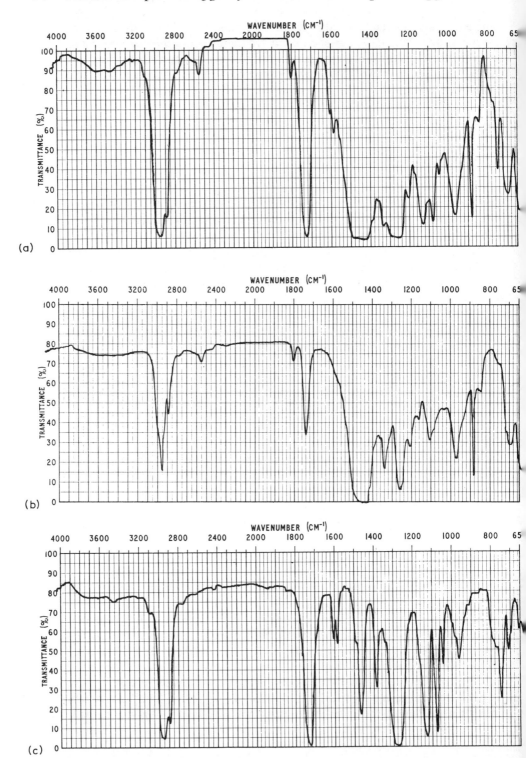

to markedly different spectra. It should be noted that structural differences between some polymer types, for example polyamides, do not necessarily give such marked differences.

As we have seen the concept of group frequencies which is being applied here, depends on the existence of only a small degree of coupling between vibrations so that particular chemical groups in different molecules give rise to absorptions in the same region of the spectrum. This empirical approach has proved to be effective by a great deal of practical experience and a number of excellent atlases of spectra and IR correlation tables is now available. In principle the same general approach can be applied to Raman spectroscopy but its role in polymer identification is limited by the paucity of reference spectra currently available. No doubt with time this limitation will be overcome as more experimental data are generated. This should prove particularly beneficial in the analysis of those polymers such as the polyamides which are difficult to distinguish by IR but whose Raman spectra show characteristic features.

(b) COUPLING EFFECTS

Although the concept of group frequencies is based on there being minimal interaction between vibrations and neighbouring groups, it is important to recognize that the coupling that does occur permits detailed structural information to be derived. In the absence of intermolecular or intramolecular interactions, it would be perfectly possible to assign particular IR absorptions to specific chemical groups, but the positions of the absorption peaks would then be largely insensitive to the molecular environment of the group. For example, the presence of an ester group in an aromatic polymer such as poly(ethylene terephthalate) would not be distinguishable from that in a long chain aliphatic polyester. Identification of the polymer would still be possible by examination of the total spectrum but the amount of structural detail would be less. In particular, evidence for different chain conformations, stereochemical arrangements or morphology would be largely absent in the spectra. Fortunately, the interactions that do occur between vibrating groups leads to shifts in the absorption frequencies which may be interpreted in terms of the molecular environment of the particular group.

The most marked interactions occur intramolecularly with adjacent groups and interchain effects are found to be slight. As a general rule coupling effects are greatest for deformational modes (bending) than for stretching vibrations and they are also reduced when the vibrating atoms have significantly different masses. It is well to recall at this point that coupling may affect both peak intensity, when the coupling is essentially electronic, or position when

FIGURE 5.7 IR spectra: (a) plasticized PVC, (b) (a) following extraction with acetone, (c) DIOP plasticizer.

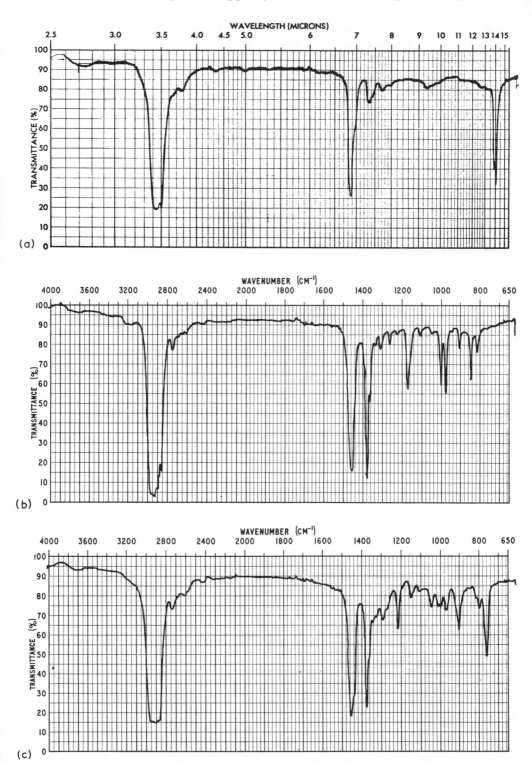

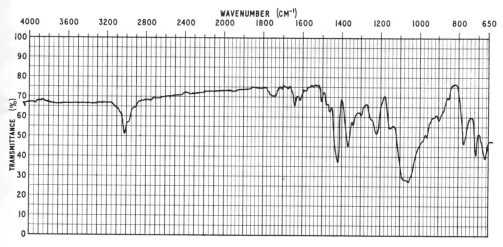

FIGURE 5.9 IR spectrum of poly(vinylidene chloride) film.

the coupling is largely mechanical. This may be seen by comparing the IR spectra of PVC (Fig. 5.7) and poly(vinylidene chloride) (PVDC) (Fig. 5.9). In addition to other spectral changes, the intensity of the $-CH_2$ stretching vibration at $2900-3000 \, cm^{-1}$ is reduced in PVDC compared to PVC and the $-CH_2$ bending absorption is shifted from 1450 to $1400 \, cm^{-1}$ as a result of the interaction with the $-CCl_2$ group. Coupling effects can also be exemplified by comparing the IR spectra of blends of polyethylene and polypropylene with that of an ethylene–propylene copolymer (Fig. 5.10). The former spectrum is the result of simple superposition of the individual homopolymer spectra whereas the copolymer spectrum differs as a result of the intrachain coupling between the different copolymer sequences.

(c) CONFORMATIONAL STUDIES

Coupling effects play a central role in characterizing polymers which differ in their stereochemical structure or spatial arrangement of the polymer chains. The most pronounced coupling effects are found to arise from conformational isomerism. This is in marked contrast to other spectroscopic techniques such as NMR where conformational effects are slight. The consequences of different conformational changes may be seen from the following. Consider first of all the different conformations which may be assumed by a disubstituted ethylene molecule. When viewed along the C–C bond and assuming that rotation is

FIGURE 5.8 IR spectra: (a) low density polyethylene, (b) isotactic polypropylene, (c) isotactic polybutene-1. In (a), the spectrum is plotted against wavelength for comparison.

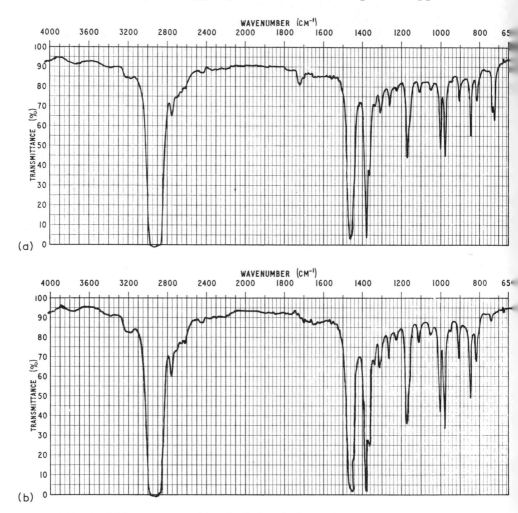

FIGURE 5.10 IR spectra: (a) polyethylene/polypropylene blend, (b) polyethylene-propylene random copolymer.

possible about this bond, there are three conformations which give minima in the potential energy versus angle of rotation curve (Fig. 5.11). These conformations are designated *trans* and *gauche*. It is not too surprising that the respective conformations influence the vibration frequency of the C–X bond such that it is split into a doublet, corresponding to the *trans* and *gauche* conformations respectively. For bonds situated in larger molecules such as polymers, the conformations of both adjacent groups need to be considered (Fig. 5.12). The C–X vibrational frequency may then differ depending on whether the bond is *trans* with respect to the C–H bonds on the two adjacent carbon atoms or *trans* to the chain carbon atom. In principle each of these

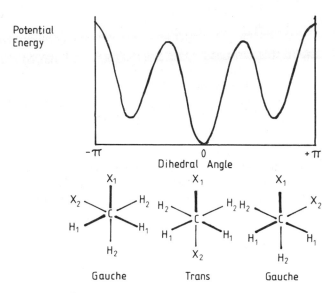

FIGURE 5.11 Potential energy curve for different conformations of di-substituted ethylenes; viewed along C–C bond.

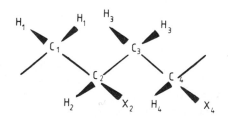

FIGURE 5.12 Different conformations of segment of chain of a substituted polyethylene.

conformational arrangements may give rise to splitting of the IR absorption peaks. Since there is likely to be a distribution of different conformations along a polymer chain, this can result in very complex spectra. Nevertheless there has been some notable success in interpreting spectra on the basis of chain conformations. The IR spectra of PVC will serve as an example. For syndiotactic PVC the all *trans* planar zig-zag arrangement, in which the C–Cl bond is *trans* to the C–H bond, the stretching vibrations are observed to occur at 603 and 639 cm^{-1}. By contrast, the *trans-gauche, trans-gauche* sequence in isotactic PVC results in absorptions at 695 cm^{-1}. Other bonds have been assigned in a similar manner. Improved spectral resolution is likely to reduce some of the problems associated with accidental overlap of lines and the distribution of chain conformers but complete analysis will not be possible in all but a few cases.

(d) STEREOCHEMICAL STUDIES

We have used the example of PVC in its different tactic forms to illustrate the effects of conformation on IR spectra. It is quite reasonable to ask whether the differences observed could be explained equally well on the basis of the stereochemical structures or possibly of differences in morphology which also occur. It is found, however, that the observed differences can be attributed almost entirely to conformational changes. However it is also true that the conformational structures are themselves dependent on and hence indicative of the basic stereochemistry of the polymers. As a consequence, IR spectroscopy is capable of being used to determine, albeit indirectly, the stereochemical structure of polymers. For example, the IR spectra of solid isotactic, syndiotactic and atactic forms of polypropylene differ in detail (Fig. 5.13) and may then be used to identify different tactic forms of the polymer. Evidence that the differences are a function of conformation is given by the spectra of the molten polymers which are essentially identical even though the tacticities are unaffected by the melting process. Despite the fact that the spectral changes are interpreted in terms of chain conformations, IR spectra are used successfully to identify different tactic forms of polymers.

(e) CRYSTALLINITY IN POLYMERS

The crystal structures of isotactic and syndiotactic polypropylene also differ, with the former forming 3:1 helix whilst the latter crystallizes in the form of either a 4:1 helix or in a planar zig-zag conformation. Atactic polypropylene on the other hand is non-crystalline. Again it is reasonable to question whether or not the observed IR spectral differences arise due to the different

---

FIGURE 5.13 IR spectra of different tactic forms of polypropylene: (a) isotactic, (b) syndiotactic, (c) atactic.

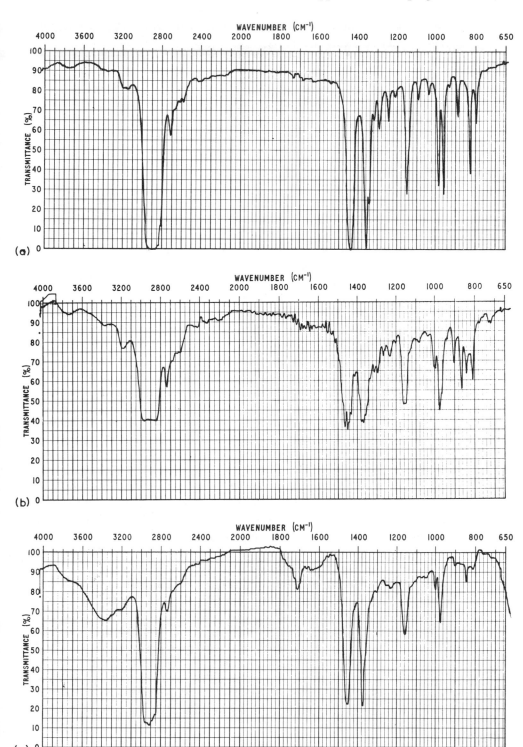

morphologies. For this to be the case it would be necessary to presume that significant intermolecular coupling occurs. It may be shown, however, that intermolecular coupling is slight and only becomes at all significant when more than one chain passes through the unit cell. In these special cases coupling between adjacent chains can give rise to splitting due to the effects of in phase and out of phase vibrations, and is termed Davidov splitting. In most cases however, although specific intermolecular splitting has been observed at low temperatures, the IR spectra can be adequately interpreted in terms of chain conformations.

(f) IR DICHROISM

Both IR dichroism and Raman polarization have proved useful in studies of molecular orientation in polymers. An essential prerequisite in applying vibration spectroscopy is an understanding of the relationship between molecular orientation and the direction of the bond dipole moments. In the simplest approach it is relatively straightforward to correlate the dichroic ratio R with the orientation of the vibrating bonds. Thus for example, in stretched samples of polyamides the measured value of $R = 0.67$ for the N–H stretching vibration is consistent with the N–H bond lying perpendicular to the chain direction. By comparison, the smaller extent of dichroism, $R = 0.8$, for the C=O stretch implies that the –C=O dipole moment is not colinear with the bond direction. A more detailed analysis is possible from the application of Group Theory, which defines the expected dichroism of particular vibrating species, preferably when applied in conjunction with Raman polarization studies.

The extent of chain orientation may also be derived by considering changes in IR dichroism with increase in draw ratio. Since this is not strictly an absolute method it should not be too surprising that differences are found between the degree of orientation predicted by IR measurements and that calculated for example, using X-ray diffraction methods. It is probable that the observed discrepancies reflect different conformations of the oriented chains. In addition, for semicrystalline polymers, differences in the degree of orientation in crystalline and amorphous regions introduce some further complication. In this context it is also necessary to distinguish between orientation of lamellar crystallites and chain orientation. For low draw ratios the chain direction may not coincide with the draw direction but with increased draw ratio fibrillar structures, in which the chain direction lies in the draw direction, are produced.

The influence of orientation and conformational changes on IR spectra consequent upon mechanical deformation also promises to be an additional tool in the investigation of mechanical behaviour of polymers. Polymer chains subjected to stress are known to give rise to shifts in peak positions, peak shape, dichroism and polarization and possibly the formation of additional absorption bands particularly as a result of chain scission. The ability to observe changes directly by stressing samples in the spectrometer, particularly using FTIR, provides the opportunity for dynamic measurements to be made.

We see therefore that in terms of polymer characterization IR spectroscopy may be employed to identify unknown polymers and by means of conformational coupling effects, permit investigations of stereochemical structures, orientation and morphology of polymers. It also follows that the effect of variables which produce changes in these properties such as temperature, processing parameters, annealing times etc. may be monitored by following changes in the vibrational spectra.

5.3.2 *Raman analysis*

Raman spectroscopy is a less common experimental technique but one which has its own particular advantages to complement IR spectroscopy. One particular advantage we have seen lies in the fewer constraints on sample size and form which means that the technique may be used for samples such as aqueous solutions not readily amenable to IR analysis. Perhaps of much more relevance are the fundamental aspects of Raman spectroscopy which permit the detection of vibrations inactive in the IR. On the other hand, some IR active vibrations are inactive in Raman and these differences provide important confirmatory evidence when making assignments of particular absorptions. Although Raman spectroscopy is less likely to be used for routine identification purposes, there is nothing in principle to preclude its use. At the present time it is limited by the lack of comprehensive reference spectra. Nevertheless the use of Raman as a specialized analytical technique for characterizing polymers will certainly increase.

Superficially IR and Raman spectra determined for the same material are quite different because of the different vibrational modes active in each and some care has to be exercised in comparing the spectra. The relatively simple example of polyethylene will serve to illustrate expected differences. The Raman spectrum obtained using the 514.5 nm exciting line of the argon ion laser is shown in Fig. 5.14. Prominent peaks are evident at $2900 \, \text{cm}^{-1}$ (doublet) and $1460 \, \text{cm}^{-1}$ which are readily assigned to the $-CH_2$ stretching and deformation modes respectively and are consistent with IR absorptions. By contrast, the IR active $-CH_2$ rocking modes (720 and $730 \, \text{cm}^{-1}$) are inactive in the Raman. Of perhaps more interest are the additional Raman lines which appear in the region of 1070 to $1300 \, \text{cm}^{-1}$ which are shown to arise from vibrations of the $-C-C-$ backbone and which are absent in the IR. These Raman-active skeletal modes may be particularly useful in characterizing polymers since they are likely to be sensitive to microstructural changes. Careful examination of the Raman scattered light having frequencies close to the excitation line frequency reveals lines which arise from characteristic backbone vibrations. Specifically these are a consequence of what are termed longitudinal acoustic modes or accordion vibrations (Fig. 5.15). There is good

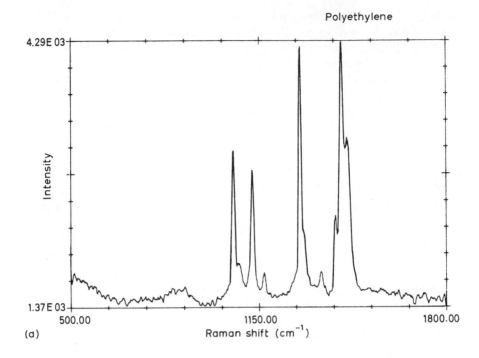

(a)

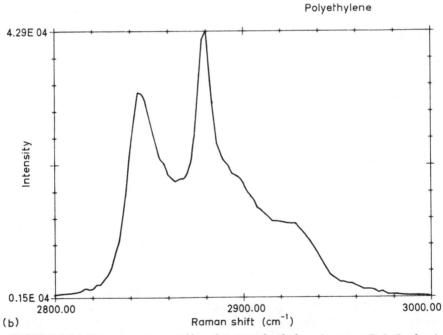

(b)

FIGURE 5.14 Raman spectrum of low density polyethylene (courtesy D. J. Gardner).

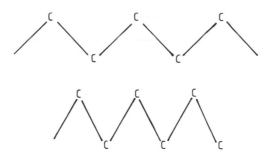

FIGURE 5.15 Main chain accordion vibrations.

evidence that these skeletal vibrations are sensitive to crystallite lamellar thickness. Using a simplified model in which the accordion vibration is represented by a vibrating spring, it is possible to estimate the length (L) of the planar zig-zag accordion from the equation

$$v = \frac{n}{2L} \frac{E}{\rho}$$

where v is the vibration frequency, n is the order of the vibration, E is Young's modulus and ρ is the density. Alternatively the approach can be used to estimate a value for Young's modulus of the extended chain. For quenched specimens the calculated L is in agreement with the long spacing calculated from small angle X-ray diffraction measurements. As expected, the estimated lamellar thickness increased with annealing time and temperature. It is likely that examination of these low frequency vibrations in Raman spectra will provide detailed information on lamellar structure even though not easy experimentally.

The skeletal vibration modes have also proved of value in the identification of polyamides. The IR spectra of different nylons (Fig. 5.16) are dominated by the absorptions of the amide groups. Comparison of the spectra reveals few characteristic differences between them such that unequivocal assignments to particular polyamides are difficult. The Raman spectra (Fig. 5.17) contain additional peaks due to the polymethylene chain segments which differ sufficiently for the different nylons to permit more definite identification.

Raman spectra are also sensitive to chain conformations in much the same way as IR spectra. In Fig. 5.18 the Raman spectra of various tactic forms of polypropylene are shown. It is instructive to compare Fig. 5.18 with Fig. 5.13 to emphasize the differences between IR and Raman spectra. The observed differences in the Raman spectra are also attributed in large part on the basis of conformational arrangements. Thus it is found that Raman spectra of the different tactic forms of polypropylene become essentially equivalent in the melt in much the same way as the previously noted IR spectra. However despite the fact that the differences between the different tacticities arise from

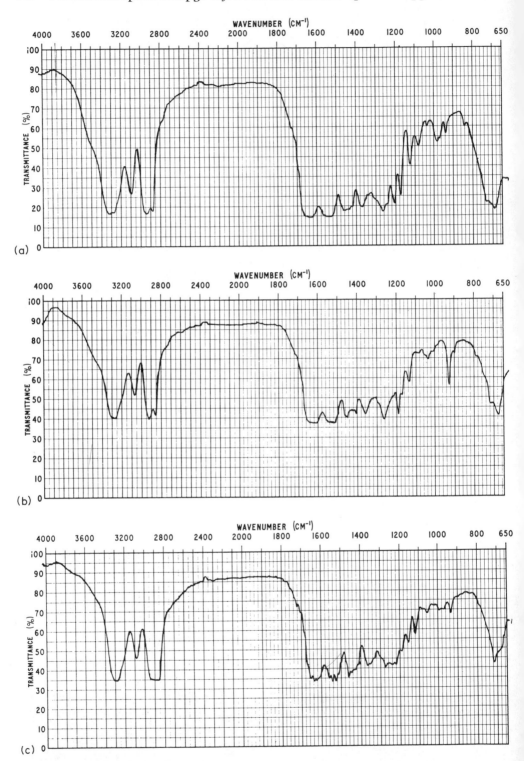

differences in chain conformation, Raman spectroscopy may be used to distinguish different stereochemical structures.

Changes in conformation which accompany phase transitions offer another field of application for Raman spectroscopy. Poly(tetrafluoroethylene) (PTFE) exhibits a phase transition at 19°C which involves a change in conformation from a 13:1 helix to a 15:1 helix, with consequent changes in the Raman spectrum. The spectra have been analysed in some detail using group theory and have been shown to be consistent with the different conformations. Depolarization studies of a transparent sample of hexafluoropropylene-tetrafluoroethylene copolymer provided confirmatory evidence for the assignments. Although not strictly comparable, data from similar structures or model compounds may, with care, be used to assist in assignments.

(a) DEPOLARIZATION STUDIES

In principle Raman spectroscopy depolarization is capable of providing detailed information on orientation in polymers. Raman spectroscopy has the additional advantage that, since only a small sample volume is necessary to give measurable spectra, examination of microtomed sections through sample thickness allows orientation distributions to be determined, rather than simply average orientation values. The approach taken is to determine the extent of depolarization ($I_\perp/I_\parallel$) with respect to the molecular axis for anisotropic Raman bands. Depolarization studies are difficult to carry out with accuracy using opaque specimens due to scattering effects, so the method is really only applicable to transparent materials.

Quantitative analysis is much more problematical when using Raman spectroscopy. In contrast to IR spectroscopy where the Beer–Lambert Law applies, the intensity of Raman bands is directly proportional to the concentration of the absorbing species. As a consequence it becomes necessary to carry out absolute intensity measurements with the attendant experimental difficulties. It is feasible, however, to carry out comparative methods using well characterized bands in the spectrum of the sample, or of internal standards. This approach has been used successfully to determine copolymer compositions by monitoring bands specific to the two components in the copolymer.

5.3.3 *Near infra-red (NIR)*

NIR is not a routine analytical technique but it has found some applications in polymer characterization. As with IR, the technique may be used to identify polymers and additives by reference to standard spectra and applying the

FIGURE 5.16 IR spectra of different polyamides: (a) Nylon 6, (b) Nylon 6.6, (c) Nylon 11.

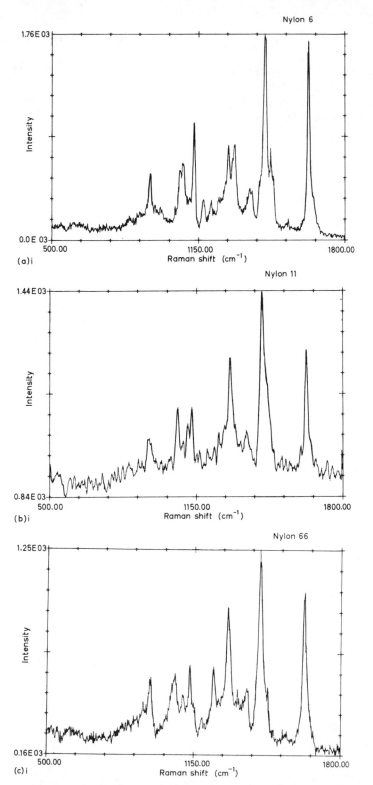

FIGURE 5.17 Raman spectra of different polyamides: (a) Nylon 6, (b) Nylon 6.6, (c) Nylon 11 (courtesy D.J. Gardner).

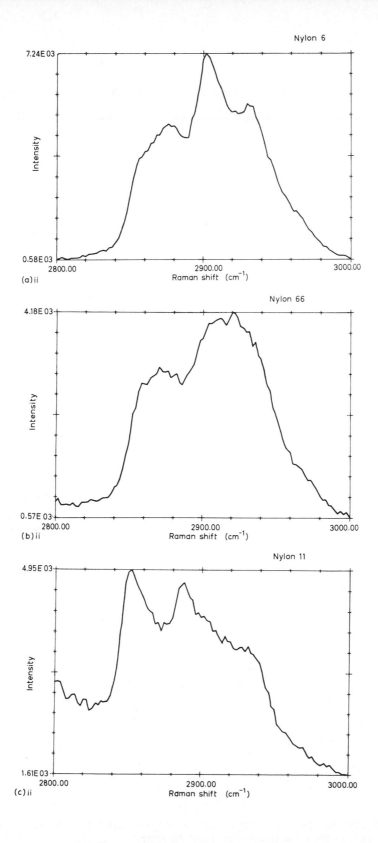

Nylon 6

(a)ii

Nylon 66

(b)ii

Nylon 11

(c)ii

81

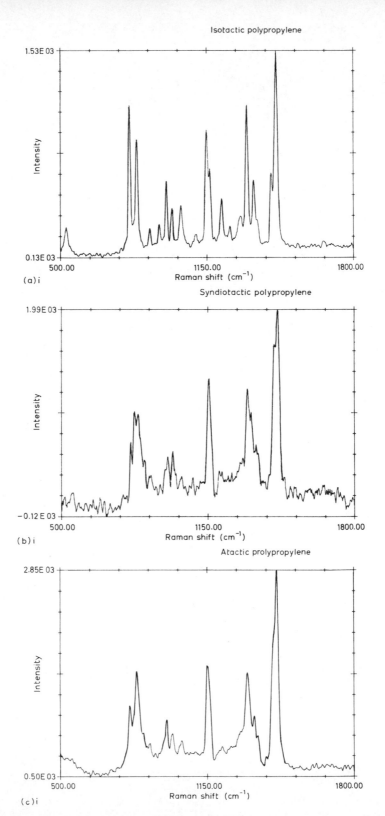

FIGURE 5.18 Raman spectra of different tactic forms of polypropylene: (a) isotactic, (b) syndiotactic, (c) atactic (courtesy D.J. Gardner).

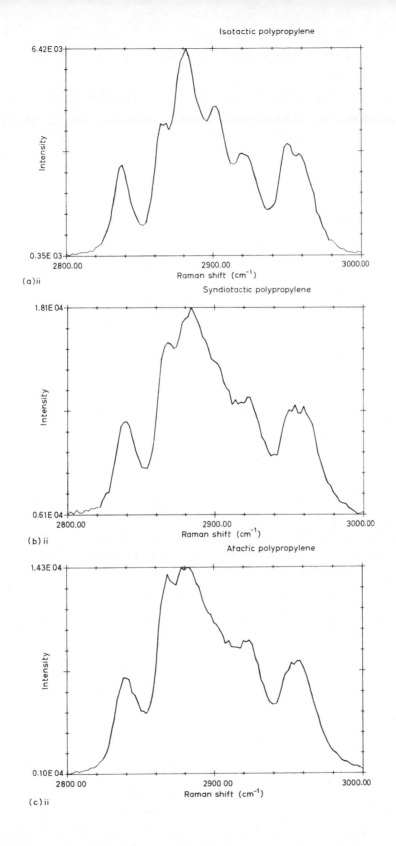

Isotactic polypropylene

(a)ii

Syndiotactic polypropylene

(b)ii

Atactic polypropylene

(c)ii

83

concept of group frequencies. Copolymer compositions may be determined by measurements of the absorption of characteristic functional groups. Measurements of dichroic ratios provide information on orientation in polymers. Kinetic studies of chemical reactions in solution have also been reported so that the technique is likely to offer a number of useful applications in the study of polymers.

Conclusions

In the foregoing some of the applications of vibrational spectroscopy to polymer characterization have been considered. There is a range of other important applications which should also be noted. Infra-red spectroscopy in particular is used to investigate the chemical reactions of polymers including oxidation and environmental degradation (weathering), thermal degradation and pyrolysis so important to understanding in-service behaviour of polymers. Developments in the use of FTIR are particularly encouraging in that they enable dynamic or kinetic measurements to be made much more readily. FTIR is also used in combination with gel permeation chromatography to permit identification and characterization of the various eluted fractions. Identification of characteristic end groups in polymer chains by means of IR is a well established technique for molecular weight determination.

References

1. Siesler, H.W. and Holland Moritz, K. (1980) *Infrared and Raman Spectroscopy of Polymers*, Marcel Dekker, New York and Basel.
2. Maddams, W.F. (1982) Infrared and Raman Spectroscopy in *Analysis of Polymer Systems*, (eds L.S. Bark and N.S. Allen), Applied Science, London.
3. Hummel, D.O. (1974) Vibration Spectroscopy in Polymer Spectroscopy, (ed. Dieter O Hummel), Verlag Chemie, Weinheim.
4. Painter, P.C., Coleman, M.M. and Koenig, J.L. (1982) *The theory of vibrational spectroscopy and its application to polymeric materials*, Wiley, New York.
5. Zbinden, R. (1964) *Infrared spectra of high polymers*, Academic Press, New York.
6. Haslam, J., Willis, H.A. and Squirrel, D.C.M. (1980) *Identification and analysis of plastics*, Heyden, London.

6

Nuclear magnetic resonance spectroscopy

6.1 Introduction

Nuclear Magnetic Resonance (NMR) spectroscopy is now well established as one of the most useful instrumental techniques for characterizing polymeric materials. Even the earliest applications to the study of solid polymers, which were limited due to line broadening effects and consequent low resolution, provided valuable information on the structure and properties of polymers. The very significant developments in instrumentation and experimental techniques which have occurred over the past 20 years have resulted in improvements in sensitivity and resolution of spectra so that NMR now provides an invaluable means of probing molecular structure. Polymer scientists and technologists have taken full advantage of these developments and important strides have been made in our understanding of the detailed chain structure of polymers and copolymers and of the morphology and transitions in the solid state. NMR has also provided a means for identifying intermediate structures formed during polymerization reactions so permitting more detailed reaction mechanisms to be proposed. When used for end group analysis, NMR also provides a useful absolute method for determining average molecular weights (Chapter 2). There is little doubt that the techniques of NMR will continue to prove to be amongst the most useful and versatile analytical methods to assist in our understanding of the structure and properties of polymeric materials.

This chapter will present an outline of the basic resonance phenomenon, including magnetic interaction, so important in structure determination, instrumentation and experimental techniques and will discuss particular examples of application to the characterization of polymers. It is not intended to give a comprehensive treatment of the theory and experimental aspects since there are many excellent texts available and the reader is referred to them [1–3]. The following is intended to provide sufficient background information to enable the reader to appreciate the ways in which NMR may be used and its potential for characterizing polymers.

6.2 Principles of magnetic resonance

6.2.1 Nuclear spin and quantization of spin states

NMR spectroscopy, in common with other forms of spectroscopy, is concerned with the detection of the absorption or emission of EM radiation by stimulated transitions between energy levels in the system under investigation. Since the energy levels are influenced by the environment of the nuclei, the resulting spectrum gives more or less direct evidence regarding the chemical nature of interacting atoms both qualitatively and quantitatively. NMR arises from the interaction of the applied electromagnetic radiation with nuclear spins when the energy levels in the latter are split by an external magnetic field. An essential prerequisite, then, is that the material contains nuclei which possess nuclear spin. It is found that atoms or isotopes whose nuclei contain either odd numbers of protons or odd numbers of neutrons, possess nuclear spins and so are detectable by NMR, whereas atoms or isotopes containing even numbers of both protons and neutrons have zero spin and consequently are not detectable by NMR (Table 6.1). Nuclear spin is specified by a non-zero value of the nuclear spin quantum number (I) which can take integral or half integral values. For the vast majority of polymeric materials, only ^1H exists in high concentration and has been the subject of the majority of applications of NMR to date*. However, important developments in the technique have enabled other important constituent elements in polymers whose magnetic nuclei only exist naturally in low abundance to be investigated, e.g. ^{13}C, ^{19}F, ^{15}N.

In an analogous manner to that of the electron spin, the nuclear spin has associated with it angular momentum which is characterized by the angular momentum vector $(\mathbf{I})$. The magnitude of $\mathbf{I}$ is related to the nuclear spin quantum number (I) by

$$\mathbf{I} = \frac{h}{2\pi}\sqrt{I(I+1)} \tag{6.1}$$

TABLE 6.1 Magnetic properties of nuclei encountered in NMR studies of polymers

Nucleus	I	g_N	γ_N	Resonance frequency (MHz) for $B = 1.0\,T$
^1H	$\frac{1}{2}$	5.5854	26.753	42.577
^2H	1	0.8574	4.107	6.536
^{13}C	$\frac{1}{2}$	1.4042	6.728	10.705
^{14}N	1	0.8072	1.934	3.076
^{15}N	$\frac{1}{2}$	-0.566	-2.712	4.315
^{19}F	$\frac{1}{2}$	5.2576	25.179	40.055
^{31}P	$\frac{1}{2}$	2.261	10.840	17.235

*Fluoropolymers are exceptions and are dealt with in section 6.3.6.

where h = Planck's constant. Since the nuclei also possess electric charge, they likewise possess a magnetic moment which may be expressed by the magnetic moment vector $\boldsymbol{\mu} = \gamma\mathbf{I}$, where γ is termed the magnetogyric ratio, and lies in the same direction as the angular momentum vector.

It is sometimes convenient to express the magnetic moment in the form

$$\mu = g_N\beta_N I \tag{6.2}$$

where g_N is a dimensionless constant, $\beta_N = eh/4\pi mc$ is the nuclear magneton in which e and m are the charge and mass of the proton respectively, and c is the velocity of light. In this form the expression is similar to that used to describe the magnetic moment of the spinning electron (Chapter 7).

In the absence of an applied magnetic field, the magnetic moment vector is randomly orientated and the spinning nuclei all possess the same energy. When the spinning nuclei are placed in a strong magnetic field, however, the magnetic moment vectors become aligned by interaction with the magnetic field. Quantum theory tells us that only certain orientations of the magnetic moment vector with respect to the magnetic field are permissible. These permitted orientations are given by the values of the magnetic quantum number m_I which takes only those values for which $m_I = I, I - 1, \ldots - I$. One way of thinking of this is that the projection of the spin angular moment vector in the field direction is only allowed to take certain values. That is to say that the angular momentum and hence magnetic moment are both quantized. The allowed values of the angular momentum in the field or z-direction, in units of $h/2\pi$ are

$$I_z = m_I h/2\pi = m_I\hbar \tag{6.3}$$

where $\hbar = h/2\pi$.

The corresponding values of the magnetic moment are then

$$\mu = \gamma m_I\hbar = g_N\beta_N m_I \tag{6.4}$$

Since most of the situations of interest in polymers involve nuclei for which $I = \frac{1}{2}$, it will simplify the discussion if we consider the resonance phenomenon by reference to such nuclei. For $I = \frac{1}{2}$ the nuclear quantum number m_I can only take the two values $m_I = +\frac{1}{2}$ or $-\frac{1}{2}$ and may be taken to represent situations where the nuclear spins are aligned respectively anti-parallel and parallel to the magnetic field direction. The alignment of the magnetic moments provides the energy separations necessary for the resonance phenomenon.

6.2.2 The resonance condition

The effect of placing an isolated magnetic moment in a strong magnetic field results in a change of energy which is derived from the torque experienced by the spinning nucleus, i.e. $E = -\mu B_0$ where B_0 is the applied magnetic field

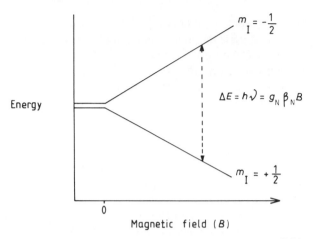

FIGURE 6.1 Energy levels of proton nuclear spins in a magnetic field.

(Fig. 6.1). In classical electrodynamics, this is equivalent to stating that the potential energy of the nucleus changes by

$$E = -\gamma \hbar m_I B_0 = g_N \beta_N m_I B_0 \tag{6.5}$$

For nuclei for which $I = \frac{1}{2}$ there are then two possible energy states corresponding to $m_I = \pm\frac{1}{2}$ with an energy difference

$$\Delta E = \gamma \hbar B_0 = g_N \beta_N B_0 \tag{6.6}$$

Transitions between the two energy states are induced by illuminating the sample with an oscillating electromagnetic field of frequency v_r, applied perpendicular to the applied field B_0 such that

$$h v_r = \Delta E = g_N \beta_N B_0 \tag{6.7}$$

For the strong fields used in NMR studies, v_r is in the radio frequency range.

Absorption will be detected if there is a net excess of nuclear spins in the lower energy state. In a collection of nuclei the spins are distributed between the two energy states according to Boltzmann's Law which expresses the ratio of the number of spins in the upper energy state (N_U) to that in the lower energy state (N_L) by

$$\frac{N_U}{N_L} = \exp\left(\frac{-\Delta E}{kT}\right) \approx 1 - \frac{\Delta E}{kT} = 1 - \frac{g_N \beta_N B_0}{kT} \tag{6.8}$$

In practice ΔE is very small. For example for the proton in a typical magnetic field of 2.3 Tesla (T), $\Delta E = 6 \times 10^{-12}$ Jmol^{-1} so that the excess of nuclei in the lower energy state is only about 10^{-5}.

In order to maintain equilibrium there must be simultaneous relaxation of the excited spins back to the ground state, which necessitates transfer of energy to the surroundings or lattice. This process is characterized by a relaxation time, termed the longitudinal or spin–lattice relaxation time T_1. The magnitude of T_1 is important in determining the saturation behaviour of NMR signals. The signal intensity is inversely proportioned to T_1, i.e. intensity = constant/$B_1 T_1$ where B_1 is the strength of the RF field. Thus long relaxation times and strong oscillating fields reduce the signal intensity by saturation.

In addition to spin-lattice relaxation, there is a second important relaxation mechanism namely spin–spin or transverse relaxation T_2 which is particularly relevant with respect to line broadening. Thus for a Lorentz shaped line, the line width $\Delta = 2/T_2$. Since measured line widths are also due to field inhomogeneity, it is usual to define a relaxation time T_2^* such that $\Delta = 2/T_2^*$.

Experimentally NMR may be carried out either by fixing the frequency (v) and varying the applied field (B) until the resonance condition is satisfied, or by varying the frequency at constant field strength. In practice both methods are employed.

6.2.3 The NMR spectrometer

The typical arrangement of an NMR spectrometer is shown in Fig. 6.2. It consists of 5 main components:

1. high field strength magnet
2. radio frequency source
3. NMR probe
4. sweep system
5. amplification and recording system

The *magnet* is required to have high homogeneity (better than 1 in 10^8) and may be a permanent magnet, an electromagnet or a superconducting magnet; each has its own particular advantages. Field strengths are

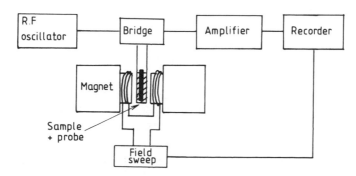

FIGURE 6.2 Schematic of NMR spectrometer.

normally in the range 1.4 to 2.3 T corresponding to operating frequencies of 60 and 100 MHz respectively. Permanent magnets and electromagnets have maximum field strengths of about 2.1 T and 2.3 T respectively. Higher field strengths require the use of superconducting solenoids which extend the operating frequency up to about 300 MHz. This has the great advantage that by operating at high field strengths both the magnitude of the chemical shifts (section 6.2.4) and the sensitivity increase.

The *radio frequency source* is a crystal controlled oscillator having high stability (1 in 10^9). Commercial instruments normally operate at 60, 100 or 220 MHz.

The *NMR probe* consists of transmitter and receiver coils which surround the sample and act to couple the incident radio frequency and the receiver signal to the sample. When the resonance condition is satisfied, the sample absorbs energy from the transmitter and the change in the coil is detected by the receiver coil and is then amplified as the NMR signal. Stability is increased by simultaneously modulating the sample by varying the field in the audio frequency range (5 kHz) and the resultant signal is detected by means of a phase sensitive detector which filters out unwanted signals.

The *sweep system* may operate by either varying the magnetic field or varying the frequency. Field sweep is achieved by means of sweep coils attached to the pole faces of the magnet and is used particularly when operating in the integral mode to obtain quantitative data. The varying frequency mode is used for spin decoupling and is achieved by varying the frequency of the audio frequency modulation field or by varying the radio frequency oscillator frequency directly. Sweep rates are typically in the range 1–2 Hz/s.

Amplification and recording system. The resonance signals may be displayed as the absorption signal, the integral or as the first or second derivatives of the absorption curve. The spectra are usually plotted on a chart recorder with the x-axis linked to the sweep system which may be precalibrated in terms of chemical shift values. The NMR signal is recorded by sweeping through the resonance position from low to high field. The samples are contained in glass tubes (about 5–12 mm diameter), located in an air driven rotating spinner or turbine to even out any local field inhomogeneities. The higher the spinning frequency, the higher the intensity of the signal and the smaller the spinning side band signals which result from sample spinning. Rates of up to about 200 rpm are used. It is common practice to locate a reference standard in a small capillary tube within the sample tube.

The use of variable temperature accessories permit NMR spectra to be recorded over temperature ranges from -150 to $+200°C$ by flowing cooled or heated nitrogen over the sample holders.

A number of specialized techniques which facilitate the interpretation of NMR spectra will be considered later in this section, following the discussion of the chemical shift and spin–spin coupling effects.

6.2.4 *Chemical shift*

In the absence of other interactions, for a given applied frequency and nuclear spin type, NMR resonance would only be observed at a single value of field strength and this would result in a single absorption peak. If this were the case, NMR spectroscopy would be very limited as an analytical technique. In fact the field experienced by the nucleus is modified due to magnetic shielding by electrons orbiting the nucleus. In effect, induced electric currents produce magnetic fields which oppose the applied magnetic field. The higher the electron density around the nucleus and the higher the applied field, the greater the extent of shielding.

The effective field experienced by the nucleus is expressed by

$$B_{\text{eff}} = B_0(1 - \sigma) \qquad (6.9)$$

where σ is termed the screening constant. The magnitude of σ depends on the particular chemical environment, since it is this which determines the electron density around the nucleus. As a consequence, the effect is referred to as the chemical shift.

The difference in energy between the levels for nuclei for which $I = \frac{1}{2}$ is now given by

$$\Delta E = g_N \beta_N B_0 (1 - \sigma) \qquad (6.10)$$

so that the positions of resonance lines can be used to identify particular chemical environments, e.g. different hydrogen atoms in organic molecules. In order to do so it is necessary to have some practical measure of the extent of shielding. We have already seen that the chemical shift is dependent on the applied field strength. However NMR spectrometers operate at different field strengths so it is clearly desirable to be able to express the chemical shift in a form that is field independent. This can be done quite easily by referring the chemical shifts to a standard substance. The accepted standard most usually employed for ^{1}H resonance is tetramethylsilane (TMS) which contains 12 equivalent protons and so gives rise to a single intense absorption line. For convenience the TMS signal can be recorded at the same time as the sample under examination. The procedure used for comparison leads to the definition of a dimensionless parameter termed the chemical shift (δ) such that $\delta = (\sigma_{\text{TMS}} - \sigma) \times 10^6$ expressed in ppm, and which is independent of the operating conditions. In practice, measurements are made of the frequencies at which the individual peaks occur and the chemical shift is then expressed by

$$\delta = \frac{\nu_{\text{sample}} - \nu_{\text{TMS}}}{\nu_0}$$

where ν_0 is the operating frequency of the spectrometer or alternatively, the frequency of the proton for which $\delta = 0$.

Chemical shifts for protons in organic molecules, including polymers, lie in the range 0–10 ppm. By comparison ^{13}C chemical shifts extend over a range of 250 ppm with respect to TMS and this has certain advantages.

6.2.5 *Spin–spin interactions*

For a collection of nuclear spins in a sample one could expect interactions to occur between them such that the effective magnetic field experienced by any nucleus would increase or decrease relative to the applied field B_0. In general the interaction between two dipoles a distance r apart and with the line joining them orientated at an angle θ to the direction of B_0, produces a change in the local magnetic field.

$$\Delta B = \pm \tfrac{3}{4} g_N \beta_N (1 - 3 \cos^2 \theta) r^{-3} \tag{6.11}$$

In the liquid and gaseous states the continuous motion of the molecules cancels out the spin–spin coupling since the time average of $(1 - 3 \cos^2 \theta)$ becomes zero. This is fortunate in that it permits high resolution of spectra in the liquid state. For polymer solutions molecular chain motions are restricted to some extent so that changes in chain conformations require longer times than the mean lifetime of the spin states and this results in broadening of the resonance lines. This can present some problems in interpreting spectra and also means that the spectra can show marked temperature dependence although this may be used to advantage.

In the solid state the occurrence of spin–spin interactions results in much greater line broadening for polycrystalline or amorphous polymers and anisotropy in single crystal spectra. Interestingly, however, the term $(1 - 3 \cos^2 \theta)$ becomes zero for $\theta = 54°44'$ and this provides a means of suppressing line broadening to some extent by rapidly rotating solid samples about an axis set at the 'magic angle' of $54°44'$ to the magnetic field. Magic angle spinning (MAS) is being used increasingly in the study of solid polymers and concentrated polymer solutions.

6.2.6 *Coupling constants*

In addition to intermolecular spin–spin coupling, and of more use in structural analysis, is the interaction which occurs between nuclei via the bonding electrons of closely spaced nuclei in molecules. If we again consider the particular case where $I = \tfrac{1}{2}$, the spin state of any nucleus can be $+\tfrac{1}{2}$ or $-\tfrac{1}{2}$ and as a consequence electrons in adjacent covalent bonds will become correspondingly polarized. As a result the effect of the particular spin states will be transmitted by means of this polarization to neighbouring magnetic nuclei and the effective field at those nuclei is increased or decreased proportionately. For $I = \tfrac{1}{2}$ this means that the absorption line of the latter nuclei will be split into two equally intense lines (since the populations of the two spin states of the former nucleus are equally probable to a first approximation). In like manner, the respective spin states of the second nucleus will produce a splitting of the absorption peaks of the first nucleus.

Spin–spin coupling is normally only effective between nuclei separated by one or two covalent bonds. The magnitude of the spin–spin coupling decreases with increasing bond separation and is characterized by the coupling constant J_{AX}. Coupling constants are expressed in units of frequency (Hz) and typical values of $^1H-^1H$ coupling in organic molecules lie in the range 2–20 Hz. The magnitude of spin–spin coupling constants depends on the chemical nature of the interacting nuclei and this sensitivity to structure is one of the principal advantages of NMR in that it provides a means of identifying particular chemical groups in molecules. Coupling constants are also sensitive to differences in conformations of molecules and consequently depend on the rates of bond rotations and so may also be used to provide information on molecular motions in polymers.

In contrast to the chemical shift, coupling constants are essentially independent of the applied field strength. Thus, measurements of spectra at different field strengths show that line separations due to spin–spin coupling remain unchanged whereas those due to chemical shift change in proportion to the change in field strength.

Where coupling occurs with more than one (n) equivalent nuclear spin, the resonance line is split into $2n + 1$ lines with the intensities of the peaks varying according to the coefficients of the binomial expansion. The peak intensities may be represented by means of the Pascal triangle (Fig. 6.3). For non-equivalent interacting nuclei the splitting is somewhat more complicated but can usually be deduced from measured coupling constants and the general shape of the spectrum.

The simple rules for line splitting just outlined only apply to situations where the differences in chemical shift (measured in Hz) are large compared to the magnitudes of the spin–spin coupling constants. The spectra are then termed first order. Since chemical shifts increase with increasing operating field

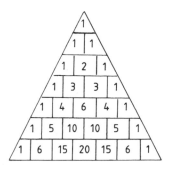

FIGURE 6.3 Pascal triangle.

strength or frequency, then working at high field strengths is more likely to result in first order spectra.

When the differences in chemical shift and spin–spin coupling constants are of comparable magnitude, additional line splittings and variations in line intensities are observed. Although these effects can be useful in the interpretation of spectra, they will not be covered further in this text and the interested reader is referred to more detailed discussions of the effect [1, 4].

6.3 Experimental techniques

6.3.1 *Fourier transform spectroscopy*

Fourier transform spectroscopy relies on the response of the magnetic nuclei to strong RF pulses, and the subsequent analysis of the signal. This may be seen by the following. The application of a sequence of pulses of a strong RF signal of frequency v_0 and pulse length t, produces side-band frequencies of width $\pm 1/t$, each separated by a frequency difference $1/t_2$, where t_2 is the pulse repetition time. In effect the pulse technique converts the monochromatic RF signal of frequency v_0 into a polychromatic signal of width $\pm 1/t$ (Fig. 6.4). For example, for a pulse width of 5 μs, the band width is 200 kHz. Provided that the resonance frequencies of the nuclei in the sample lie within this frequency range, they will all be excited to their upper energy state. This is equivalent to simultaneous illumination of the sample with a range of frequencies. The relaxation of the excited spin states back to their ground state causes an alternating RF output voltage to be produced which is detected by the receiver coils. The sinusoidal signal decays exponentially to zero with the time constant T_2^* and is termed the free induction decay (FID) signal (Fig. 6.5).

If we consider for simplicity a resonance spectrum consisting of a single line we can see how the pulse technique is employed to generate the NMR spectrum. For the single absorption line, the output signal frequency v_1 will be given by the difference between the pulse frequency (v_0) and the resonance frequency of the particular nucleus (v_r), i.e. $v_1 = v_0 - v_r$. It is then possible to obtain from the FID signal all of the information required to construct the NMR

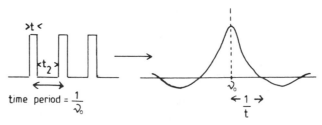

FIGURE 6.4 Representation of pulsed NMR experiment.

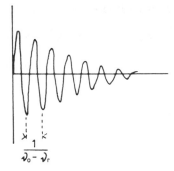

FIGURE 6.5 Example of free induction decay (FID) signal.

signal; v_1 gives the resonance frequency and T_2^* enables the line width to be determined. The same is true for more complicated spectra in that the single decay curve is a complete representation of the NMR spectrum in terms of decay time. What is then required is the conversion or transformation of the time dependent decay curve into the field or frequency dependent absorption curve. We will not go into detail here but simply state that this is achieved by means of the mathematical Fourier transform (FT) operation.

In terms of time

$$F(t) = \int_{-\infty}^{+\infty} F(v) \exp(-i2\pi vt)\, dv \tag{6.12}$$

and in terms of frequency

$$F(v) = \int_{-\infty}^{+\infty} F(t) \exp(+i2\pi vt)\, dt \tag{6.13}$$

so that what is needed is the transformation

$$F(t) \rightarrow F(v) \tag{6.14}$$

This is achieved by digitizing the output signal and carrying out the FT by means of a digital computer using standard procedures.

The particular advantage of FT spectroscopy is the reduction in time required to develop the spectrum. A computer scan may be carried out in as little as 0.5 s. By recording possibly several hundred scans and then averaging to increase the signal to noise ratio, much better resolutions can be achieved using the pulse method in times comparable to those needed to carry out a single scan using continuous wave techniques. This greatly improved sensitivity has found application in the analysis of spectra from nuclei of low concentration or low sensitivity particularly ^{13}C, and has been applied fairly widely in the study of polymers.

6.3.2 Double resonance techniques

(a) THE NUCLEAR OVERHAUSER EFFECT (NOE)

Double resonance techniques find use in enhancing the intensities of spectral lines. The enhancement arises due to changes in relaxation behaviour of one nuclear transition as a consequence of the saturation of other nuclear transitions. There is a necessary requirement, therefore, that dipole–dipole coupling be responsible for the relaxation phenomenon. This means that the nuclei in question are located sufficiently close together since the magnitude of dipole–dipole coupling falls off as $1/r^6$. The result of saturating one spin type is to disturb the normal Boltzmann distribution of the other spin, when they are coupled, such that the frequency of spin transitions of the latter nucleus are significantly increased.

A particular application of NOE to polymers is in recording ^{13}C spectra where the intensity of the otherwise low intensity spectra is increased when using broad band decoupling (section 6.3.2(c)).

Since the effect is sensitive to the magnitude of spin–spin separation distances, NOE is also useful in structure determination by enabling estimates of relative distances between different nuclei to be made. The nuclear spin which produces the greatest signal enhancement of the reference nucleus is closest to that nucleus.

(b) SPIN DECOUPLING

Spin decoupling is widely used in the interpretation of complex spectra and operates by selectively removing interactions with specific nuclei so determining which resonance lines show splitting from which nuclei. This is achieved by simultaneously irradiating the sample with the resonance frequency of the nucleus to be decoupled whilst sweeping through the entire spectral region with the variable frequency. It is brought about electronically by means of side band modulation of the central band of the RF oscillator. It is also possible to apply a third resonance frequency to produce additional decoupling giving rise to triple resonance.

The use of the frequency sweep mode to produce decoupling is experimentally much simpler in that the full decoupled spectrum can be produced in a single sweep whereas separate experiments are required for each pair of nuclei to be decoupled when using the field sweep mode.

(c) BROADBAND DECOUPLING

There are some situations where it is necessary to decouple nuclei whose resonance absorptions occur over a range of δ values. The method of broadband decoupling is used in these cases. This is achieved by irradiating with a band of frequencies covering the relevant range and centred on the central absorption frequency, e.g. 100 MHz, such that the whole spectrum of the nucleus is decoupled without saturating the signal of interest. This technique finds particular application in the study of ^{13}C and ^{19}F spectra in polymers where the spectra are obtained with broadband decoupling of ^{1}H nuclei. At the same time, enhancement of the ^{13}C and ^{19}F nuclear resonance lines may be observed due to NOE.

Interpretation of complex spectra may also be facilitated by selective deuteration of protons in the material under investigation.

6.3.3 Deuteration

The spin quantum number of the deuterium isotope is 1 so that coupling to protons results in splitting of resonance lines of the latter nuclei into $1:1:1$ triplets. The magnitude of the coupling between protons and deuterons is much less (about one-seventh) than that between protons alone because of the much smaller magnetogyric ratio of deuterium (Table 6.1). Also, since the resonance frequency of deuterium nuclei differs from that of the proton, the result of deuteration of a molecule (i.e. replacement of ^{1}H by ^{2}H) is effectively to remove those lines from the spectrum arising from the substituted protons. This can provide a useful way of selectively simplifying spectra. It is also fortunate that deuterium substitution has little effect on the chemical shifts of adjacent protons. Where line broadening does result from spin–spin coupling with many deuterium nuclei in a deuterated sample, it is possible to use deuterium decoupling to eliminate the line broadening.

(a) DEUTERIUM NMR

In considering deuteration, it is worth noting in passing the more recent development of deuterium NMR which permits detection of molecular motion not readily detectable by proton NMR. The technique will not be discussed in any detail but reference is made to work that has been reported into the investigation of molecular motions in flowing concentrated polymer solutions.

6.3.4 Broad line NMR

Most applications of polymeric materials necessarily involve the solid state and hence characterization of polymers in the solid state is of direct interest. However, as was noted in section 6.2, the existence of internuclear dipolar interaction leads to line broadening in solids and to a lesser, but still significant extent, in polymer solutions. Consequently the spectra obtained are essentially featureless broad lines and interpretation is somewhat more difficult and the information that can be derived from them is limited. Nevertheless, many of the early investigations of polymers involve broad line NMR and valuable information on polymer structure can be obtained from such studies.

The width of the NMR absorption peak may be considered to arise from dipolar interactions between immobile magnetic nuclei with the spatial distribution of the spins determining the shape of the curve. What is needed then is some means of relating the line shape to the distribution of the magnetic interactions.

The resonance line shape may be expressed in a general form by a normalized line shape function $f(B)$ such that

$$\int_{-\infty}^{+\infty} f(B)\, dB = 1 \tag{6.15}$$

A characteristic of the curve which can be used to indicate the distribution of interactions is the second moment of the curve. This is defined as the mean square width of the line measured from the centre of the resonance or alternatively, the mean value of the square of the internal field. For a symmetrical spectrum the centre of the resonance line (B_c) is given by

$$B_c = \int_{0}^{\infty} B f(B)\, dB \tag{6.16}$$

and is the average value of the magnetic field. The second moment of the line is then defined by

$$\Delta B^2 = \int_{-\infty}^{+\infty} (B - B_c)^2 f(B)\, dB \tag{6.17}$$

In practice, the second moment may be determined graphically from the observed line shape [5] with some correction necessary for effects such as broadening due to modulation amplitude and sweep rate. In some cases the line width can be calculated by means of Van Vleck's formula [6], although not widely applicable, permitting comparison of experimental with predicted values of second moments. For most cases, however, the variations in second moments are only taken to be indicative of changes in molecular structure and molecular mobility rather than providing detailed structural information.

6.3.5 ^{13}C *Resonance*

Naturally occurring carbon consists of approximately 98.9% of the ^{12}C isotope for which $I = 0$ and only 1.1% of the ^{13}C isotope having spin $I = 1/2$ which is detectable by NMR. The sensitivity of ^{13}C in NMR is also significantly lower than ^{1}H so that the intensities of resonance absorption lines for ^{1}H are more than 5000 times greater than those for ^{13}C. In addition, the spin–lattice relaxation times for ^{13}C are generally much longer than those for ^{1}H so that saturation of the resonance signals is more likely to occur.

It is not surprising therefore, that successful detection of ^{13}C resonance has presented experimental difficulties. Fortunately, despite these limitations, developments in experimental techniques made possible by the advent of more readily available computers have enabled ^{13}C spectroscopy to become established as an almost routine procedure and one which finds particular application in the characterization of polymers. Early developments included the use of computers of average transients (CAT) which enabled a number of sweeps to be made about the resonance position and averaging out noise so improving the signal to noise ratio. However, FT spectroscopy is now the established technique for the study of the ^{13}C spectra.

Experimentally the procedure usually involves the use of ^{1}H broadband decoupling (section 6.3.2(c)) to eliminate line splittings from protons in the polymer. This serves to simplify the spectra. However, the method suffers from the disadvantage that potentially useful structural information is thereby lost. The use of gated decoupling can be employed to retain some of this structural information. Gated decoupling enables the two effects of spin-decoupling and the nuclear Overhauser enhancement to operate. This is possible because spin–spin decoupling occurs almost instantaneously, whereas NOE, involving relaxation processes, takes a finite time (typically of the order of seconds). In FT spectroscopy, the saturation field B_2 may be switched off whilst the RF pulse is applied so that data collection occurs without decoupling but still retaining some of the nuclear Overhauser enhancement. Conversely the saturation frequency is only turned on at the same time as the RF pulse so that complete saturation does occur although spin decoupling takes place.

In contrast to ^{1}H resonance, ^{13}C spectra occur at much lower resonance frequencies. For example, in a field of 2.3 T the respective resonance frequencies are 100 MHz and 25 MHz. The chemical shift ranges also differ substantially being approximately 250 ppm and 10 ppm respectively for ^{13}C and ^{1}H (relative to TMS). Together with the fact that ^{1}H–^{13}C spin–spin coupling constants are about 150 Hz, this means that the resolution of ^{13}C spectra is much improved.

The chemical shift ranges for different chemical groupings have been tabulated so that many assignments may be made by reference to model compounds.

6.3.6 ^{19}F Resonance

Although there are fewer reported investigations of ^{19}F NMR for polymers, the interest in fluorinated polymers and the relative ease of recording ^{19}F spectra means that the technique is of some importance. ^{19}F spectra for unscreened nuclei lie in the same general region as ^{1}H. At an applied field of 2.3 T the resonance frequency is approximately 94 MHz which compares with 100 MHz for protons. They differ more markedly in terms of their chemical shift values which extend over a range of about 500 ppm, very much greater than ^{1}H or ^{13}C. The sensitivity of ^{19}F is also comparable to that of ^{1}H and since ^{19}F is the natural isotope of fluorine, there are fewer experimental problems in detecting NMR absorptions. Nevertheless, as with ^{1}H spectroscopy, spin decoupling techniques are required to permit more detailed analysis of the spectrum.

It is also important to note that hyperfine interactions between hydrogen and fluorine nuclei in the same molecule provide very useful structural information for fluorinated polymers.

6.4 Applications of NMR to polymers

6.4.1 *High resolution NMR of polymers*

In order to achieve high resolution of NMR spectra, it is necessary to work with dilute solutions of polymers and generally to make measurements at high temperatures where molecular motions are sufficiently rapid to average out dipolar interactions. Solution concentrations of the order of 5–40% and temperatures up to about 150°C are typical values, but optimum conditions need to be determined for each system.

The early applications of high resolution NMR to polymers were concerned almost exclusively with ^{1}H resonance determined at 60 and 100 MHz (field strengths of 1.4 and 2.3 T respectively), and such studies yield a great deal of information on molecular chain structures. With the development of super-conducting magnets, better resolution is achievable enabling more positive assignments to be made and has extended the utility of NMR in characterizing polymers. More recent developments in experimental techniques, notably FT spectroscopy, have simplified detection of ^{13}C nuclei to such an extent that many current NMR studies of polymers concentrate on ^{13}C resonances.

High resolution NMR is employed particularly in determination of details of chain structure in polymers. This is, of course, central to any understanding of the properties of polymeric materials and hence the continual interest in the subject. Differences in chain structure arise during polymerization depending on the reaction conditions and may be controlled to some extent to give desired features. Before we consider the application of NMR it will be helpful to consider some of the structural variations that can occur during polymeriz-ation of both homopolymers and copolymers that are amenable to investig-ation by NMR.

Homopolymers:
1. mode of addition (head-to-head, head-to-tail etc.)
2. stereochemical addition (tacticity)
3. isomerism of diene polymers
4. chain branching
5. chain end groups

Copolymers:
1. copolymer composition
2. monomer sequences
3. configurational sequences
4. reaction mechanisms

If we take as an example vinyl monomers of the general structure

$$CH_2=C\diagdown{}^{X}_{Y}$$

where X and Y are different chemical substituents, one of which may be H, we can see how different structure variations arise. During addition polymerization monomer units successively add to the growing chain but can do so in several ways, e.g.

1. Head-to-tail: $-CH_2-\underset{Y}{\overset{X}{C}}-CH_2-\underset{Y}{\overset{X}{C}}-$

2. Head-to-head: $-CH_2-\underset{Y}{\overset{X}{C}}-\underset{Y}{\overset{X}{C}}-CH_2-$

3. Tail-to-tail: $-\underset{Y}{\overset{X}{C}}-CH_2-CH_2-\underset{Y}{\overset{X}{C}}-$

In most cases steric considerations dictate that head-to-tail addition predominates and this simplifies matters. Where X and Y substituents are different, the carbon atom to which they are attached (α-carbon) is pseudoasymmetric and this gives rise to different stereochemical structures.

(a) DETERMINATION OF STEREOCHEMICAL CONFIGURATIONS

There are two distinguishable ways in which the monomer units may add to the chain and unit, termed meso and racemic configurations.

meso   *racemic*

Groupings of two monomer units like this are termed dyads. When a third monomer unit is added, this may also add meso or racemic with respect to the adjacent monomer unit. Sequences of three monomer units or triads may be represented as mm, (mr, rm), and rr where m = meso and r = racemic. These triad configurational sequences are also more commonly termed isotactic, heterotactic and syndiotactic respectively. The existence of these structural arrangements is particularly important in determining crystallization of thermoplastics and is therefore of considerable interest. It is fortunate that these configurational sequences give rise to distinguishable magnetic environments for the nuclei in the monomer units and hence different NMR absorption frequencies.

For the meso placement the two protons attached to C_3 will experience different extents of shielding from the X and Y substituents and hence will have different chemical shifts leading to a four line spectrum. By contrast, for the racemic placement the C_3 protons experience on average the same shielding due to X and Y substituents and so would be expected to have equivalent chemical shifts provided that there is sufficiently rapid rotation about the bonds in the chain (i.e. when the time for rotation about the bonds is small compared to the reciprocal of the line separation in Hz). The result is then a single line absorption.

When the shielding effects of the X and Y substituents are sufficiently strong, the $-CH_2$ resonances may be able to distinguish longer configurational sequences. For reasons of symmetry, the β-CH_2 proton resonances reflect even numbered monomer sequences, i.e. tetrads (e.g. mmm, mrm etc.). We shall see later that ^{13}C resonances can also be used to distinguish configurational sequences.

For vinyl monomers in which X is H the α-proton resonance is also sensitive to configurational arrangements in the chain. In this case the proton resonance is dependent on the relative positions of the substituents in both adjacent (vicinal) α-carbon atoms. That is to say that the chemical shielding of the proton attached to the α-carbon may differ depending on whether the sequences are isotactic, heterotactic or syndiotactic triads. Again, provided that the shielding effect of the Y substituent is sufficiently strong and the resolution of the spectra sufficiently good, the effects of longer sequences of monomer units may be distinguishable. In this case it is odd numbered sequences of monomer units that are distinguishable by the αC–H proton resonance, i.e. triads and pentads. It is also probable that protons in the substituents can give resonance absorptions that also reflect different configurational sequences. This is the case for PMMA which we shall consider in some detail to show how the information may be derived.

(i) Configurational sequences in poly(methylmethacrylate)

Specific application of NMR in determining stereoisometric sequences in polymers can be illustrated by reference to poly(methylmethacrylate) (PMMA)

$$-CH_2-\underset{(\alpha)}{\overset{\displaystyle \nearrow CH_3}{C}}\underset{COOCH_3}{\diagdown}$$

(β) (α)

which has been subject to detailed investigation. The polymer can be synthesized by several different methods which may be shown by e.g. X-ray diffraction techniques to give different tactic forms. Predominantly isotactic PMMA consists essentially of meso–meso triad groupings whereas syndiotactic PMMA consists predominantly of racemic–racemic groupings. However, since the synthesized polymers are never 100% isotactic or syndiotactic, there are also likely to be mr and rm heterotactic sequences present.

(i) ^1H *resonance.* In general, the α-CH$_3$ protons are expected to be equivalent due to rapid rotation about the C–CH$_3$ bond. However, differences in chemical shift would be expected for different tactic sequences in much the same way as for the αC–H proton resonances. The β-CH$_2$ protons are non-equivalent for mm isotactic polymers and should give rise to quartet absorptions, whereas single lines are expected for the equivalent protons in racemic placements.

Typical spectra for different tactic forms of PMMA are shown in Fig. 6.6. Considering the spectrum of predominantly isotactic PMMA, it is possible to assign the peaks based in part on evidence from model compounds. With increasing value of chemical shift (relative to TMS) the protons would appear in the order α-CH$_3$, β-CH$_2$, and α-COOCH$_3$. Thus assignments can then be made in the following manner.

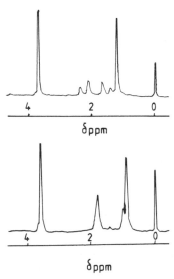

FIGURE 6.6 60 MHz proton NMR spectra of different tactic forms of poly(methylmethacrylate): (a) isotactic, (b) syndiotactic (from Johnsen, U. (1961) *Kolloid. Z* **178**, 161).

α-CH$_3$ singlet at $\delta = 1.20$ ppm
α-CH$_2$ symmetrical quartet centred at $\delta = 1.9$ ppm
α-COOCH$_3$ protons give a strong singlet at $\delta = 3.6$ ppm

The spectrum in Fig. 6.6(b) was obtained from an 85% syndiotactic (i.e. predominantly rr) polymer. Three predominant singlet peaks are evident and are assigned thus

α-CH$_3$ at 0.91 ppm
β-CH$_2$ at 1.9 ppm
α-COOCH$_3$ at 3.6 ppm

which are entirely consistent with the earlier discussion.

Additional peaks are also evident at 1.04 and 1.20 ppm which are assigned to α-CH$_3$ protons in mr and mm sequences which are present since the structure is not completely syndiotactic. The experimental conditions under which these spectra were obtained were insufficient to distinguish longer configurational sequences.

A spectrum recorded at 220 MHz for a 10% solution of syndiotactic PMMA in chlorobenzene is shown in Fig. 6.7. The spectrum covers the range 1–1.5 ppm and the peaks are assigned to the α-CH$_3$ protons with splittings arising from pentad monomer sequences. With increasing chemical shift, it is possible to assign the peaks to mrrm, mrrr, rrrr, (rmrm + mmrm), (rmrr + mmrr), rmmm and rmmr. The β-CH$_2$ protons also show additional lines, this time due to tetrad monomer sequences which, although quite complex, have been largely assigned satisfactorily. The details will not be presented here.

For the predominantly isotactic polymer pentad splittings of the α-CH$_3$ and tetrad splittings of the β-CH$_2$ have also been resolved at high field strength and fairly detailed assignments have been proposed. Again, it is not the intention to

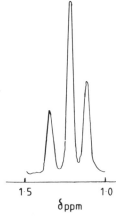

1·5   1·0
δ ppm

FIGURE 6.7 220 MHz proton NMR spectrum of poly(methylmethacrylate) (from Ferguson, R.C. (1969) *Macromolecules*, **2**, 237).

present any detailed discussion of these effects. Suffice to say that the high sensitivity of NMR clearly lends itself to detailed study of intramolecular effects in polymer chains, which may be interpreted in terms of chain structure.

The relative fractions of isotactic and syndiotactic sequences in PMMA may be obtained by comparing the intensities of the α-CH$_3$ peaks at 1.20 ppm (isotactic) and 0.91 ppm (syndiotactic). However, a fuller understanding of the distribution of the various stereoregular sequences in polymers requires some knowledge of the statistics of monomer unit placements. Detailed consideration is beyond the scope of this text, and can be found elsewhere [7], but some brief comments may be helpful.

When, during the course of polymerization, the placement of the monomer units is entirely dependent on the nature of the polymerizing chain end group then Bernoullian statistics are found to apply. When the placement is also dependent on the penultimate chain end group, then the placements correspond to Markovian statistics and may be described by conditional probabilities of the first order. (A Markovian chain of second order is found when the placement is determined by neighbouring dyads; similarly for higher order dependences.) It is possible to calculate the Bernoullian and Markovian probabilities of the different stereochemical sequence lengths for comparison with experimentally determined distributions which assists in assigning the individual peaks.

So far we have considered some aspects of NMR studies of PMMA. The technique is similarly applicable to other disubstituted ethylenes, although when the size of the α-C substituent is larger than the methyl group, the spectra are correspondingly more complex and less amenable to analysis. For some polymers such as poly(methacrylic acid) and poly(acrylonitrile) the stereochemical structure can be investigated more conveniently by converting the polymers to PMMA chemically since the assignment of its spectral lines has been established with greater certainty.

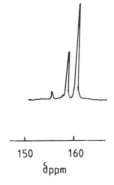

150   160
δppm

FIGURE 6.8 ^{13}C NMR spectrum of α-CH$_3$ syndiotactic poly(methylmethacrylate) (from Inoue, Y. *et al.* (1971) *Polymer J.*, **2**, 535).

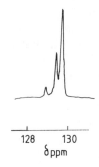

128   130
δ ppm

FIGURE 6.9 ¹³C NMR spectrum of α-C atoms in poly(methylmethacrylate) (from Inoue, Y. *et al.* (1971) *Polymer J.*, **2**, 535).

(ii) ¹³C *Resonance.* Although a later development, the application of ¹³C resonance to the study of configurational sequences in polymers has increased considerably in the last decade. ¹³C resonance has proved particularly suitable for the study of vinyl monomers where proton–proton coupling gives rise to complex ¹H spectra. Experimentally, as we have seen, this can involve broadband decoupling of the ¹H interactions and using FT techniques (section 6.3.1). Since PMMA has been most widely studied, it is sensible to use this polymer to illustrate applications of ¹³C NMR. The ¹³C spectrum of the α-CH_3 group in predominantly syndiotactic PMMA is shown in Fig. 6.8. The three peaks evident in the spectrum are assigned respectively to rr, (mr + rm), and mm sequences with decreasing values of chemical shift. This is entirely consistent with the ¹H spectra. In addition, ¹³C resonance is able to resolve spectra of the in-chain or backbone carbon atoms. Figure 6.9 shows the ¹³C resonance from the in-chain α-C atom and once again chemical shift differences corresponding to rr (rm + mr) and mm sequences are clearly resolvable.

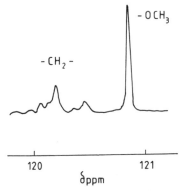

120   121
δ ppm

FIGURE 6.10 ¹³C NMR spectrum showing β–CH_2 and –O–CH_3 carbon resonances in syndiotactic poly(methylmethacrylate) (from Inoue, Y. *et al.* (1971) *Polymer J.*, **2**, 535).

The ^{13}C spectra for the β carbon atom and for the $>C=O$ carbon also show resolvable splitting due to tetrad monomer sequences (Fig. 6.10), whereas the $-OCH_3$ carbon exhibits only a single line absorption.

(ii) Configurational sequences in vinyl polymers

For vinyl monomers of the form

$$-CH_2-CH-$$
$$\underset{X}{|}$$

the NMR spectra tend to be more complex than for PMMA depending on the extent of coupling from the X substituent, and also on the magnitude of the coupling between the α-CH (methine proton) and the β-CH$_2$ groups. For polymers such as PVC, polystyrene and poly(methylacrylate), where coupling from the substituent is essentially absent, interpretation of the spectra involves consideration of the coupling only between the two methylene protons themselves (geminal coupling) and that between the methine protons and neighbouring methylene protons. Even in the simplest case the methine proton resonance will be a quintet due to interaction with two pairs of equivalent methylene protons. However, as we have already seen, different stereochemical sequences result in non-equivalence of β-CH$_2$ protons so that the methine resonances are correspondingly more complicated. Simplification of the spectra can be brought about by making use of the various approaches outlined in section 6.3 such as working at higher field strengths, selective spin decoupling and deuteration. Reference to spectra of model compounds can also assist in confirming possible assignments.

Fairly detailed investigations have been carried out on PVC and can serve to show what information on molecular structure can be derived. The ^1H NMR solution spectrum of PVC recorded at 120°C (Fig. 6.11), shows two main

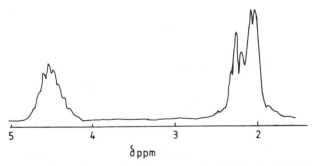

FIGURE 6.11 100 MHz proton NMR spectrum of poly(vinyl chloride) in solution (from Sieloff, G. and Hummel, D.O. (1974) *Makromol Chem*, **175**, 1561).

groups of absorption lines which can readily be assigned to the α-CH and β-CH$_2$ protons respectively by virtue of their overall intensities and relative positions based on expected chemical shifts. Depending on the distribution of different configurational sequences, it is not always possible to make detailed assignments of all of the absorptions due to accidental superposition. However, by selectively decoupling the α-CH interaction the multicomponent β-CH$_2$ proton resonances can be reduced to two peaks. Selective deuteration of the α-CH proton produces a similar effect. This might be taken to imply that there are only two main configurational sequences but it has been shown, by additional deuteration of the β-CH$_2$ protons, that longer sequences (tetrads of monomer units) need to be considered in making detailed assignments. Calculation of the expected statistical probabilities of placements can also assist in assigning peaks of the deuterated material.

Interpretation of the α-CH absorption peaks also requires the use of spin decoupling which reduces the complex multi-line spectrum in Fig. 6.11 to three lines, which have been assigned to rr, (mr + rm), mm triads. The intensities of the peaks are found to be consistent with expected Bernoullian statistics which adds confidence to the assignments. The use of deuterated samples provides evidence of interaction due to longer configurational sequences. Thus specially synthesized samples in which the β-CH$_2$ protons are substituted by deuterium and for which the spectrum of the α-CH proton is recorded with simultaneous ^{1}H-^{2}H decoupling permits resolution of splittings from longer range interactions.

The greater chemical shifts of ^{13}C resonances are also particularly useful in differentiating effects of longer sequences and are being used more frequently to confirm assignments.

Interpretation of spectra from vinyl polymers in which additional spin–spin interactions arise from the substituent X is likely to be more difficult. For example, polypropylene –CH$_2$–CH(CH$_3$)– has been studied fairly extensively and although many of the features of the spectra have been positively assigned by reference to highly tactic forms and using selectively deuterated samples, there is still uncertainty about some fine detail in the spectra. In spite of this the value of NMR spectroscopy in providing detailed structural information on vinyl monomers should not be underestimated.

6.4.2 NMR studies of copolymers

Probably the largest proportion of polymeric materials in use are copolymers of one sort or another, and so it is clearly of importance to gain as much understanding as possible of their structure. From what has already been discussed, NMR would appear to be a suitable technique for characterizing copolymers and it has in fact, been used to a great extent in structure determination. The information that may be derived ranges from determination of the overall comonomer compositions in the copolymer through to detailed descriptions of the arrangement of monomer sequences in the chain and, where applicable, determination of configurational sequences in the chain. This type of structural information is particularly important in understanding the mechanisms of copolymerization and the influence of reaction conditions on the copolymer structure.

The ability of NMR spectroscopy to provide structural information such as the distribution of monomer units along the chain and the length of monomer sequences is dependent, in much the same way as that discussed earlier in determining configurational sequences, on the influence of neighbouring monomer units via chemical shifts and spin–spin coupling on the ^{1}H and ^{13}C resonances. Again, it may be stated that in general for vinyl monomers, the αC–H, ^{1}H and ^{13}C resonances reflect interactions within odd numbered monomer sequences and βC–H$_2$, ^{1}H and ^{13}C resonances show dependences due to even numbered monomer sequences.

In chain addition copolymerizations, the order of addition of monomer units to the propagating chain can be expressed in terms of monomer reactivity ratios

$$r_1 = \frac{k_{11}}{k_{12}} \quad \text{and} \quad r_2 = \frac{k_{22}}{k_{21}}$$

where k_{ij} are the kinetic rate constants for the propagation reaction

$$M_i{}^{\boldsymbol{\cdot}} + M_j \xrightarrow{\;k_{ij}\;} M_j{}^{\boldsymbol{\cdot}}$$

The instantaneous copolymer composition is then dependent on the monomer reactivity ratios and on the monomer feed composition. Implicit in this statement is the assumption that the order of addition of monomer units is highly dependent on the nature of the terminal group of the growing copolymer chain and hence may be expressed in statistical terms. The advantage of these statistical relationships is that they may provide a means of testing possible assignments of the spectral lines to particular interactions.

(a) STRUCTURE OF VINYL COPOLYMERS

Many of the applications of NMR to copolymers have been concerned with chain-addition vinyl copolymers, including for example ethylene-propylene, ethylene-vinyl chloride, methylmethacrylate-methyacrylic acid etc. The following examples will serve to indicate how the spectral information may be used to characterize copolymers.

(i) Vinylidene chloride-co-isobutylene

Interpretation of the NMR spectra is simplified when there are no configurational isomers to complicate matters. This is essentially true for the copolymer formed by reaction between vinylidene chloride and isobutylene which has been the subject of a number of studies. The spectra recorded for the two homopolymers are, as expected, relatively simple. Poly(vinylidene chloride) gives a sharp singlet due to the equivalence of the β-CH$_2$ protons, whilst polyisobutylene gives a doublet spectrum with separate absorptions from the β-CH$_2$ and the α-C–CH$_3$ protons, there being little resolvable interaction between the methylene and methyl protons. The simplicity of the spectra is entirely consistent with predominant head-to-tail addition.

The ^{1}H spectra of the copolymer are more complex, multicomponent and vary with copolymer compositions. It is reasonable to conclude that the additional lines arise from interactions of different monomer sequences in the copolymer chains, and may therefore be used in describing the chain structure. The approach involves the recognition that the methylene proton resonances can be influenced by dyad and tetrad monomer sequences and that the possible monomer sequences may be determined by particular probabilities, e.g. may be fitted to first order Markovian statistics and should be consistent with known monomer reactivity ratios. In practice, it may be found that the spectra are sensitive to longer monomer sequences so that it may be necessary to examine the possibility of second order (or higher) Markovian statistics being applicable.

(ii) Stereochemical sequences in vinyl copolymers

The possibility of configurational variations in the chain adds complications to the interpretation of the spectra but can be taken into full account in many examples. The problem is simplified somewhat when the comonomer units have the same tacticity, i.e. co-isotactic or co-syndiotactic. Particular examples are those of isotactic and syndiotactic forms of PMMA which have been partially hydrolysed to give what are co-isotactic or co-syndiotactic forms of methylmethacrylate-methacrylic acid copolymers. This is obviously an atypical example, but it can serve to show how the monomer distribution affects the ^{1}H and ^{13}C NMR spectra. As we have already seen, the ^{1}H spectrum of syndiotactic PMMA consists of three predominant absorptions assigned with increasing field to –CH$_3$, –CH$_2$, –OCH$_3$ protons, with some additional splitting of the methyl proton line due to triad monomer sequences. In the hydrolysed copolymer, the methyl spectral region consists of six lines which may be shown to arise from six different triad compositional sequences in the chain. This ability to distinguish so clearly different monomer sequences promises to be most useful in characterizing copolymer chain structures.

Broadly similar effects are observed for hydrolysed isotactic PMMA but in this case, only four components are evident in the CH$_3$ peaks due to some accidental line overlap. Nevertheless the information gained in this type of study should enable more definite assignments to be made of the spectra recorded for other copolymers formed by radical polymerization in the presence of the two monomers.

6.4.3 *Studies of polymerization reactions*

High resolution ^{13}C NMR has also been applied to the study of the reactions of, for example, phenol formaldehyde novolak and resole resins and related model compounds. The composition of the polymerizing resins is complex but it is possible to identify in the NMR spectra contributions from hemiacetal monomers, dimers and trimers. The method promises to be particularly useful in providing confirmatory evidence for predictions from computer simulations of the reaction mechanism of use in designing continuous reactor systems. Another example of kinetic studies is that of the regeneration of cellulose which has been reported recently.

6.5 NMR of polymers in the solid state

Structural information that may be derived from NMR studies of polymers in the bulk state depends on the form of the sample. In principle, single crystal studies should provide the most detailed information which can directly complement that obtained by X-ray and electron diffraction. Unfortunately, single crystals are rarely available and in any event their properties do not necessarily reflect directly those of bulk polymers.

For oriented polymers, useful information can be deduced from the anistropy evident in the spectra. Changes in line shape with angle can be interpreted in terms of the degree of molecular orientation, as well as temperature dependent effects which arise due to changes in the degree of molecular orientation.

For the majority of bulk polymers which are unorientated (polycrystalline or

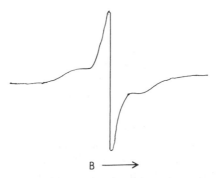

B $\longrightarrow$

FIGURE 6.12 Typical broad line NMR spectrum of a semicrystalline polymer. This is the first derivative curve obtained by modulating the sweep field as shown in Fig. 7.6.

amorphous), the resolution of the spectra is much poorer and consequently the amount of information obtainable is correspondingly limited. Nevertheless, they are still very useful to our understanding of the nature and properties of polymeric materials.

6.5.1 Broad line NMR line shapes

In order to see how NMR may be usefully employed to study bulk polymers, let us first of all consider the most general case, that of unorientated polymers and the spectral information which may be derived. A typical line shape is shown in Fig. 6.12. It can be seen to consist of two main components: a broad absorption with a narrower central component superimposed. Differences in line shape can arise from several different causes

1. The presence of different magnetic nuclei having different chemical shifts;
2. Spin–spin interactions of different chemical groupings in the polymers;
3. The presence of different phases as e.g. in polycrystalline polymers, alloys and network polymers;
4. The presence of low molecular mass additives such as monomer or absorbed water.

It is necessary therefore to distinguish these different effects.

The chemical shifts for protons in organic molecules lie in the range 0–10 ppm relative to TMS. Such values correspond to ΔB values of about 10^{-6} T which are very small compared to typical line widths in broad line NMR of polymeric solids. For ^{19}F nuclei, however, chemical shifts are of the order of 500 ppm which means that the ΔB values for different nuclei are typically more than an order of magnitude greater and may well lead to line splitting in the solid state spectra. Thus, for example, spectra of copolymers of vinylidene fluoride with hexafluoropropylene show separate components from both fluorine and hydrogen nuclei in the solid state.

In general it is found that for solid polymers spin–spin interaction is only resolved at low temperatures when molecular motions in the polymer are effectively frozen out. Even then resolution of the spectra is difficult but may be improved by working with orientated polymer samples.

Assignment of spectra to different phases is usually more straightforward although not without its problems. Consider a two phase system in which the degree of molecular mobility differs in the two phases as, for example, in a semicrystalline polymer above T_g. The higher molecular chain motions in the amorphous regions produce some averaging out of the dipolar interactions. This results in narrowing of the line due to that phase giving a sharper central component in the spectrum. The low mobility in the crystalline regions gives rise to the broad spectral component. It would, of course, be expected that with decreasing temperature the amorphous component would broaden as the molecular motions are gradually frozen in and in fact this is observed.

It follows that the temperature dependence of NMR spectra provides a means of distinguishing interactions in polymers that arise from either different phases and from different chemical shift effects.

6.5.2 *Determination of degree of crystallinity*

The presence of the narrow spectral component arising from the amorphous regions in the polymer offers a more or less direct method for determining the degree of crystallinity in semicrystalline polymers. This has been confirmed for a number of polymer systems where very good agreement has been found with X-ray diffraction, IR and specific volume measurements. However care is required in applying the NMR technique to determine the degree of crystallinity. In essence the technique relies on the fact that molecular chain motion and the consequent line widths differ in the crystalline and amorphous regions. More formally the requirement is that the correlation time for molecular chain rotation should be less than about 10^{-4} s for line narrowing to occur at NMR frequencies. The situation is more complex in practice since for any polymer, there is likely to be a distribution of correlation times corresponding to different levels of molecular motion in the polymer chains. In addition, since molecular motions are temperature dependent, a knowledge of the activation energies for molecular motions is necessary to permit a full interpretation of the NMR data. If this is not taken into account there appears to be a variation in degree of crystallinity with temperature, particularly for branched chain polymers where molecular motions in the branch chains differ from those in the polymer backbone.

Other factors such as crosslinking can also substantially reduce chain mobility in amorphous regions and hence result in broadening of the resonance lines. Obviously measurements at temperatures below T_g also give broader lines for amorphous regions. Effects due to molecular chain orientation in polymers are likely to complicate the spectra with additional lines arising from orientation in otherwise amorphous regions.

Although it has been assumed that the narrow component in polycrystalline polymers arises from the amorphous regions, it may sometimes be due, in part, to defects in the crystalline regions of the polymers. A number of studies of solution grown polymer single crystals have demonstrated that heating and subsequent cooling leads to the development of a narrow spectral component, the intensity of which is a function of the maximum heat treatment temperature. The defect regions giving rise to the narrow component evidently form reversibly since they can be annealed out by careful heat treatment. Complementary X-ray and electron diffraction studies may be used to confirm that structural changes are occurring in the crystalline region, brought about by these heat treatments. This is one example of the ways in which NMR investigations can provide useful confirmatory evidence for detailed microstructural changes in polymer crystals. In another example, defect regions produced by high energy irradiation are detectable by NMR.

6.5.3 Molecular motions in polymers

Implicit in the discussion of uses of NMR to detect different phases in polymers was the assumption that differences in line width or second moment of the absorption line arise from differences in molecular chain motions. It follows that NMR may provide a means of investigating molecular motions and factors which influence chain rotations. In order to appreciate how the technique may be used, it is necessary to consider just how the shape and width of the lines depend on molecular motions.

Recalling the earlier discussion, the line width of the NMR line arises in solids due to the distribution of dipole–dipole interactions between many interacting nuclei, i.e. the local magnetic field experienced at the nucleus varies due to the distribution of magnetic moments according to

$$\Delta B = \pm \tfrac{3}{2}(3 \cos^2 \theta - 1)r^{-3}\frac{\mu_0}{4\pi} \qquad (6.18)$$

In the solid state these interactions are not averaged out, giving rise to the observed line broadenings. Increased molecular motion would be expected to cause a reduction in line widths as some averaging of the dipolar interaction takes place. The simplest application of NMR to studies of molecular motion is the determination of line width as a function of temperature. Consider first spectra of amorphous polymers where it is found that typically the line width varies stepwise with increasing temperature (Fig. 6.13). This evidently arises from different thermal transitions in the polymer. The transitions involve rotations and vibrations of pendant groups and side chains or movement in the backbone of the polymer. For semicrystalline polymers, additional line narrowing may arise from transitions in the crystalline region, e.g. folded chain region and internal defects.

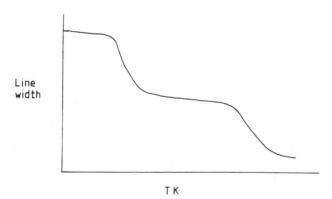

FIGURE 6.13 Variation of broad-line NMR line width with temperature showing transition regions.

The effects of line broadening may be expressed either in terms of the line width or as the second moment of the line (or even higher moments). The latter is usually more sensitive to internal changes and is the preferred method for expressing line broadening. It is also possible to calculate second moments for rigid structures if certain assumptions about the structure are made [5]. It may then be possible to correlate the temperature dependence of the second moment with specific molecular motions in polymers. For complex spectra it may also be possible to calculate second moments for the individual components and compare with experimentally determined values. Thus, for example, different phases or different elements may be distinguished.

6.5.4 ^{13}C *spectra of solid polymers*

So far we have considered broad line NMR applied particularly to ^1H nuclei which was the subject of most of the early applications of NMR to polymers. However as in the case of high resolution spectroscopy applications to ^{13}C resonance, making use of spin decoupling and MAS techniques, have become established techniques for investigating molecular motions in polymers. Comparisons of static broad line NMR with MAS have revealed the existence of internal motions in polyolefins that are spatially anisotropic. The technique therefore should enable more detailed investigations to be carried out. In addition, whereas broad line NMR can provide information on what might be termed gross molecular motions, the temperature dependence of spectra produced with MAS show that line width variations can be associated with small scale conformational motions. For example, investigations of molecular motions in solid 1, 2-polybutadiene, using FT techniques to determine spin–lattice relaxation times, has demonstrated that micro-transitions can occur in the region of T_g which may be taken to represent the macroscopic transitions. It would appear that with the improved sensitivity offered by ^{13}C, FT spectroscopy will provide a very sensitive probe of molecular transitions in solid polymers.

^{13}C resonance spectroscopy is, of course, even more widely applicable in the study of solid polymers. It has been used to provide information on the effects of chain conformation, crystal packing and temperature dependence of polycrystalline polyolefins. Thus in polycrystalline syndiotactic polypropylene splitting of methylene protons has been shown to arise from different conformations of the polymer in the solid state.

6.5.5 *Chain branching*

The effects of chain branching on the morphology and properties of polymers, in particular of polyolefins, continues to be of importance in polymer technology. ^{13}C and ^1H NMR have proved of value to such investigations. Measurements of spin relaxation times of 1-alkene copolymers have demonstrated that there are essentially three spectral components which decay at room temperature corresponding to crystalline, interfacial and amorphous regions. In order to identify the specific influence of the side chains, it is necessary to suppress the backbone resonances. This can be achieved by synthesizing copolymers having –CD– units in the backbone but leaving the side chains undeuterated. ^{13}C resonances then enable the side chain resonances to be investigated. It is usual when determining ^{13}C resonances, as with other nuclear resonances which are present in low concentration, to use cross-polarization, i.e. polarization of the dilute spins by energy transfer from the abundant nuclei so that the ^{13}C resonance is observed with simultaneous decoupling of the high concentration spins.

6.6 Two dimensional NMR

This chapter would be incomplete without some reference to the important developments in 2D NMR which is now coming to be applied to polymeric materials. As the term implies, the technique provides a two dimensional

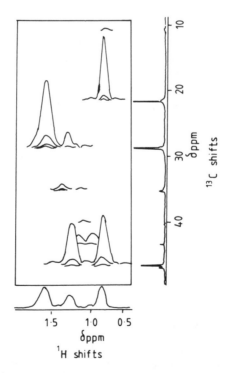

FIGURE 6.14 2D NMR spectrum of poly(propylene butylene) copolymer (Cheng H.N. and Lee, G.J. (1987). *J. Polymer Science, Polym. Phys. Ed.* **25**, 2355).

representation of the NMR spectrum and there are several variants on the basic principle. Compared to conventional, one dimensional NMR, 2D spectra have the advantage that the NMR parameters such as chemical shift and spin–spin coupling are observed separately. This means that overlapping spectra can be well resolved thus considerably simplifying assignments.

The experimental technique involves illuminating the sample with a sequence of RF pulses, the order in which they are utilized determining the form of the 2D representation. In general, 2D NMR experiments comprise four different periods: preparation, evolution, mixing and detecting. In other words, the nuclear transitions are stimulated by the pulse sequence and the decay signals are transformed (FT) in various ways to represent the different interactions.

The technique has been applied to a number of polymer systems. For example, the NMR spectra of some propylene copolymers have been determined using $^{13}C-^{1}H$ shift correlations. In this method the 2D representation is the line-to-line correspondence between ^{1}H and ^{13}C splitting. The ^{1}H NMR spectra of polypropylene copolymers contain contributions from primary methyl ($- CH_3$), secondary methylene ($-CH_2-$) and tertiary methine ($\supset CH-$) protons. Similarly, the ^{13}C spectra comprise contributions from the corresponding carbon atoms. The peaks in the 2D spectra are due to correlated absorptions involving simultaneous ^{1}H and ^{13}C splitting from the same group (Fig. 6.14).

It is expected that, with the increased availability of commercial spectrometers incorporating pulse sequence and FT facilities, 2D NMR will be applied to characterize a wider range of polymeric materials.

6.7 Conclusions

In this Chapter we have attempted to provide sufficient fundamental theory to enable the reader to appreciate the ways in which NMR may be used in the characterization of polymers. Inevitably, much detail has been omitted. The more important applications of NMR to polymers have been discussed to show the wealth of information that may be derived. It is certain that in any detailed investigation of polymers NMR spectroscopy will play a central role. Although still specialized techniques, the various means of determining NMR spectra will find more general use for routine analytical work with the more ready availability of the instruments.

References

1. Pople, J.A., Schneider, W.G. and Bernstein, H.J. (1959) *High Resolution Nuclear Magnetic Resonance*, McGraw Hill, New York.
2. Slichter, C.P. (1978) *Principles of Magnetic Resonance*, Springer-Verlag, Berlin.

3. Harris, R.K. (1983) *Nuclear Magnetic Resonance Spectroscopy*, Pitman.
4. Carrington, A. and McLachlan, A.D. (1967) *Introduction to Magnetic Resonance*, Harper, London.
5. Slonim, I. Ya and Lyubimov, A.N. (1970) *The NMR of Polymers*, Plenum Press.
6. See Ref. 2, Chapter 7.
7. Klesper, E. (1974) in *Polymer Spectroscopy* (ed. D.O. Hummel), Verlag Chemie, Weinheim, Chapter 3.

Exercises

6.1 Calculate the population difference for ^{1}H nuclear spins in fields of 1.5 and 12 T at 300 K.

6.2 Calculate the resonance frequencies for the following nuclei in a field of 2.3 T.
(i) ^{1}H; (ii) ^{2}H; (iii) ^{19}F; (iv) ^{31}P.

6.3 In a certain magnetic field protons are found to resonate at 63.855 MHz.
(i) Calculate the magnetic field strength.
(ii) What magnetic field strength would be required to observe the NMR spectrum of (a) ^{19}F and (b) ^{13}C nuclei using the same radiofrequency?

6.4 For a molecule containing only two interacting protons, NMR absorption peaks are observed at $+296.3, 297.7, 399.5$ and 400.9 Hz relative to the TMS peak for a spectrometer operating at 60 MHz.
(i) Calculate (a) chemical shifts for the two protons; (b) the hyperfine coupling constant.
(ii) What would be the resonance frequencies (relative to TMS) for a spectrometer operating at 100 MHz?

6.5 The ^{1}H NMR spectrum for a sample of polybutene consists of two sharp lines which are located relative to TMS at $+64.8$ and 87.6 Hz for a spectrometer operating at 60 MHz. The peak heights are in the ratio 3:1. There are two possible isomers of polybutene

(A) $[(CH_3-CH_2)CH-CH_2-]_n$ and (B) $[-(CH_3)_2C-CH_2-]_n$.

(i) Explain how you would assign the spectrum to one of the isomers.
(ii) Calculate the chemical shifts for the different protons. Why don't the lines show additional splitting? Under what conditions could additional splitting be observed?

7

Electron spin resonance spectroscopy

7.1 Introduction

The first application of electron spin resonance spectroscopy (ESR) to problems in polymer science was probably the observation of the spectra of trapped radicals in X-irradiated poly(methylmethacrylate). Since that time the techniques of ESR have been used in many areas of polymer science in which paramagnetic species are involved. In principle ESR may be used to examine any system possessing unpaired electrons including atoms, free radicals, radical ions and transition metal salts. Although restrictive in that sense, it is fortuitous that macro-radicals are fairly readily produced and trapped in polymers and are hence amenable to study.

Polymer scientists and technologists are of necessity greatly concerned with the various degradation reactions of polymers and with ways of combating degradation. In many instances degradation reactions involve free radical intermediates and ESR spectroscopy provides a very direct means for determining their structure and investigating their reactions. This has been particularly true of degradation induced by high energy irradiation (γ-rays, X-rays, fast electrons, protons, neutrons) and ultraviolet photolysis where fairly detailed reaction mechanisms have been elucidated. Detection of chain scission radicals during mechanical degradation of polymers and their subsequent reactions have also enabled more detailed theories of strength of materials to be formulated and tested.

Under suitable experimental conditions, propagating radicals may also be detected during radical and ion initiated addition polymerization and copolymerization reactions. The high sensitivity of modern ESR spectrometers which can detect radical concentrations as low as 10^{-9} molar permits examination of systems not readily accessible by other techniques. In many cases hyperfine structure, i.e. splitting of the resonance signal, in the spectra provides means of determining details of molecular structure of the radicals which also facilitates greater understanding of reaction mechanisms. Another area of interest is that concerned with Ziegler–Natta catalysts which possess unpaired electrons and are therefore detectable by ESR and the relevance of the paramagnetic species to catalytic activity can be investigated.

119

In general, studies of radical accumulation and decay in different circumstances provide information concerning reaction mechanisms and indirectly
on the effects of microstructure on reactivity and molecular mobility and
transitions in polymers.

This chapter will attempt to cover the basic theory of ESR at a level that
should enable students to appreciate the information that can be derived from
ESR experiments although no detailed treatment is intended and the reader
should refer to other texts for more comprehensive treatment. The remainder
of the chapter is concerned with consideration of application of ESR to
polymers with emphasis on the information that may be obtained and its
relevance to the understanding of the properties of polymers.

7.2 Basic theory

7.2.1 *Electronic structure and orbital motion*

ESR, in common with other spectroscopic techniques, is concerned with the
detection of stimulated transitions between different energy states in
systems excited by incident EM radiation of the correct frequency (v) such
that

$$hv = \Delta E \qquad (7.1)$$

where h is Planck's constant and ΔE the difference between the energy levels.

In order to understand the origin of the transitions observed in ESR, it will be
convenient to consider some of the properties of electrons and their relevance
to the resonance phenomenon. In atoms and molecules, the electrons move
around the nucleus in well defined orbitals and hence possess angular
momentum which is quantized. The azimuthal quantum number I may be
either zero or an integral positive value. The unit of angular momentum is
$h/2\pi$ and the orbital angular momentum may be represented by the orbital
angular momentum vector $\mathbf{p}_1$ of length l and magnitude $lh/2\pi$ (more precisely,
the value of the length of the vector is given from quantum mechanics as
$\sqrt{l(l+1)}$).

Since the electron possesses electric charge, the orbital motion produces a
magnetic moment ($\boldsymbol{\mu}_l$) acting along the axis of rotation. It may be shown from
classical considerations that the magnetic moment vector is related to the
orbital angular momentum vector by

$$\boldsymbol{\mu}_l = \frac{-e}{2mc}\mathbf{p}_l \qquad (7.2)$$

where e = charge on the electron, m = mass of the electron, and c = velocity of
light. The negative sign implies that the direction of the magnetic moment
vector is opposite to that of the angular momentum vector. The magnitude of

the magnetic moment vector is given by

$$\mu = \frac{eh}{4\pi mc} \tag{7.3}$$

The quantity $e/4\pi mc$ is defined as the Bohr magneton (β), having a numerical value of $9.2731 \times 10^{-24}\,\mathrm{JT}^{-1}$, and may be thought of as being the unit of magnetic moment. The magnetic moment of an orbital electron is thus an integral multiple of Bohr magnetons.

In addition to orbital angular momentum, electrons also possess intrinsic angular momentum or spin. The spin angular momentum is also quantized and may only take certain values characterized by the spin quantum number (S) with values of $\pm\frac{1}{2}$. The magnitude of the spin angular momentum vector ($\mathbf{p}_s$) is given by

$$\mathbf{p}_s = \frac{Sh}{2\pi} \tag{7.4}$$

and may be represented by a vector of length $\sqrt{S(S+1)}$ in units of $h/2\pi$. The spin angular momentum also gives rise to an associated magnetic moment ($\boldsymbol{\mu}_s$). The magnitude of the magnetic moment due to spin cannot be derived in an analogous manner to that of the orbital motion from classical considerations. In order to obtain agreement with experiment, it is found necessary to assign to the spin angular momentum the value

$$\boldsymbol{\mu}_s = \frac{-e}{mc}\mathbf{p}_s \tag{7.5}$$

That is to say the factor relating the spin angular momentum to the magnetic moment is twice that found for the orbital motion. The precise reason for this difference need not concern us. Rather, for the present purposes, we can be content with the assurance that the relationship has been clearly shown to be correct by many experimental observations.

The total angular momentum of a single electron is the vector sum of the orbital and spin angular moments and may be represented by the vector $\mathbf{j}$ such that

$$\mathbf{j} = \mathbf{l} + \mathbf{s} \tag{7.6}$$

It follows that the magnitude of the total angular momentum is $j(h/2\pi)$, (or more precisely $(j(j+1))^{1/2}(h/2\pi)$).

7.2.2 Magnetic quantization

We shall now consider the effect of placing the system of electrons in a uniform magnetic field (H). Because of their magnetic moments, the electrons will experience torques tending to align their magnetic moment vectors along the field direction in an analogous manner to the behaviour of bar magnets.

However, because of the quantization of the magnetic moments, the vector may only take up certain orientations with respect to the field. These are characterized by the magnetic quantum number M_l having possible values of $+l, l-1, \ldots 0, \ldots -l$ (i.e. $(2l+1)$ sublevels). The torque due to the magnetic field causes the angular momentum vector to precess about the direction of the magnetic field with angular frequency (ω) given by the Larmor theorem

$$\omega = \gamma H \qquad (7.7)$$

where γ, the magnetogyric ratio, relates magnetic moment to angular momentum.

The interaction energy between the electron magnetic moment and the applied field is

$$E = -\mu H \cos \theta \qquad (7.8)$$

where θ is the angle between the axis of the dipole and the magnetic field direction. The different orientations denoted by the different magnetic quantum numbers M_l correspond to different energy states of the orbital electrons.

Similarly, the spin angular momentum vector **S** for a single electron can only take up two possible orientations in the magnetic field. In other words, quantum theory demands that, in the presence of a magnetic field, the allowable spin states are quantized and the energy is split into $(2S+1)$ sublevels. For an electron with $S = \frac{1}{2}$, the magnetic spin quantum number M_s (the projection of the angular momentum vector in the magnetic field direction) can only take the two values of $+\frac{1}{2}$ and $-\frac{1}{2}$. The total angular momentum vector **j** is also restricted in the possible values it may take, represented by values of M_j. For systems of more than one electron, it is necessary to consider the resultant angular momentum vector **J** obtained by the vector sum of the individual electron contributions to **L** and **S**.

The total magnetic moment associated with an electron will depend on the way in which the spin and orbital moments couple, and may be derived by considering the vector diagram (Fig. 7.1) to be

$$\mu_J = -\frac{eh}{4\pi mc} g\sqrt{J(J+1)} = -g\beta\sqrt{J(J+1)} \qquad (7.9)$$

where

$$g = 1 + \frac{J(J+1) + S(S+1) - L(L+1)}{2J(J+1)} \qquad (7.10)$$

The corresponding energy levels are given by

$$E_J = \mu H = g\beta M_J H \qquad (7.11)$$

g is the Landé factor and is the ratio of the magnetic moment of the system in Bohr magnetons to angular momentum in units of $h/2\pi$.

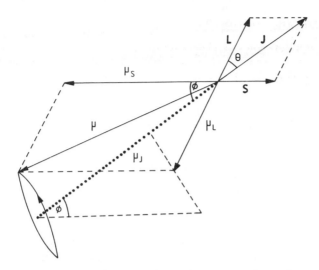

FIGURE 7.1 Vector diagram showing relation between total angular momentum of an atom and the individual angular momentum vectors.

ESR spectroscopy is concerned predominantly with situations in which there is little spin–orbit coupling. When the magnetic moment is derived solely from the spin angular momentum, i.e. $L = 0$, then the magnetic moment is given by

$$\mu = -2\frac{eh}{4\pi mc}M_s = -2\beta M_s \tag{7.12}$$

and the corresponding energy levels are

$$E_{M_s} = 2\beta H M_s \tag{7.13}$$

In practice, the degree of spin–orbit coupling is usually not known and it is found convenient to express the energy and the magnetic moment of the electron in terms of an experimental g-factor or spectroscopic splitting factor such that

$$\mu = -g\beta M_s \tag{7.14}$$

$$E_{M_s} = g\beta H M_s \tag{7.15}$$

Measured g-values for free radicals are approximately equal to 2 which is consistent with there being little spin–orbit coupling. For completely free electrons, the g-factor has been shown to be 2.002322. Deviations from the free spin value are a useful indication of the extent of spin–orbit coupling in the paramagnetic species and hence provide information on the molecular environment of the unpaired electron.

7.2.3 Resonance condition

The foregoing has laid the basis for an understanding of the resonance condition. For free radicals (i.e. molecules possessing one unpaired electron with the spin and orbital motions largely decoupled)

$$M_s = M_J = \pm\tfrac{1}{2}$$

which shows that only two possible energy states exist for the electron

$$E_1 = +\tfrac{1}{2}g\beta H$$
$$E_2 = -\tfrac{1}{2}g\beta H$$

so that

$$\Delta E = g\beta H \tag{7.16}$$

It is possible to induce transitions between these states by stimulated emission and absorption when the system is illuminated with EM radiation of the correct frequency such that the resonance condition

$$h\nu = g\beta H \tag{7.17}$$

is satisfied.

The stimulated transitions are achieved by applying electromagnetic radiation of frequency ν such that the magnetic field component (H_1) is perpendicular to the main magnetic field. It is the interaction of the oscillating field with the magnetic dipole (unpaired electron) which induces transitions between the different energy levels.

Equation (7.17) shows that there is interdependence of the magnetic field strength (H) and the incident radiation frequency (ν) at resonance. In principle, it ought to be possible to observe transitions at any field strength assuming the resonance condition is satisfied. As will be seen later, greater sensitivity is achieved by working at as high a frequency as possible. However, practical and partly historical reasons have resulted in most spectrometers operating in the X-band microwave region with $\nu = 9.5$ GHz (3 cm wavelength) which for spins with $g = 2$ necessitates magnetic fields H of approximately 0.33 T. Other spectrometers used less frequently operate in the K-band ($\nu = 23$ GHz, $H = 0.8$ T) and Q-band ($\nu = 34$ GHz, $H = 1.2$ T) microwave regions.

7.2.4 Transition probabilities

In order to achieve detectable absorption or emission, there must be a net population difference between the two energy states since the transition probabilities for stimulated emission $(P\downarrow)$ and absorption $(P\uparrow)$ are equal.

For a macroscopic assembly of N spins placed in a magnetic field such that there are N_α spins in the upper state and N_β spins in the lower state, the distribution of spins at thermal equilibrium is determined by Boltzmann's law.

$$N_\beta/N_\alpha = \exp\left(\frac{\Delta E}{kT}\right) = \exp\left(\frac{g\beta H}{kT}\right) \tag{7.18}$$

At normal temperatures, $\Delta E \ll kT$ so that equation (7.18) reduces to

$$N_\beta/N_\alpha = 1 + \frac{g\beta H}{kT} \tag{7.19}$$

and consequently, there is a slight excess of spins in the lower energy state at thermal equilibrium and one should be able to detect absorption of energy so long as the net population difference remains.

When the sample is illuminated with EM radiation, the change in the spin population is given by

$$\frac{dn}{dt} = (-2N_\beta P\uparrow) + (2N_\alpha P\downarrow) \tag{7.20}$$

where $n = N_\beta - N_\alpha$.
Since $P\uparrow = P\downarrow = P$

$$\frac{dn}{dt} = -2P(N_\beta - N_\alpha) = -2Pn \tag{7.21}$$

On integrating

$$n = n_0 \exp(-2Pt) \tag{7.22}$$

where n_0 = population difference at thermal equilibrium at $t = 0$. This implies that eventually n will reduce to zero, i.e. the energy states will be equally populated, there will be no net absorption of energy and the signal is then said to be saturated. Clearly there must be mechanisms other than stimulated emission which act to dissipate the energy of the spins in the upper energy state in order to maintain a net excess population in the lower energy state.

7.2.5 Relaxation times

When the radiation is switched off, the spin system will return to its equilibrium state. The rate at which the system returns to thermal equilibrium is determined by the mechanism through which the spins in the upper energy state lose their excess energy and this depends on the degree of coupling between the spin system and the lattice and is characterized by relaxation times. This may be seen more clearly by the following.

Let $W\uparrow$ and $W\downarrow$ be the lattice-induced transition probabilities and since the lattice is at thermal equilibrium $W\uparrow \neq W\downarrow$.

By analogy with equation (7.20)

$$\frac{dn}{dt} = -2N_\beta W\uparrow + 2N_\alpha W\downarrow \tag{7.23}$$

Since $N_\beta = (N+n)/2$ and $N_\alpha = (N-n)/2$

then

$$\frac{dn}{dt} = -(N+n)W\uparrow + (N-n)W\downarrow \tag{7.24}$$

$$= (W\downarrow + W\uparrow)\left(\frac{N(W\downarrow - W\uparrow)}{(W\downarrow + W\uparrow)} - n\right) \tag{7.25}$$

At equilibrium $dn/dt = 0$
therefore

$$n_0 = N\frac{(W\downarrow - W\uparrow)}{(W\downarrow + W\uparrow)} \tag{7.26}$$

Hence

$$\frac{dn}{dt} = (n_0 - n)(W\downarrow + W\uparrow) \tag{7.27}$$

Which may be written as

$$\frac{dn}{dt} = \frac{n_0 - n}{T_1} \tag{7.28}$$

since $1/(W\downarrow + W\uparrow) = T_1$ has the dimensions of time.

(a) SPIN–LATTICE RELAXATION

T_1 is termed the spin–lattice relaxation time and is the time for the spin population to complete $(1 - 1/e)$ of its return to equilibrium following some perturbation. We may now see the effect of this relaxation mechanism in the presence of the exciting radiation by combining equations (7.21) and (7.27)

$$\frac{dn}{dt} = -2Pn - \frac{(n - n_0)}{T_1} \tag{7.29}$$

Equation (7.21) denotes the situation where the spins are essentially isolated from the lattice, i.e. T_1 is very long. At equilibrium $dn/dt = 0$

therefore $$n = \frac{n_0}{1 - 2PT_1} \tag{7.30}$$

Provided that $2PT_1 \ll 1$ then $n \approx n_0$ which is the population difference at thermal equilibrium so that one is able to observe net absorption of energy from the radiation field continually. The transition probability P is proportional to the microwave field amplitude H_1 and so, in practice, low power levels are used to avoid problems of saturation. Saturation–recovery studies do, however, provide a valuable means of determining relaxation times as we shall see later.

The linewidth of the resonance line is inversely proportional to the spin–lattice relaxation time (T_1). That is, for very short T_1, the linewidth is determined by the lifetime of the spin state according to the Heisenberg uncertainty principle since reduction in the lifetime of the spin state causes a corresponding uncertainty in the energy level. Absorption will then occur over a range of field strengths at constant frequency. T_1 may be lengthened either by increasing ΔE or by working at low temperature.

(b) SPIN–SPIN RELAXATION

In addition to spin–lattice relaxation, there is another important relaxation mechanism, spin–spin relaxation, which also produces line broadening. Unpaired electrons can interact with other electronic dipoles in their neighbourhood. This is not in itself a means of dissipating excess energy, but rather it provides a means whereby spin–lattice relaxation becomes more effective. The fields experienced by each electron tend to blur the energy levels and it is this which leads to additional line broadening. Although there is no direct effect on the lifetime of the spin state, it is still convenient to define a spin–spin relaxation time T_2^1.

The experimental linewidth is then proportional to $1/T_2$ where

$$\frac{1}{T_2} = \frac{1}{2T_1} + \frac{1}{T_2^1} \tag{7.31}$$

7.2.6 Nuclear hyperfine interactions

(a) ISOTROPIC INTERACTIONS

From what has been discussed so far, it would appear that an assembly of spins placed in a magnetic field would give rise to a single absorption line at resonance characterized by a certain g-factor and with line width determined by the various relaxation mechanisms. If this were to be the full story, ESR would be extremely limited in its usefulness as an analytical technique.

Undoubtedly the most valuable and informative features of ESR spectra are those which arise from interaction of the unpaired spins with neighbouring nuclei possessing nuclear spin quantum numbers $I > 0$. Table 7.1 lists the most common isotopes encountered in ESR studies of polymers.

Associated with the nuclear spin there is also a corresponding magnetic moment which interacts with the magnetic field (H) and with the electron magnetic moment. The nuclear spin is also quantized and in the magnetic field there are $(2I - 1)$ allowed orientations of the spin magnetic moment vector characterized by the nuclear magnetic quantum number M_I which may take values of $I, I - 1, \ldots 0, \ldots - I$.

Interaction of the nuclear spin magnetic moment with the electron spin results in additional splitting of the electron energy levels. The magnitude of the splitting is dependent on the strength of the interaction between the

TABLE 7.1 Isotopes encountered in ESR studies of polymers and their nuclear spin quantum numbers

Isotope	I	Isotope	I	Isotope	I
H^1	$\frac{1}{2}$	D^2	1	C^{12}	0
C^{13}	$\frac{1}{2}$	N^{14}	1	O^{16}	0
N^{15}	$\frac{1}{2}$	O^{17}	5/2	O^{18}	0
F^{19}	$\frac{1}{2}$	Cl^{35}	3/2		
		Cl^{37}	3/2		

electron and nuclear spins. This is known as hyperfine interaction. Interaction of an unpaired electron with a nuclear spin I leads to splitting of the electron energy levels into $(2I + 1)$ sublevels. For example, interaction with a proton for which $I = \frac{1}{2}$ produces splitting into two sublevels (Fig. 7.2). Similarly, a deuterium nucleus ($I = 1$) produces splitting into three sublevels corresponding to the quantum numbers $+ 1, 0, - 1$.

The resonance frequency for a proton in a field of 0.3 T is approximately 14 MHz so that nuclear spin transitions are not observed in an ESR experiment. The permissible transitions are given by the selection rules $\Delta M_S = \pm 1$, $\Delta M_I = 0$. Since, to a good approximation, all the nuclear spin levels are equally populated at normal temperatures, the observed hyperfine lines arising from a single proton are of equal intensity. The separation between lines is characteristic of the particular interaction and is determined by the hyperfine splitting constant.

Additional interaction of the electron spins with other magnetic nuclei causes further splitting of the energy levels in a similar manner. Fig. 7.3 shows the expected energy level diagram for interaction with two equivalent nuclei having $I = \frac{1}{2}$, and the corresponding line intensities. Since the nuclei are equivalent, the magnitude of the separation due to the second nucleus is identical to that of the first and results in overlap of lines for which the total nuclear quantum number $M_I = 0$. The resultant spectrum is then a triplet of equally spaced lines with intensity ratios 1:2:1. For the general case of interaction with n equivalent nuclei with $I = 1/2$, the spectra consist of $n + 1$ equally spaced lines with binomial intensity distribution.

Interaction with n non-equivalent nuclei gives rise to 2^n lines of equal intensity although accidental overlap of lines usually results in complex spectra being observed.

(b) ANISOTROPIC INTERACTIONS

Although the above discussion may suggest that the hyperfine interaction is

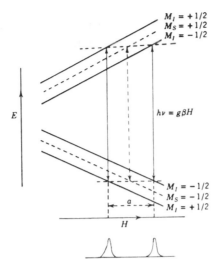

$M_I = +1/2$
$M_S = +1/2$
$M_I = -1/2$

E

$h\nu = g\beta H$

$M_I = -1/2$
$M_S = -1/2$
$M_I = +1/2$

a

H

FIGURE 7.2 Splitting of electron energy levels by interaction with magnetic nuclei having $I = \frac{1}{2}$.

isotropic, i.e. the spectra are independent of the orientation of the radicals with respect to the magnetic field, studies of radicals in the solid state show that in many cases the spectra show orientation dependence, i.e. they are anisotropic. Some understanding of the nature of anisotropic interactions therefore, is important in structure determination. Since in polymers we are concerned primarily with organic π-radicals, the following discussion will be confined to such systems.

The interaction between a nuclear spin I and an electron spin S may be shown to be composed of two parts: the isotropic and the anisotropic contributions. The isotropic part arises from the Fermi contact interaction and implies that isotropic hyperfine splitting will be observed only if there is a finite probability of finding unpaired electron spin density at the interacting nucleus. The anisotropic part of the hyperfine interaction is the angular dependent electron spin–nuclear spin, dipole–dipole interaction and is only observed in solids. In liquids and gases, rapid tumbling of the radicals averages these interactions to zero.

The anisotropic interaction can be expressed most conveniently in terms of components resolved along selected principal axes in the crystal. The principal components may be determined from single crystal spectra observed as a function of rotation about three mutually perpendicular axes. In suitable cases it is also possible to derive this information from polycrystalline samples. The spectra of randomly orientated solids are the result of contributions from radicals having different orientations with respect to the magnetic field and this can lead to difficulties in assignment. It is fortunate that in polymers,

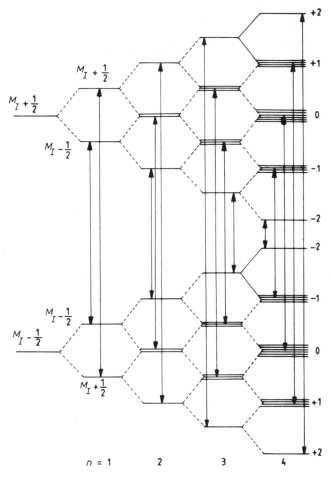

FIGURE 7.3 Splitting of electron energy levels by interaction with n magnetic nuclei having $I = \frac{1}{2}$.

despite the lack of orientation, reasonable interpretations can be made since in many cases radicals are trapped in specific sites in crystalline regions.

Some general rules have been formulated to describe the expected hyperfine coupling constant in organic radicals which have proved useful in polymer applications. For radicals of the form

$$
\begin{array}{c}
\text{R}_1 \\
\diagdown \\
\quad \dot{\text{C}}\text{–H} \\
\diagup \\
\text{R}_2
\end{array}
$$

the coupling constant for the hydrogen attached to the carbon atom containing the unpaired electron, is given by the McConnell relation $a_\alpha^H = Q\rho_a$

which expresses the dependence of the isotropic α-proton coupling constant a_α^H on spin density ρ_a. The proportionality Q has a value of 2.3×10^{-3} T. Hyperfine interaction with protons attached to the adjacent carbon atom, β-protons, i.e.

$$-CH_2-\overset{\cdot}{C}\overset{\nearrow R_1}{\searrow R_2}$$

may be expressed by

$$a_\beta^H = B_1 + B_2 \cos^2 \theta \tag{7.32}$$

θ is the angle between the plane of the C_α–C_β and the C_β–H bonds and the axis of the $p_z\pi$ orbital. $B_1 = 0.4 \times 10^{-3}$ T, $B_2 = 5 \times 10^{-3}$ T so that β-proton coupling constants are found to be larger than the α-proton coupling constants.

The ESR spectra of radicals in the solid state are described formally by means of the hyperfine tensor **T** which couples the electron and nuclear spin vectors and contains contributions from both isotropic and anisotropic interactions.

$$\mathbf{S \cdot T \cdot I} = A_x S_x I_x + A_y S_y I_y + A_z S_z I_z \tag{7.33}$$

A_x, A_y, A_z are the principal values of the hyperfine coupling constants with respect to orthogonal axes in the crystal and are related to the isotropic (a) and anisotropic tensor elements (t_x, t_y, t_z):

$$A_x = t_x + a$$
$$A_y = t_y + a$$
$$A_z = t_z + a$$

Since the trace ($t_x + t_y + t_z$) vanishes a is given by the average of the principal hyperfine coupling constants

$$a = \frac{A_x + A_y + A_z}{3} \tag{7.34}$$

The principal components of the hyperfine tensor for a typical organic π-radical derived from malonic acid have been obtained from single crystal spectra to be

$$A^1_x = -29 \text{ MHz}$$
$$A^1_y = -91 \text{ MHz}$$
$$A^1_z = -61 \text{ MHz}$$

from which $a = -60$ MHz, $t_x^1 = +31$ MHz, $t_y^1 = -31$ MHz and $t_z^1 = -1$ MHz. The hyperfine coupling constants may be expressed in Tesla (T) or MHz where 1×10^{-4} T $\equiv 2.8$ MHz. Conventionally coupling constants have been expressed in MHz for single crystal studies, although in polymer studies they have been reported almost exclusively in Tesla.

7.3 Experimental considerations

7.3.1 The ESR spectrometer

Absorption of energy from the oscillating microwave magnetic field will occur when the resonance condition $h\nu = g\beta H$ is satisfied. It is found most convenient to operate at constant frequency and to observe ESR signals by varying the magnetic field (H) about the resonance position.

In principle, it ought to be possible to observe transitions at any field strength assuming the resonance condition is satisfied. It is found that greater sensitivity is achieved by operating at as high a frequency as possible. The principle components of ESR spectrometers (Fig. 7.4) are

1. source of microwave radiation (Klystron valve)
2. absorption cell or resonant cavity
3. electromagnet
4. detection system

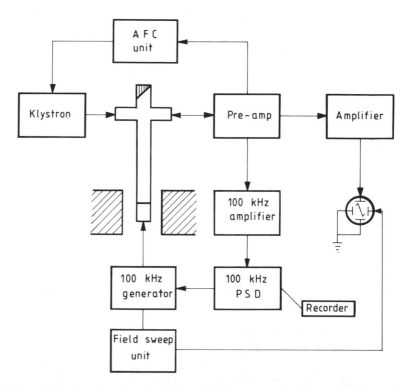

FIGURE 7.4 Schematic of typical X-band ESR spectrometer with reflection cavity.

The output from the Klystron valve may range from $0.001 \ \mu W$ up to several hundred mW. The constant frequency is stabilized by means of an automatic frequency control (AFC) which controls the voltage to the Klystron giving stabilities of the order of 1×10^{-6}. For X-band spectrometers, the electromagnetic radiation is transmitted to and from the cavity via a rectangular waveguide of wavelength 3 cm.

The simplest form of resonance cavity, a transmission cavity, consists of a section of the waveguide an integral number of wavelengths long (usually 2) designated (TE102) blanked off at each end with iris plates which have small iris holes drilled in the centre whose function is to couple the microwave radiation into and out of the cavity. A standing wave is set up in the cavity and this serves to concentrate the microwave energy on to the sample which is located in the region of maximum magnetic field vector (Fig. 7.5). However, most commercial spectrometers employ reflection cavities in conjunction with some form of balanced bridge network which gives greater sensitivity. Reflection cavities have only one iris hole for coupling with the microwave energy and may be either rectangular (TE102) or cylindrical (TE 011), the latter, having cylindrical symmetry, being particularly useful for single crystal studies. The efficiency of cavities is described by the Q-factor which is the ratio of the energy stored in the cavity to the rate of energy loss per cycle. Q-factors for commercial spectrometer cavities are typically in the range 3000–7000. The Q-factor and hence the sensitivity of cavities is reduced substantially by dielectric absorption by the sample or sample holder. In particular the presence of water in the cavity has a very marked effect on spectrometer sensitivity and should be avoided as far as possible, although as will be described later special aqueous cells can be employed for aqueous systems. Absorption of microwave

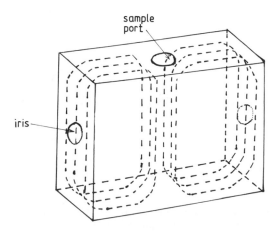

FIGURE 7.5 TE102 rectangular resonant cavity (---- = microwave magnetic field).

energy by the sample at resonance is usually detected by a semiconductor crystal rectifier as a reduction in crystal current.

In practice, the main magnetic field is set at a value corresponding to the centre of resonance and additional modulation coils attached to the pole faces of the electromagnetic are used to sweep the magnetic field about the resonance position. Greater sensitivity may be achieved by simultaneously applying a high frequency, low amplitude, modulation field whilst sweeping the main field. The result of applying the modulation field is to produce an AC signal at the detector proportional to the slope of the absorption curve. This is shown diagramatically in Fig. 7.6. The AC signal is passed into a phase sensitive detector (PSD) where it is mixed with a reference signal of the same amplitude but of variable phase to yield a DC output signal which is, in effect, the first derivative of the absorption curve. The signal to noise ratio and hence the overall sensitivity is significantly increased by the use of the phase sensitive detector since only input signals of the same frequency as the reference signal contribute to the output. The derivative curve has the additional advantage in being much more sensitive to changes in the shape of the absorption curve and hence hyperfine structure is resolved more clearly. Even better resolution is achieved by application of a second modulation signal together with a second phase sensitive detector to give the second derivative of the absorption curve. This has the additional effect of preferentially suppressing broad lines.

The sensitivity of ESR spectrometers is usually expressed in terms of minimum numbers of detectable spins times the signal linewidth (ΔH). Commercial spectrometers have sensitivities of the order of $1.5 \times 10^{10} \Delta H$.

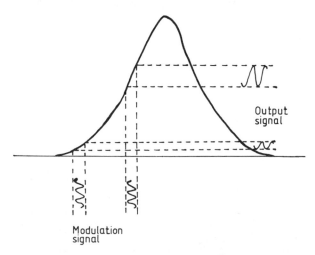

FIGURE 7.6 Modulation of absorption signal with high frequency low amplitude signal. The derivative curve reverses sign as it passes through the resonance position (peak) resulting in an appearance as shown in Fig. 6.12.

7.3.2 Sample preparation

Macroradicals in polymers react readily with oxygen to give peroxy radicals and consequent radical decay and it is usually essential to exclude air from samples by sealing either in an inert atmosphere or under a vacuum. However, there are situations where it is desirable to study behaviour of peroxy radicals. Glass or quartz tubes may be used with high purity quartz being the preferred material owing to its lower dielectric loss and because glass is more likely to contain paramagnetic impurities. High energy irradiation of quartz also gives rise to paramagnetic colour centres but these may be annealed out by flame treatment, care being taken to prevent heating of the polymer sample possibly by immersing the section of the sample tube containing the sample in liquid nitrogen. Quartz sample tubes have the additional advantage of being transparent to ultraviolet radiation and so permit direct photolysis in the cavity, provided that the cavity has a suitable window.

Variable temperature studies, in particular the effects of thermal transitions, are of great value when investigating radical reactions. These may be carried out conveniently by flowing gas at controlled temperatures through a double walled Dewar vessel in the cavity. Cooling to low temperatures can also increase the overall sensitivity of the experiments since the signal amplitude is inversely proportional to temperature.

Although polymer single crystals of sufficient size to permit orientation dependence experiments to be carried out are not available, a certain degree of molecular chain orientation of bulk polymers can be achieved by uniaxial stretching of samples. In some instances orientated single crystal mats have also been used to give information on anisotropic interactions. Cylindrical cavities are best for examining orientation dependence but satisfactory results can be obtained using rectangular cavities. Depending on the size and geometry of the samples rotation in rectangular cavities is likely to produce some signal distortion due to the sample being located in different regions of the microwave magnetic field so that retuning of the cavity is necessary.

Although it is feasible to determine absolute radical concentrations from knowledge of the various instrumental parameters, it is more usual to use comparative methods. The integrated absorption signal intensity which is proportional to the spin concentration is compared to the signal from a standard sample obtained under the same experimental conditions. The primary standard most commonly used is freshly recrystallized α, α-diphenyl picryl hydrazyl (DPPH) either as a powdered solid or in solution in benzene. Peroxylamine disulphonate ion radical (Fremy's salt) ($K_2NO(SO_3)_2$) in aqueous sodium carbonate solution is also used both as primary standard for concentration measurements and for field sweep calibration. The stability of the primary standards is not high so that signal decay occurs with time. In practice it is frequently more convenient to make use of secondary standards whose spin concentrations remain more or less constant. One commonly used secondary standard is powdered pitch having a known spin concentration which can be prepared to give other known concentrations by dilution in an inert powder.

Calibration of the magnetic field sweep is also most conveniently carried out by using standard reference signals. Mn^{2+} salts have six line spectra having line separations of 8.4×10^{-3} T giving a total spread of 42×10^{-3} T. Peroxylamine disulphonate gives 3 equally spaced lines of 1.31×10^{-3} T separation. DPPH in solution may also be used.

The use of a dual cavity obviates many of the problems met in trying to ensure that the experimental conditions for sample and reference standard are the same when determining spin concentrations. Determination of g-values to a high degree of accuracy can be achieved by measurement of the microwave frequency using frequency counters and accurate determination of the resonance field using an NMR probe. Alternatively, a comparative method may be employed by simultaneously displaying the spectra of, for example, DPPH or Fremy's salt whose g-values are known accurately.

When making ESR measurements, care must be taken to avoid distortion of the signal by, for example, increasing the modulation field amplitude too much which may cause line broadening and eventual decrease in signal amplitude or by working at too high microwave power which may cause saturation. For maximum sensitivity, the modulation field amplitude should be approximately one-tenth of the line width.

7.4 Polymerization studies

7.4.1 Solution polymerization

One of the most important applications of ESR in polymer science has been to the study of chain addition polymerization and copolymerization reactions involving radical and radical ion propagating species. Although radical concentrations in most solution polymerization systems are usually of the order of 10^{-8} M, it is possible to generate radical concentrations in the range of 10^{-5} M in flow systems which is well within the range of most spectrometers. Even higher radical concentrations may be trapped during solid state polymerizations and graft copolymerizations. Identification of the propagating species can then provide information on, for example, the molecular chain structure, the influence of reaction conditions on chain transfer reactions and reactions of radicals and monomers in copolymerizations.

The spectra of propagating radicals generated in solution are usually well resolved due to the averaging out of anisotropic interactions so that detailed information about their structure, conformation and spin density distribution may be derived. The latter may in turn provide some measure of reactivities relative to other monomers in copolymerizations.

The flow technique most frequently employed involves production of hydroxyl radicals by the redox reaction of hydrogen peroxide with Ti^{3+} ions in acidified aqueous solution:

$$H_2O_2 + Ti^{3+} \rightarrow Ti^{4+} + OH^- + OH\cdot$$

The OH· acts as the initiating species and in the presence of vinyl type monomers $(CH_2{=}CH(X))$ will initiate polymerization. Reaction of $TiCl_3$ with hydroxylamine $(NH_2OH)_2$ gives rise to $\dot{N}H_2$ radicals and similarly reaction with CH_3COOH gives methyl radicals $(\dot{C}H_3)$ and both have been used to initiate polymerization. Other initiator systems which have been investigated include t-butyl peroxide $+ SO_2$, which produces $HO\dot{S}O_2$ and t-butyl $O\dot{S}O_2$ radicals. $\dot{S}O_3^-$ radical ions may also be produced by reaction of OH· with SO_3^{2-} ions.

The reactants are thoroughly mixed just prior to passing through the resonant cavity. An aqueous solution cell must be used to minimize dielectric losses. This consists of a flat quartz cell with sample width of approximately 0.25 mm which is located in a rectangular cavity in the region of minimum electric field vector, i.e. with the broad face parallel to the direction of the main magnetic field.

By monitoring changes in spectral shape with increasing monomer concentration, it is possible to distinguish the primary species $(HO{-}CH_2{-}\dot{C}H(X))$ and the propagating radical $(HO{-}(CH_2{-}CH(X))_n{-}CH_2\dot{C}H(X))$. The differences in the corresponding spectra can be explained on the basis of differences in conformations of the β-CH_2 protons. Accurate measurement of coupling constants have enabled the spin densities on the α-carbon atoms to be related to the structure of the substituents X.

The effects of substituents on radical reactivities may also be related to spin density variation. Substituents may act partly to stabilize propagating radicals by resonance stabilization and so reduce reactivity. In general, ESR studies have demonstrated that the higher the spin density on the α-carbon atom, the higher the reactivity of the monomer radical. However, care needs to be taken when deducing radical reactivities from spin density measurements since, for certain monomers, steric effects become significant in determining reactivities.

The relative reactivities of monomers and radicals in copolymerizations may also be investigated using the above techniques. Reaction between radicals of high reactivity (M_1) and monomers of lower reactivity (M_2) give spectra of $M_2^\cdot$ with increasing concentration of the second monomer. Likewise, reaction of a monomer of relatively high reactivity with different monomer radicals would result in the less reactive (more stable) radical predominating in the spectra. The method then permits the relative order of the reactivities of different monomers and radicals to be assessed which may be compared with known reactivity ratios.

The importance of steric effects may also be examined by measurement of β-proton coupling constants which, as we have seen, depend on the angle of the C–H bonds with respect to the π-electron orbital. In this way, it may be possible to determine the influence of steric effects in the radical on the polymerizability of some monomers. The ESR spectra obtained from flow systems show marked asymmetry which is dependent on the magnitude of the hyperfine coupling constants, and on monomer concentration. The asymmetry arises from there being a non-Boltzmann distribution of the various energy populations. This is due to the production of uncorrelated radical pairs by the flow system with an initial over-population of the upper nuclear hyperfine levels for which $M_I = +1/2$. This results in reduced absorption and even emission in the low field lines and to enhanced absorption in the high field line. Although thermal equilibrium is restored within 10^{-6} to 10^{-4} s, this is just the order of lifetime of radicals in the flow system. It should be noted that by considering the kinetics of radical formation and decay in flow systems, it is deduced that only those initiating radicals which are produced and react actually within the cavity are detected. Virtually none of the radicals generated before the upper end of the cavity survives long enough to be detected.

For some polymerization systems, the use of redox initiation presents a number of difficulties. In particular there is the limitation in the number of solvents that may be used and the possibility of secondary reactions with components of the redox system. These limitations may be minimized in part by the use of photolytically generated initiator radicals. Direct photolysis can be carried out in the cavity itself in the presence of initiators such as azobisisobutyronitrile (AZBN) in a continuous flow system such that a steady state situation is achieved. The use of non-aqueous solvents reduces the problems associated with the dielectric loss of water.

The occurrence of chain transfer reactions during solution polymerizations may also be deduced in some systems from identification of additional radicals formed, for example, by hydrogen abstraction from different molecules in the system.

7.4.2 Solid state polymerizations

ESR is capable of offering particular advantages in studies of solid state polymerizations of crystalline monomers which involve radical intermediates. Specifically the use of single crystals should permit determination of the structure with greater precision from the angular dependence of the coupling constants. Initiation by means of high energy irradiation at low temperature is found to be a satisfactory means of trapping propagating radicals whilst some studies have been carried out using UV initiation. In solid state polymerizations, it is not infrequently found that the reactions occur preferentially in defect regions in the crystal where molecular mobility is presumably higher. Changes in radical concentrations with time and temperature may be associated with secondary transitions in the crystalline regions and may be used to provide additional information on the transitions.

Measurements of spin–spin relaxations and their temperature dependence may also be used to give direct information on radical distributions in the solid. This approach will be considered in section 7.5 on general radiation effects.

7.4.3 *Ionic polymerizations*

High energy irradiation of diene and vinyl monomers which polymerize by cationic mechanisms has been shown to give rise to trapped ion radicals. Because of the effects of water on the reaction mechanism, it is essential to dry the system vigorously before irradiation. Cationic polymerization may also be initiated by means of strong electron acceptors such as BF_3 and $SnCl_4$ and again vigorously dried systems are necessary to ensure observation of the radical ions. In these cases flow systems may be used.

Strong electron acceptors such as tetracyanobenzene may initiate polymerization of some vinyl monomers by a cationic mechanism when photolysed by UV light. Ionic polymerizations in solution show strong solvent effects and these will tend to influence the ESR spectra observed. Anionic polymerizations may also give rise to ESR spectra. The polymerizations may be initiated by interaction of monomers with electron transfer agents such as metallic sodium, lithium or potassium. Samples for ESR examination are prepared in vacuum systems to ensure exclusion of oxygen and water. Anionic radicals have also been observed in some monomer systems, produced by high energy irradiation.

7.4.4 *Other polymerization systems*

Useful information may be derived concerning the mechanisms of curing of unsaturated polyesters using vinyl monomers. In particular, the relative importance of chain transfer reactions to the overall curing mechanism has been investigated. A number of studies of radiation-induced graft copolymerization of vinyl monomers to natural and synthetic polymers have also been reported.

7.4.5 *Ziegler–Natta catalysts*

ESR signals are detectable from paramagnetic species in Ziegler–Natta catalysts and also for other catalyst systems and some correlations have been established between spin concentrations and catalytic activity. The ESR signals have been observed both in solution and in solid precipitates. For a representative system produced by reducing $TiCl_4$ with $(Al(C_2H_5)_2Cl)_2$, the spectrum is singlet with $g = 1.925$ and $\Delta H_{msl} = 3.5 \times 10^{-3}$ T with no clearly resolved hyperfine interactions (ΔH_{msl} is the line width at maximum slope). The signal intensity depends on the Al/Ti ratio, and the signal is attributed to Ti^{3+} with the unpaired electron in the $3d^1$ energy level. Catalytic activity appears to be reflected in the intensity of the singlet spectrum which increases in the presence of olefin monomers. For $TiCl_3$, two absorptions have been observed and catalytic activity has been correlated with the peak having a g-valve of 1.97. The spectra observed for reaction of VCl_4 with aluminium alkyls show eight line spectra in the liquid phase due to interaction of the vanadium nucleus for which $I = 7/2$, but little hyperfine structure is observed in the solid state.

7.5 Radiation effects in polymers

7.5.1 High energy irradiation

The majority of ESR studies of polymeric materials have been concerned with radicals produced in solids by either high energy ionizing radiation or ultraviolet photolysis, their subsequent reactions and relevance to the overall radiation effects. This is perhaps not too surprising since most radiation induced reactions involve radical or radical-ion intermediates which may be trapped in the polymer matrix under suitable conditions, and ESR has contributed significantly to the understanding of radiation effects and photolytic degradation.

When considering the radiation processes involved, distinction must be made between the primary species produced by high energy radiation which may not be sufficiently stable under the experimental conditions employed and the more stable secondary species which are more readily trapped in the polymer matrix. Most early radiation studies were concerned with determining the structure of stable trapped radicals by interpretation of the hyperfine structure, although a number of kinetic studies were reported. Lack of spectrometer sensitivity and the failure to appreciate fully effects of, for example, temperature and morphology limited these early investigations, valuable though they undoubtedly were in establishing the techniques. Later work has confirmed the utility of the method to observe and study primary species including trapped electrons, radical ions and radical pairs and the effects of morphology, environment, temperature etc. on radical reactions.

By far the most direct method for observing primary species is by irradiation in the cavity where short-lived species may be produced in sufficient concentration to be detected. However, direct irradiation in the ESR cavity presents a number of experimental problems and has been undertaken in only a small number of cases and will not be considered to any extent here. The method is more applicable to liquid systems where, by the use of a flow system, steady state conditions can be obtained.

For many polymer systems it has now been demonstrated that electrons and ions which are sufficiently stable to be detected may be generated and trapped by γ-irradiation at 77 K. It is not intended to review previous work in the area of radiation effects but rather to consider the general utility of ESR to study radiation effects in polymers. It will be helpful, however, to consider particular

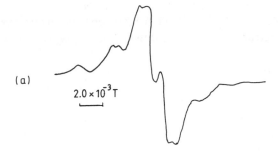

(a)

2.0×10^{-3} T

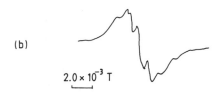

(b)

2.0×10^{-3} T

FIGURE 7.7 First derivative ESR spectra of polyethylene γ-irradiated and observed at (a) 77 K, (b) RT.

examples which illustrate the techniques and the information that may be derived and the limitations of the techniques.

As a case in point, let us consider the study of radiation effects in polyethylene which has been the subject of a large number of investigations. Typical first derivative ESR spectra obtained with unorientated polyethylene by irradiation and examination at 77 K and at room temperature are shown in Fig. 7.7. It has been well established that the predominant species trapped at 77 K is the alkyl radical I, the sextet spectrum arising from interaction with five protons having hyperfine splitting constants $a_H = 3.3 \times 10^{-3}$ T. The species formed at room temperature is assigned to the allyl radical II with the seven lines arising from interaction with six approximately equivalent protons with coupling constants $a_H = 1.4 \times 10^{-3}$ T. Prolonged irradiation yields a singlet spectrum due to polyene radicals III, the electron being delocalized to interact with a large number of equivalent protons giving unresolved hyperfine structure.

I	$-CH_2-\dot{C}H-CH_2-$	(sextet $9 = 3.3 \times 10^{-3}$ T)
II	$-CH_2-\dot{C}H-CH=CH_2-$	(septet $9 = 1.4 \times 10^{-3}$ T)
III	$-CH_2-\dot{C}-(CH=CH)_n-$	(singlet)

It has been shown that the allyl radicals II may be converted to alkyl radicals by photolysis at 77 K with light of wavelength 254 nm.

Electrons are trapped by high energy irradiation at 77 K and are sufficiently stable to be observed. The ESR signal from trapped electrons superimposed on the radical spectrum, is a sharp singlet, $\Delta H_{msl} = 4$ Gauss, which readily saturates at even moderate power levels. Typically, microwave power levels of < 0.004 mW are required to give maximum signal intensities. Problems of saturation, no doubt, were responsible for the failure to detect electrons in earlier studies. The trapped electrons decay on exposure to infra-red radiation or by gentle warming above 77 K so that care must be exercised to prevent decay. The yield of trapped electrons is found to be higher the greater the degree of crystallinity. In addition to the trapped electron and free radical signals, additional spectral components are detectable which are attributable to radical ions. The presence of deliberately added electron or positive ion scavengers influences both the trapped electron and the ionic spectra and it is generally believed that the ionic contribution is due to anion radicals, although there is some contradictory evidence.

Alkyl radicals I possess an α-carbon atom proton and the spectra should be anisotropic. The use of stretched polyethylene film or fibres, or orientated single crystal polymer mats permits orientation dependence to be examined from which the principal components of the anisotropic tensor can be determined. Spectra observed with the magnetic field respectively perpendicular (10 lines) and parallel (6 lines) to the c-axis, can be reconstructed by assuming $a_\beta^H = 33$ Gauss, $a_{\alpha\perp}^H = 18$ Gauss and $a_{\alpha\parallel}^H = 34$ Gauss. For orientated single crystal polymer mats, electron and X-ray diffraction studies show that the c-axis orientation is well-defined being perpendicular to the plane of the mat but that the a and b axes are distributed more or less randomly in the plane of the mat. In such cases, in order to analyse the spectra as a function of rotation angle, it is necessary to take into account the distribution of the a and b axes in the plane of the mat. For a random distribution of the C–H bonds about the c-axis, the α coupling constant (A) has been shown to be linearly dependent on $\cos^2 \phi$ where ϕ is the direction between the c-axis and the applied field and the following equation has been found to hold.

$$A(\phi) = a - \tfrac{1}{2}(C + D)[1 - 3\cos^2 \phi] \qquad (7.35)$$

from which the Fermi-contact contribution (a) is determined to be 24 Gauss and $(C + D) = 10^{-3}$ T where C and D are constants depending on the spin density of the α-carbon atom.

The values of the principal components of the hyperfine tensor A_x, A_y, A_z were found to be 2.59×10^{-3}, 3.41×10^{-3} and 1.2×10^{-3} T respectively, where A_x corresponds to $\theta = 90°$, $\phi = 0$; A_y, $\theta = 90°$, $\phi = 90°$; A_z, $\theta = \phi$. θ is the angle between the direction of the c-axis and the field H. It is also possible, in principle, to determine the principal components of the hyperfine coupling tensor directly from the spectra of polycrystalline samples. For example, the alkyl radical spectrum (Fig. 7.8) obtained by UV photolysis of allyl radicals at 77 K gave the following values: $A_x = 2.6 \times 10^{-3}$, $A_y = 4.06 \times 10^{-3}$,

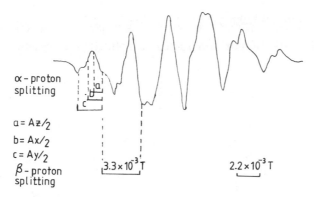

α - proton
splitting

$a = Az/2$
$b = Ax/2$
$c = Ay/2$
β - proton
splitting

3.3×10^{-3} T

2.2×10^{-3} T

FIGURE 7.8 First derivative ESR spectrum of alkyl radicals in γ-irradiated polyethylene showing derivation of anisotropic coupling constants.

$A_z = 1.36 \times 10^{-3}$ T, which are in general agreement with the values obtained from orientated samples.

Anisotropic interactions of the evidently simple allyl radicals have been investigated but the spectra have proved to be too complex for complete interpretation. This is regrettably one of the major problems encountered in ESR studies of solid polymers. Even relatively simple radicals can give rise to complex spectra due to the effects of chain conformation, orientation and morphology.

A particular case in point is that of the radicals trapped by γ-irradiation of poly(methylmethacrylate) which have been the subject of numerous investigations. The familiar $5 + 4$ line spectrum (Fig. 7.9) has been variously considered to arise from the superposition of two spectra or from a single radical species, the propagating radical. The weight of evidence points towards

$$
\begin{array}{c}
\text{CH}_3 \\
| \\
-\text{CH}_2-\text{C}\cdot \\
| \\
\text{COOCH}_3
\end{array}
$$

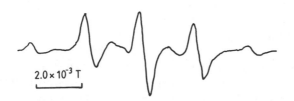

2.0×10^{-3} T

FIGURE 7.9 First derivative ESR spectrum of γ-irradiated poly(methylmethacrylate).

the latter interpretation, although there is evidence for underlying spectra of other radical species. The difficulties in the interpretation arise due to the local environment of the chain end radical which gives a distribution of the conformation angle of the β-proton bonds about the chain axis. The β-proton coupling constants depend on the conformation of the β-proton relative to the unpaired electron orbital. It may be shown that the observed spectra can be adequately reconstructed by assuming Gaussian distribution of conformation angles about a most probable value $\theta_1 = 55°$, $\theta_2 = 65°$ and with half height width of 5–6°. Although this interpretation is very reasonable, it is unfortunate that other independent evidence is not available as confirmation.

One area in which ESR may be used to provide valuable information to complement what are more conventional studies, is that concerned with radical formation and decay kinetics which may be related to molecular relaxation processes. Radical decay behaviour in polyethylene may be used to demonstrate how this may be usefully employed. Alkyl radical decay which occurs on warming from 77 K has been shown to correlate with steric conformations and molecular motions for samples of varying crystallinity. For both high- and low-density polyethylene (HDPE and LDPE), three temperature regions at which more rapid decay occurs have been identified as 120, 200 and 250 K respectively, with the decay rates in each of the regions following second order kinetics. The quantity $(k[R \cdot])^{-1}$ where k is the relevant second order rate constant, and $[R \cdot]$ the corresponding radical concentration has the dimensions of time and represents the average duration of encounter between two radicals, or in other words, the mean life-time of the radicals. This may be considered as being equivalent to the time constant τ for molecular motion of the polymer chain segments assuming that radical decay is governed by chain mobility. Comparison with relaxation times derived from dynamic mechanical experiments show some differences so that although the temperature regions correspond to the β and γ relaxations in polyethylene, the correspondence does not appear to be direct.

From variations in decay rates with temperature, activation energies may be calculated for the three temperature regions. For HDPE radical decay in the region of 200 K is found to be dependent on the degree of crystallinity and the activation energy (12 kcal mol^{-1}) is in agreement with molecular motions at lamellar surfaces. For LDPE, on the other hand, decay in this region does not appear to depend on the degree of crystallinity and the calculated activation energy (25 kcal mol^{-1}) more closely corresponds to motion in amorphous regions rather than lamellar surfaces.

Radical decay at the lower temperatures could not be directly related to any relaxation processes evident in dynamic mechanical tests. It has been presumed that since the fraction of radicals decaying in this region was proportional to the amorphous content, the mechanism probably only involves very small scale movement of molecules in the amorphous regions, not detectable in mechanical tests. This appears to be one example where ESR techniques may prove to be very sensitive for probing localized molecular motion in polymers.

Differences in spectral shape which are observed for radicals decaying in the different temperature regions indicate differences in the trapping sites in the polymer. The influence of polymer morphology on the efficiency of radical trapping has been of considerable interest over the years and ESR studies should be a means of providing specific information on morphological effects. Thus it has been shown that radicals tend to be produced fairly uniformly throughout the polymer but then migrate to the crystalline lamellar surfaces ultimately to decay and form crosslinks.

There is also evidence that the radicals may be trapped preferentially within crystal defects so that ESR provides additional evidence to complement NMR, electron microscopy and viscoelastic measurements on, for example, the increase in the mobile fraction in crystalline polymers on annealing. Radical decay at room temperature has also been shown to be dependent on different morphological features. Activation energies for slow radical decay ($22 \, \text{kcal mol}^{-1}$) are much smaller than those expected for the crystalline dispersion but can be explained by assuming that radical decay only requires movement of shorter chain segments than those responsible for the α_c transition in the chemical relaxation studies. This is very likely a general effect and is found, for example, in polypropylene where activation energies for radical decay above T_g in crystalline regions for both isotactic and syndiotactic polymers are only about half those determined for α_c by viscoelastic techniques.

Since the ESR spectral shape and decay kinetics are sensitive to the environment of the radicals, the technique may be used in conjunction with mechanical treatments which are expected to modify the internal molecular structure of polymers. For example, fatigue of samples subjected to cyclic stresses results in some increase in disordered regions in the bulk polymer. γ-Irradiation of fatigued samples at low temperature and ESR examination can thus be used to give information on the extent of disruption produced which can again be used in conjunction with more conventional density and X-ray measurements.

Radiation effects as evidenced by ESR studies are also influenced by stereoisomerism in tactic polymers. Thus for different tactic forms of polypropylene although the eight line spectra, which arise from the mixture of two radicals ($-CH_2-\dot{C}(CH_3)-CH_2-$ and $-CH_2-CH(\dot{C}H_2)-CH_2-$) are superficially the same, resolution of the hyperfine structure increases with increasing crystallinity in the order $iPP > sPP > aPP$. For the stereo regular forms the spectra can be adequately reconstructed by considering different conformational angles of the β-protons. Power saturation studies, which will be covered more fully later (section 7.8) can also show differences in radical environment which arise from differences in molecular conformations.

It is sometimes more convenient and informative to study the properties of peroxy radicals which are formed by chemical interaction of oxygen with carbon radicals. For example, γ-irradiation of poly(tetrafluoroethylene) in air produces peroxy radicals

$$(-CF_2-\underset{\underset{\displaystyle O-O\cdot}{|}}{CF}-CF_2-)$$

Peroxy radicals have characteristic spectral shapes (Fig. 7.10) due to g-factor anisotropy and the temperature dependence of spectral shape may be used to probe molecular motion in the polymer. A K-band spectrometer may be used here since that is more sensitive to g-factor variations. At 77 K, the spectrum arises due to principal values of the g-factor tensor $g_1 = 2.038$, $g_2 = 2.007$ and $g_3 = 2.002$. With increase in temperature, the peroxy radicals undergo free rotation resulting in partial averaging of the g-tensor such that at 238 K g_1 and g_2 coalesce into a single line due to motional averaging of the tensor elements whereas g_3 is essentially unchanged.

At room temperature, the spectrum can be reconstructed by assuming $g_{\parallel} = 2.006$ (parallel to chain axis) which is equivalent to g_2, and $g_{\perp} = 2.022$ (perpendicular to chain axis) where $g_{\perp} = (g_1 + g_3)/2$. The molecular motion of the chain end peroxy radicals $(-CF_2-CF_2-OO\cdot)$, which are also produced by radiation-induced chain scission reactions, may be considered to be more or less random in three dimensions and the spectra coalesce to a sharp singlet above 240 K due to averaging of the principal elements of the g-tensor.

7.5.2 *Radical pairs*

High energy irradiation may also lead to the production of radicals in pairs, possibly in a single event, which are located sufficiently close together so that

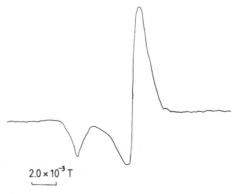

2.0×10^{-3} T

FIGURE 7.10 First derivative ESR spectrum of peroxy radical produced by mechanical degradation of polyethylene in air at 77 K.

their spins couple to give the triplet state for which $S = 1$. Radical pairs may, of course, also be produced on a statistical basis but the concentration of such pairs would be extremely small.

The interaction of the electrons in a radical pair produces characteristic ESR spectra which, in ideal cases, permit identification of the radical species, their separation and relative orientation in the solid. It will be helpful to consider briefly the interaction between radicals I and II which are sufficiently close together (distance r apart) and for which rapid exchange of the unpaired electrons occurs.

The resonance field strengths for the $\Delta M = 1$ transitions are given by

$$H = H_0 \pm \tfrac{1}{2}d - \tfrac{1}{2}\sum (A_i^I M_i^I + A_i^{II} M_i^{II}) \tag{7.36}$$

where d = spectral splitting due to the zero field spin–spin interaction. This is anisotropic and is expressed by

$$d = d_\perp (1 - 3\cos\theta) \tag{7.37}$$

and

$$d_\perp = \frac{3g}{2r^3}\cdot\beta \tag{7.38}$$

where θ is the angle between $\mathbf{r}$ and $\mathbf{H}$. The positive sign corresponds to the $0 \leftrightarrow -1$ and the negative sign to the $0 \leftrightarrow +1$ transitions and these give rise to a main doublet spectrum. The equation illustrates the important features of radical pair spectra. The hyperfine coupling involves interaction of the unpaired electrons with magnetic nuclei in both radicals of the pair and the coupling constants are consequently half those of isolated radicals. Secondly, the $(1 - 3\cos\theta)$ term shows the marked anisotropy of the doublet spin–spin interaction. The doublet lines actually cross over on rotation between $d_{//}$ and $d_\perp$ with respect to θ. If the crystal structure is known, it is then possible to determine the separation as well as the orientation of the radical in single crystals.

In addition it is possible to observe $\Delta M_s = 2$ transitions which are usually forbidden for isolated radicals and arise due to the simultaneous flipping of the two electron spins. The absorption is observed at $g = 4$ and the resonance field values are given by

$$H = H_0 - \tfrac{1}{2}\sum (A_i^I M_i^I + A_i^{II} M_i^{II}) \tag{7.39}$$

The ratio of the transition probability for $\Delta M_s = 2$ to that of $\Delta M_s = 1$ is given by

$$\frac{I_2}{I_1} = \frac{1}{2}\left[\sin^2 2\theta \left(\frac{d_\perp}{H}\right)^2 \right] \tag{7.40}$$

which for randomly orientated solids reduces to

$$\frac{I_2}{I_1} = \frac{4}{15}\left(\frac{d_\perp}{H}\right)^2 \tag{7.41}$$

The intensity of the $\Delta M_s = 2$ transitions are, therefore, approximately 1000 times weaker than the corresponding $\Delta M_s = 1$ transitions and are difficult to detect. Nevertheless, it is the observation of the $\Delta M_s = 2$ transitions which has provided the most direct evidence for radical pair formation in polymers.

Studies of radical pairs in organic single crystals have shown that the two radicals may be produced with specific orientations and separations in the crystal which suggests a specific mechanism for their formation. It is probable that the radicals are produced either on adjacent molecules or by the decomposition of a single molecule.

$\Delta M_s = 2$ transitions have been observed for a range of polymers irradiated by γ-rays or high energy electrons at 77 K and where the structure is sufficiently well resolved, the hyperfine interactions have been found to be about half those for the $\Delta M_s = 1$ transitions. As might be expected, irradiation with high energy electrons gives higher concentrations of radical pairs than samples irradiated with γ-rays as judged from the $\Delta M_s = 2$ peaks. This is a consequence of the higher rate of energy deposition or linear energy transfer (LET). However, since the intensity of the $\Delta M_s = 2$ peak is also dependent on the inter-radical distances, it is conceivable that the weaker signals observed following γ-irradiation may be due to differences in the nature of the radical pairs. The spectra observed at $g = 2$ are composed of $\Delta M_s = 1$ transitions of the radical pairs as well as those of isolated radicals and it is generally difficult to identify unequivocally the radical pair contribution. Changes in the relative intensities of $\Delta M_s = 1$ and $\Delta M_s = 2$ peaks with increasing radiation dose will usually show that radical pairs are not due simply to accidental overlap of adjacent pairs but are formed by specific mechanisms.

If orientated samples are available, these will provide better resolved spectra and, in addition, the orientation dependence permits estimates of inter-radical distances to be made. With this information and knowledge of the crystal structure from, for example, X-ray diffraction measurements, it may then be possible to identify specific radical pairs which is particularly useful in understanding some of the primary radiation events. A particular area of interest is the role that radical pairs might play in radiation-induced crosslinking reactions and also their possible involvement in solid state polymerizations where radical pairs have been detected.

7.6 Mechanical degradation

Radicals may be produced in stressed polymers when the magnitude of the applied stress is sufficient to produce homolytic scission of covalent bonds. Provided that the radicals are produced in sufficient concentration, ESR may then be used to give information on the fracture processes.

In general, radicals may be produced by grinding or milling of polymers when the radicals are produced mainly on the exposed fracture surfaces, or by subjecting the polymer to tensile, torsional, compressive or impact forces when radicals are produced up to the point of fracture of the samples. Fairly large radical concentrations are produced by grinding and milling at low temperatures due to large fracture surface areas but less success is achieved by tensile stressing of polymers. Fracture of moulded or cast samples does not usually produce measurable radical concentrations, but the use of drawn fibres leads to an increase of the order of 10^5 in the number of radicals produced by tensile stresses. The use of radical traps such as hydroquinone, duraquinone or chloranil, which are dispersed in the polymer prior to stressing and react with mechanically generated radicals to give more stable radicals, may be necessary for systems where the primary scission radicals are unstable and difficult to detect directly. Because of the sensitivity of free radicals to oxygen, it is essential in studying mechanical degradation to ensure that oxygen is excluded from the apparatus. This may involve the use of vacuum systems or inert gas atmospheres.

Information on fracture mechanisms may be derived from studies of radical decay in air which gives some indication of the location of the radicals. By comparison with the calculated number of chains passing through the fracture surface, it has been shown that for spherulitic partially crystalline polymers, many of the detected radicals lie at some distance from the fracture surface. This is taken to imply that fracture takes selective paths via less densely packed tie molecule amorphous regions which involve less bond rupture than if plane fracture occurs through the spherulites. For possibly less obvious reasons, the fracture of amorphous glassy polymers also appears to produce radicals at some distance from the fracture surface. It is likely that craze formation in the bulk of the polymer is responsible for these effects.

Grinding of polymers can lead to significantly high static charge build-up if the system is sufficiently well insulated. There is some evidence that deliberate earthing of specimens can increase radical decay rates on subsequent annealing. The reasons for this are not clear, but it does serve to highlight potential problems that may arise.

Possibly more valuable information can be obtained by direct tensile stressing of samples in the ESR cavity. Ideally provision should be made for direct measurement of load and extension and for evacuation of the sample chamber. Controlled introduction of oxygen to induce radical decay can then be carried out. The method of stressing can be either constant load or stepwise constant strain. For constant strain experiments, it is essential that the loading equipment is sufficiently rigid so that the strain can be maintained constant during stress relaxation experiments.

The particular value of ESR has been to demonstrate the influence of microstructure on deformation mechanisms. The spectra of radicals observed are essentially isotropic and evidently they are trapped in disordered amorphous regions. The sensitivity of the radicals to oxygen is also indicative of the radical location in the more accessible amorphous regions.

Although it is clear that the trapped radicals result from chain scission it should not necessarily be assumed that the numbers of radicals observed represent the number of fractured chains. The detection of chain end groups by infra-red spectroscopy has shown that chain scission may proceed by a radical initiated chain reaction mechanism. Kinetic studies of radical production indicate that mechanical degradation is a thermomechanically activated process, and activation energies may be determined. It follows from this that photolysis of stressed polymers could perhaps lead to enhanced radical formation and this has been demonstrated for a number of polymers. Ultraviolet light of different wavelengths is likely to be needed to produce preferential radical formation in different polymers.

7.7 Thermal degradation

Although of great importance technologically, thermal degradation of polymers has not proved particularly amenable to ESR investigation. The ESR spectra of thermally treated PVC and PVDC give singlet spectra of polyene radicals produced at high levels of degradation. The particular problem encountered is that at degradation temperatures radicals show rapid decay. It is possible to quench thermally treated samples in, for example, liquid nitrogen but only limited direct information on degradation processes is likely to be obtained. However, ESR in conjunction with, for example, DTA and TGA studies should provide useful additional information on reaction mechanisms.

7.8 Relaxation studies

Relaxation effects have already been introduced and from the earlier discussion it should perhaps be evident that in principle it should be possible to derive some understanding of the environment of radicals which determine spin–lattice and spin–spin relaxation times and hence ESR linewidths. There are various ways in which relaxation studies may be made but all involve some form of power saturation of the ESR signal. At this point, it will be helpful to distinguish between homogeneous and inhomogeneously broadened lines. Homogeneous broadening arises directly from the relaxation processes introduced earlier. When spin–spin relaxation is the predominant mechanism the spectrum approximates Gaussian line-shape and the relaxation time T_2^1 may be obtained from the linewidth between points of maximum slope $\Delta H_{msl} = 2/T_2$. For pure spin–lattice relaxation the line shape is Lorentzian and $\Delta H_{msl} = 2/\sqrt{3}T_2$.

Inhomogeneous broadening, which is more likely to be found in polymer systems, arises due to effects such as unresolved hyperfine interactions, anisotropy broadening, dipolar interactions between unlike spins or even inhomogeneities in the magnetic field. In these cases, the overall line may be composed from spin packets each experiencing local magnetic fields and having Lorentzian line shapes but giving an overall Gaussian line shape.

Relaxation times may be determined from saturation studies in which the signal intensity or output voltage, V_R, is measured as function of increasing microwave power, H_1. For homogeneously broadened lines

$$V_R \propto \frac{H_1}{1 + \gamma^2 H_1^2 T_1 T_2}$$

whereas for inhomogeneously broadened lines

$$V_R \propto \frac{H_1}{(1 + \gamma H_1^2 T_1 T_2)^{1/2}}$$

The saturation behaviour of polymers is found to be intermediate between homogeneous and inhomogeneous broadening with the curve of V_R versus H_1 passing through a maximum.

From experimental saturation curves, the following parameters may be defined

1. the intersection of the linear unsaturated region with the tangent to the maximum $(H_1(1/2))$
2. the value of $H_{1/2}$ which makes V_R half its maximum at high values of H_1
3. the value of H_1 which gives $V_{R/2}$ at low values of $H_1(H_1^1(V_{R/2}))$

The ratio $2:3$ may be shown to depend on $a = \Delta H_L / \Delta H_G$ where $\Delta H_L = 1/\gamma T_2 . L$ and G refer to Lorentzian and Gaussian line shapes respectively. T_2 may then be determined from

$$T_2 = \frac{\sqrt{2}}{\gamma \Delta H_{msl} a} \tag{7.42}$$

Hence T_1 may then be calculated since

$$H_1(\tfrac{1}{2}) = \frac{1}{\gamma (T_1 T_2)^{1/2}} \tag{7.43}$$

7.9 Conclusions

We have seen that, although electron spin resonance spectroscopy is restricted to samples containing paramagnetic species, there are important applications in the characterization of polymers. The technique provides direct evidence of reaction intermediates in polymerization and degradation reactions and, in many cases, permits these to be identified unequivocally. The insight this provides into reaction mechanisms greatly improves understanding of stereochemical polymerizations and relative reactivities and chain sequences in copolymerizations.

The influence of morphology on degradation reactions can be deduced more or less directly from ESR studies and assists our understanding of polymer stabilization. Structural information derived from ESR also complements X-ray and electron diffraction techniques. Since radical reactions are dependent, in part, on chain segmental motion the technique also provides a useful method for investigating thermal transitions in polymers. Thus, despite the fact that ESR provides indirect evidence on structure, selective application of the technique provides valuable information not accessible by other methods.

Further reading

1. Ayscough, P.B. (1967) *Electron Spin Resonance in Chemistry*, Methuen, London.
2. Carrington, A. and McLachlan, A.D. (1967) *Introduction to Magnetic Resonance*, Harper and Row, New York.
3. Ranby, B. and Rabek, J.F. (1977) *ESR Spectroscopy in Polymer Research*, Springer Verlag, Berlin.
4. Fischer, H. (1974) in *Polymer Spectroscopy*, (ed. Dieter O Hummel), Verlag Chemie, Weinheim, chapter 4.
5. Poole, C.P. (1967) *Electron Spin Resonance, Treatise on Experimental Techniques*, Interscience, New York.

Exercises

7.1 Calculate the ratio of the population of electron energy levels in magnetic fields of (a) 0.3 T and (b) 1.0 T at 100 K and 300 K. Comment on the significance of the results to ESR measurements.

7.2 Deduce the expected hyperfine pattern for radicals containing the following interacting nuclei:

   (i)   3 equivalent protons
   (ii) 3 equivalent deuterons
   (iii) 2 equivalent protons (A) and 3 equivalent protons (B) where $a_H^A = 2a_H^B$.
   (iv) 2 equivalent nuclei having $I = \frac{3}{2}$.

7.3 Photolysis of the free radical initiator AZBN (azobisisobutyronitrile) yields the radical

$$(CH_3)_2\dot{C}-CN$$

Assuming that interaction occurs with each of the methyl protons and with the nitrogen atom, construct the expected hyperfine pattern for the radical in solution.

7.4 The ESR spectrum of γ-irradiated, stretched poly(ethylene terephthalate) film shows marked anisotropy. With the magnetic field aligned parallel to the plane of the film, the spectrum is an octet of lines of equal intensity due to interaction with 3 protons, H_1, H_2, H_3, having coupling constants of 3.4, 2.9 and 1.8×10^{-3} T respectively. On rotation about the orientation axis, two of the coupling constants (a_{H_2}, a_{H_3}) remain essentially constant whilst the third (a_{H_1}) decreases in such a manner that when the field is perpendicular to the plane of the film, the spectrum is a sextet with intensity ratios $1:2:1:1:2:1$.

(i)   What is the value of the anisotropic coupling constant a_{H_1} with the field perpendicular to the film?
(ii) Show how the observations can be explained on the basis of the radical

(iii) By reference to equation (7.32), estimate the conformation angle of the β-CH_2 bonds.

7.5 Given that the peak to peak separation of the first derivative of an ESR absorption line is 10^{-5} T and that only spin–lattice relaxation takes place, calculate the spin–lattice relaxation time T_1 for the radicals.

What would be the relaxation time if only spin–spin relaxation occurred? (Assume $g = 2$.)

8

X-ray diffraction

8.1 Introduction

X-ray diffraction is the traditional method of crystallographic structure determination, and the standard techniques and analysis procedures can be used in the study of crystalline polymers. The most straightforward application is in the determination of crystal lattice spacings, but with refinement complete structure determination can be achieved, including the position of pendant atoms or groups (e.g. the hydrogen atoms in polyethylene crystals). Other information that can be obtained by X-ray diffraction includes crystal size and perfection, the 'long period' in lamellar polymers (= lamellar thickness + interlamellar region), the crystallinity (of semicrystalline polymers), the degree of preferred orientation in polycrystalline samples, and, in some recent studies, the conformation of chains in amorphous polymers.

There are numerous ways in which the X-ray source, the sample and the X-ray detection system (often a photographic film) can be arranged. The one chosen is dictated by the structural characteristics of the sample chosen (e.g. whether it is a single crystal or an aggregate of randomly oriented crystals) and by the information sought. A brief survey of the principles involved is included here; more detailed descriptions are to be found in standard texts. Some general diffraction theory is introduced, and this will serve also the chapter on electron diffraction and microscopy. Emphasis is placed on those aspects that are of most value in the analysis of polymers. Some of these, such as determination of crystallinity, are often ignored in general texts on X-ray methods.

8.2 Generation and properties of X-rays

X-rays are produced by bombarding a metal target with high energy electrons. The electrons are emitted from a heated filament held at a high negative potential (usually 20–60 kV) and are accelerated towards the target held at earth potential. These components are assembled into an evacuated tube to avoid scattering of the electrons by gas atoms in their path from the filament to the target. When a high energy electron passes into the target it is scattered many times and photons are produced as it decelerates. A large fraction of these have energies within the X-ray range, and form a continuous distribution of energies (and hence wavelengths).

However a second process, which is more important for the purpose of generating X-rays for analytical purposes, occurs when the bombarding electron ejects an electron from an inner shell of a target atom. An electron from a higher shell then falls into the lower depleted shell to restore its equilibrium structure accompanied by the emission of an X-ray photon. The X-ray photon has an energy equal to the difference in energy between the two participating shells and is thus discrete. Hence a substantial portion of the total X-ray spectrum emitted by the target will have particular discrete energies (and hence wavelengths) that are characteristic of the element(s) present in the target.

Ideally, the X-ray source should contain one wavelength only ('monoenergetic') so targets are made from a single element, popular ones being copper, iron, cobalt and chromium. The most intense X-ray component is that produced by electrons falling from the second (L) shell into the first (K) shell. This is called the K_α X-ray and can be separated from the rest of the spectrum by using filters to absorb the unwanted components; if it is necessary to remove components that are too close to the K_α wavelength to be removed by filters, a crystal oriented such that it diffracts (only) the chosen wavelength can be employed, and the diffracted radiation used as the source.

The X-rays escape from the tube through a window which is transparent to the required component.

8.3 Diffraction theory

The first objective here is to examine the characteristics of X-ray diffraction when a monoenergetic parallel X-ray beam impinges on a crystal. Essentially, the task is to add together the contributions scattered from each of the atoms of the crystal. Although scattering of X-rays occurs primarily by interaction with the electrons within the target (sample), the atomic (nuclear) positions can be treated as the scattering centres. The scattering from an individual atom can be represented by its own particular 'scattering factor' to represent the scattered amplitude, which will be a function of direction, $f(\theta)$, where 2θ is the angle between the direction in which the amplitude is measured and the direction of the incident radiation.

After dealing with crystal diffraction consideration is given to the influence of crystal defects, to scattering from non-crystalline samples, and to scattering at small angles from features on a larger scale such as microvoids and stacks of crystal lamellae.

8.3.1 *The Laue conditions*

Derivation of the Laue conditions can be found in standard texts on X-ray diffraction and an abbreviated version is presented to introduce the nomenclature used here and in Chapter 9. In Fig. 8.1(a) scattering centres at P and Q are in the path of a parallel coherent beam of monoenergetic radiation with wave vector **k**, which we will choose to have length $|\mathbf{k}| = 1/\lambda$ where λ is the wavelength. The plane wave travelling in the z-direction (parallel to **k**) can be represented by a function of the form

$$\psi = \psi_0 \exp(2\pi i k z).$$

Radiation will be scattered in all directions by both P and Q and for the direction represented by the wave vector **k'**, the path difference between the waves scattered from P and Q (Fig. 8.1a) is:

$$AQ - PB = \frac{\mathbf{k} \cdot \mathbf{r}}{|\mathbf{k}|} - \frac{\mathbf{k'} \cdot \mathbf{r}}{|\mathbf{k'}|} = \lambda(\mathbf{k} - \mathbf{k'}) \cdot \mathbf{r} \qquad (8.1)$$

The corresponding phase difference is

$$2\pi(\mathbf{k} - \mathbf{k'}) \cdot \mathbf{r} = -2\pi \mathbf{K'} \cdot \mathbf{r} \qquad (8.2)$$

where **K'** is represented diagrammatically in Fig. 8.1(b), and has length $|\mathbf{K'}| = (2\sin\theta)/\lambda$. It will be noted that $|\mathbf{k'}|$ has been set equal in length to $|\mathbf{k}|$ ($= 1/\lambda$) implying that the scattered radiation has the same wavelength as the incident radiation and that our considerations concern only scattered radiation which has suffered no energy change ('elastic scattering'). Generally there will be in addition some inelastic scattering involving energy loss, but this does not interfere seriously with the measurement of the elastically scattered component.

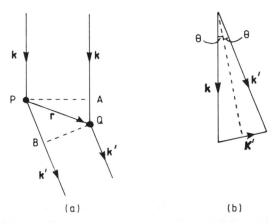

(a)   (b)

FIGURE 8.1 (a) An incoming X-ray beam, wave vector **k**, impinging on scattering centres at P and Q. A particular scattering direction represented by the wave vector **k'** is chosen for consideration; (b) vector diagram, defining the scattering vector **K'**.

From this it can be shown that the total amplitude scattered from an assembly of n atoms is

$$\psi_{tot} = \sum_{n} f_n(\theta) \exp(-2\pi i \mathbf{K}' \cdot \mathbf{r}_n) \tag{8.3}$$

where $f_n(\theta)$ is the scattering factor for the nth atom located at $\mathbf{r}_n$. In the case of a crystal it is now convenient to decompose r_n in the following manner:

$$r_n = m_x\mathbf{a} + m_y\mathbf{b} + m_z\mathbf{c} + p_x\mathbf{a} + p_y\mathbf{b} + p_z\mathbf{c} \tag{8.4}$$

where $\mathbf{a}$, $\mathbf{b}$ and $\mathbf{c}$ are the lattice vectors defining the crystal, m_x, m_y and m_z are integers that locate the unit cell within which the nth atom is located, and p_x, p_y, p_z locate the particular atom within the unit cell, and may therefore take fractional values (Fig. 8.2). At this point we have to introduce the concept of the reciprocal lattice. This idea is discussed in standard texts on crystal diffraction and we will choose simply to give the mathematical form of the definition of the reciprocal lattice. Thus the reciprocal lattice corresponding to the crystal lattice with lattice vectors $\mathbf{a}$, $\mathbf{b}$ and $\mathbf{c}$ is represented by vectors $\mathbf{a}^*$, $\mathbf{b}^*$ and $\mathbf{c}^*$ where $\mathbf{a}^* \cdot \mathbf{a} = \mathbf{b}^* \cdot \mathbf{b} = \mathbf{c}^* \cdot \mathbf{c} = 1$ and $\mathbf{a}^* \cdot \mathbf{b} = \mathbf{b}^* \cdot \mathbf{c} = \mathbf{c}^* \cdot \mathbf{a} = 0$.

It is convenient to express $\mathbf{K}'$ with reference to the reciprocal lattice as

$$\mathbf{K}' = u\mathbf{a}^* + v\mathbf{b}^* + w\mathbf{c}^* \tag{8.5}$$

If equations (8.4) and (8.5) are substituted into equation (8.3) then the total scattered amplitude becomes

$$\psi_{tot} = \left\{ \sum_{\text{sets of } p} f_n(\theta) \exp(-2\pi i(p_x u + p_y v + p_z w)) \right\}$$
$$\times \left\{ \sum_{\text{set of } m} \exp(-2\pi i(m_x u + m_y v + m_z w)) \right\} \tag{8.6}$$

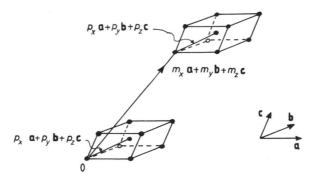

FIGURE 8.2 Unit cell with one corner located at the position chosen to be the origin, 0. Atoms within the unit cell are located at positions $p_x\mathbf{a} + p_y\mathbf{b} + p_z\mathbf{c}$. A second unit cell is shown, with the corresponding corner located at $m_x\mathbf{a} + m_y\mathbf{b} + m_z\mathbf{c}$.

The first summation relates simply to the repeat unit (unit cell) and the second one sums the contributions from all unit cells. The second summation is a maximum when $(m_x u + m_y v + m_z w)$ is a whole number, giving the condition that contributions from all unit cells are in phase. Since m_x, m_y and m_z can take all even and odd integer values in all combinations the only way to fulfil this condition is to put $u = h$, $v = k$, $w = l$ where h, k and l are all integers. These are the Laue conditions, and the particular value of $\mathbf{K}'$ at which these conditions are met is normally called the diffraction vector $(\mathbf{g})$. Each set of lattice planes gives rise to a different diffraction condition and is represented by its own particular diffraction vector, often written $\mathbf{g}_{hkl}$. Thus, at a diffraction maximum

$$\mathbf{K}' = h\mathbf{a}^* + k\mathbf{b}^* + l\mathbf{c}^* = \mathbf{g}_{hkl} \qquad (8.7)$$

It can be shown that $g_{hkl} = 1/d_{hkl}$ where d_{hkl} is the crystal lattice spacing of the hkl planes. Thus when the Laue conditions are fulfilled we have

$$|\mathbf{K}'| = |\mathbf{g}| = 1/d = (2\sin\theta)/\lambda \qquad (8.8)$$

This is the familiar Bragg law, which can be derived more directly without also establishing the Laue conditions.

It is now convenient to introduce a useful geometrical representation known as the Ewald sphere. The Ewald sphere is constructed with radius $1/\lambda$ and in Fig. 8.3 two particular radii are shown as $\mathbf{k}$ and $\mathbf{k}'$, which lie in the incident beam direction and the diffracted beam direction respectively (cf

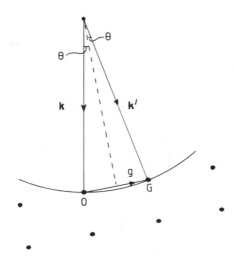

FIGURE 8.3 Ewald sphere construction, showing the reciprocal lattice points of which one, at G, lies on the sphere at the particular orientation shown. The direction of the incident beam is given by $\mathbf{k}$ and that of the diffracted beam by $\mathbf{k}'$.

Fig. 8.1b). The vector **k** intersects the sphere at the origin (O) of the reciprocal lattice and if the Laue conditions are fulfilled for diffraction in the direction **k′** then **k′** intersects the sphere at a point (G) in the reciprocal lattice of the crystal. Other reciprocal lattice points are shown and it is clear that only a small number can simultaneously lie on the surface of the sphere, confirming that for a single crystal illuminated with a parallel monoenergetic beam of X-rays very few reflections will be obtained in any particular setting (usually none or one, but multiple reflections can be obtained for certain special settings).

8.3.2 Structure factor

If we return to equation (8.6) and examine the first summation on the right hand side, this can now be modified by replacing u, v and w by integers h, k and l, for the second summation goes to zero when this condition is not met. Thus the first summation becomes

$$F = \sum_{\substack{\text{sets} \\ \text{of } p}} f_n(\theta) \exp\left(-2\pi i(p_x h + p_y k + p_z l)\right) \qquad (8.9)$$

This is known as the structure factor and gives an indication of the relative intensity of diffraction from different sets of planes within the same crystal. Sometimes as the result of the symmetry and composition of a crystal the structure factor may be zero for a particular set of planes. Thus certain sets of (hkl) planes do not give observable diffraction ('forbidden' reflections). This is known as the structure factor rule. Crystals with different symmetries have different structure factor rules, and this can be an aid to structure identification.

To evaluate the structure factor and derive the structure factor rule from equation (8.9) the first task is to choose the set of p_x, p_y, p_z values to represent the unit cell. It is essential to give proper weighting to each atom. For example, in a cubic structure each of the eight corner atoms in the representative repeat motif is shared by eight neighbours, and only one should be chosen for the structure factor calculation.

With polymer crystals the structure can be very complicated with many atoms per unit cell and at least two atomic species, more often three or four, (e.g. the polyamides ('nylons') often crystallize, and contain C, H, O and N). Thus the structure factor summation contains many more terms and there is much less likelihood of systematic absences than with say cubic structures, though large intensity differences can still occur between different reflections and act as a guide to structure identification. Standard procedures can be used to identify the crystal class and to index the reflections. Measurements of the angular positions of the (indexed) reflections then permit calculation of the interplanar spacings within the crystal by the application of Bragg's law. However, this does not give a complete determination of the location of every atom, especially those present in side groups. It is not a simple task to determine the 'setting angle' of the zig-zag backbone in polyethylene (Fig. 8.4)

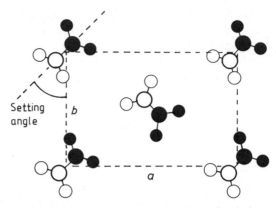

FIGURE 8.4 Projection of the unit cell of orthorhombic polyethylene along the c-axis. The molecule has an all-*trans* conformation, with the carbon atoms forming a planar zig-zag in a plane perpendicular to the diagram. Solid circles lie in planes above the adjacent open circles. Solid circles at the corner sites are all coplanar, but those in the molecule passing through the centre are displaced along the c-axis with respect to those at the corners.

though this is probably the simplest of all polymer crystal structures. To refine the crystal structure determination it is necessary to make very accurate intensity measurements of as many different reflections as possible, then to compare their relative magnitudes with those computed from the structure factor expression for selected structures. The structure chosen must conform with the measured lattice parameters, but this still leaves open the possibility of polymer main chain rotation, and a variety of side chain configurations.

(a) THE DEBYE FACTOR

In the preceding analysis it is assumed that the atoms are at rest at their lattice positions. In reality they execute thermal vibrations about their mean positions, and to take account of this each term in the structure factor summation must be multiplied by the Debye (or Debye-Waller) factor

$$\exp\left(-16\pi^2\bar{u}^2\sin^2\theta/\lambda^2\right)$$

where $\bar{u}^2$ is the mean square displacement of the atom normal to the reflecting planes. This causes the intensities of the diffraction peaks to fall as the temperature increases.

8.3.3 Line broadening

If an infinite perfect crystal is irradiated by monoenergetic parallel X-rays then diffracted rays will appear only when the Laue conditions are exactly met, and the diffracted rays will themselves be parallel and appear in discrete directions. In practice these ideal conditions are never met. Crystals are usually limited in size and contain defects, and although it is usually arranged that the X-ray beam is monoenergetic it will have a finite angular divergence. Let us consider the effect of these factors one by one, starting with those associated with the X-ray source.

Firstly, since the source of X-rays is a transition between discrete atomic energy levels, the X-ray energy is fixed, though it is sometimes necessary to take precautions to isolate the particular X-ray energy required when several lines may be available from the target. Collimation using slits and/or focusing using curved crystals can be used to produce a narrow beam with minimal angular divergence ('parallel'). Any remaining divergence in the incident X-ray beam will be carried through into the diffracted beam. This is often called 'instrumental broadening' because it does not depend in any way on the sample. When the diffracted wave is broadened by sample-related factors in addition to the instrumental broadening it is desirable to separate the two components of broadening. The instrumental broadening contains no information about the sample, but should be subtracted from the total broadening before any quantitative assessment of the sample-related factors is attempted.

Removal of an instrumental component such as this is a common problem in optics and spectroscopy (and in many other branches of physics). In mathematical terms, the way the instrumental broadening and the other sources of broadening interact is by multiplying together to form a 'convolution product'. To see how this works, consider the broadened profile shown in Fig. 8.5(a), which represents the diffracted beam ('signal') as it would appear if free from instrumental broadening, but broadened by the sample. The instrumental broadening profile is shown in Fig. 8.5(b), which shows the shape of the signal that would be obtained from an infinite perfect single crystal in place of the actual sample, and with no other alterations to any of the settings of the diffraction equipment. The task now is to multiply ('convolute') these two profiles. What happens is that each part of the profile shown in Fig. 8.5(a) is 'broadened' by the profile shown in Fig. 8.5(b). Examples are shown in Fig. 8.5(c), where selected thin strips from Fig. 8.5(a) have been represented broadened in this way. This process now has to be expanded until the whole of the profile shown in Fig. 8.5(a) has been dealt with in like manner, then the several contributions are added together to give the final result, which should represent the observed profile, containing both sample related broadening and instrumental broadening (Fig. 8.5d). Thus the observable profile is that represented in Fig. 8.5(d) and the one required is that containing information about the sample (Fig. 8.5a).

The multiplication process (convolution) outlined above now has to be reversed ('deconvolution') to recover Fig. 8.5(a). To do this the instrumental profile, Fig. 8.5(b), is required. This will be the same for all samples and can be determined experimentally with the help of a specially prepared calibration sample consisting of large crystals with minimal defect content, sometimes a naturally occurring mineral. With such a sample crystal size and defect

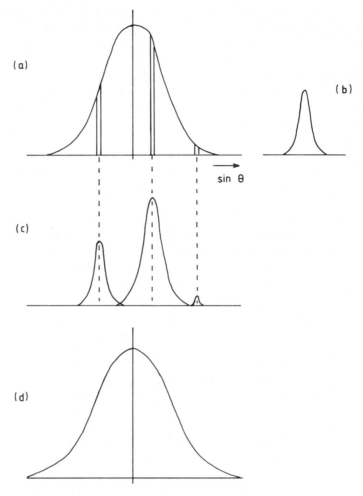

FIGURE 8.5 Convolution: (a) intensity distribution in a diffracted X-ray beam, free from instrumental effects; (b) instrumental broadening profile, as would be recorded for large perfect crystals; (c) instrumental broadening applied to selected, vanishingly narrow, portions of profile (a); (d) summation of broadened narrow strips covering the whole of (a). This is the convolution product, and is what is observed.

broadening are negligible so that the observed broadening can be assumed to be instrumental broadening only. The deconvolution process can be fairly laborious even with the help of a computer and for many purposes it is often sufficient to apply a simple approximation formula to correct for instrumental broadening and recover a measure of the peak breadth. A popular measure of broadening is the breadth (b) of the diffracted intensity curve at half the peak height (Fig. 8.6) and the most widely used correction formula is $b^2 = b_o^2 - b_i^2$, where b is the breadth of the corrected curve, b_o is the breadth of the observed curve and b_i is the breadth of the instrumental profile.

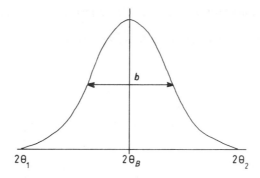

FIGURE 8.6 Schematic diffraction peak profile defining b, the peak breadth at half the peak height.

Turning now to sample-related sources of X-ray line broadening, the one that is most straightforward to analyse is the small crystal size effect. This is best understood by considering what happens if a crystal is oriented close to, but not exactly at, the Bragg position. As has already been established, if the crystal is large and perfect there will be no diffracted intensity at this setting, the consequence of destructive interference. In any chosen observation direction the phases of contributions from neighbouring scattering centres are slightly different, and from a large crystal at any instant in time all phases will be represented equally, positive and negative, thus causing mutual cancellation and leaving a net amplitude of zero. On the other hand, if the crystal is small the range of phase variation is limited, complete cancellation of all contributions is no longer possible and a residual amplitude remains even when the crystal setting departs significantly from the Bragg position. The second source of broadening is from defects. Crystals always contain defects and in most polymer crystals the defect concentration is probably quite large for reasons discussed below. Scattering from defects will clearly have different phase components to those which would occur if the perfect crystal had remained uninterrupted in the vicinity of the defect.

(a) SMALL CRYSTAL SIZE BROADENING

Several different treatments of the effect of crystal size on diffraction line broadening lead to the Scherrer equation:

$$t = \frac{K\lambda}{B \cos \theta} \tag{8.10}$$

where B is the breadth in radians, θ is the Bragg angle and t represents the extent of the crystal measured in the direction normal to the planes to which the diffraction peak corresponds. The coefficient K normally takes a value close to 0.9, but depends to some extent on the shape of the crystals. If the diffracted intensity is presented as a function of $((2 \sin \theta)/\lambda)$ instead of θ, then the breadth (β) becomes $(B \cos \theta_B)/\lambda$, whence $t = K/\beta$.

(b) DEFECT AND DISTORTION BROADENING

Broadening of the diffraction peaks may occur when the atoms are displaced from their ideal crystal lattice positions. Before considering this in some detail let us dispense with two cases in which broadening does not occur.

1. If the crystal lattice is strained uniformly the peak positions alter to correspond to the new lattice dimensions but no broadening occurs. (Note that bending the crystal does not produce uniform lattice strain and that bending does lead to broadening.)
2. If the displacements of the atoms from the true lattice positions is random with a Gaussian distribution then no broadening occurs. This is the case with thermal vibrations but is also true if a similar displacement distribution is 'frozen' into the crystal whence the main effect is a reduction of intensity as with thermal vibrations.

In both of these examples, the memory of the unstrained perfect crystal lattice is maintained, and when this is true the distortion is described as a 'lattice distortion of the first kind'. However some lattice distortions of the first kind do produce broadening. Examples are:

1. point defects, such as vacancies and interstitials
2. line defects, such as dislocations

In the case of polymer crystals, the vacancy is less easily understood than in, say, a metallic crystal, for the covalently bonded molecules from which the crystals are assembled do not readily permit the absence of an atom. Interstitials are much more easily accommodated and will take up positions in between polymer chains. Atoms or small molecules such as oxygen or water take up such positions and may diffuse through the crystal by hopping from one interstitial site to another. Other types of point defects that might occur in polymer crystals are associated with side group substitution or with chain ends. The atomic displacements around a point defect will depend on its size and will be strongly influenced by the stiffness anisotropy in polymer crystals. The strains around a dislocation are likewise determined by the (anisotropic) elastic properties of the crystal. It should be noted that if a polymer chain terminates within a crystal (as at A in Fig. 8.7) then the defect looks similar in plan to an edge dislocation, but is confined to a single plane. The strains will, of course, be three dimensional but are quite different from those obtained with an edge dislocation.

If the distortions are such that long range order is destroyed they are described as 'lattice distortions of the second kind'. This can occur if low angle

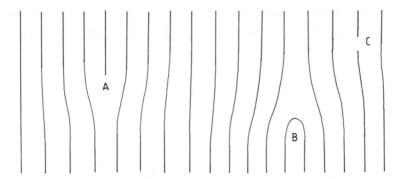

FIGURE 8.7 Examples of defects in polymer crystals: A, terminating chain; B, buried loop; C, two terminating chains in close proximity.

boundaries are present, producing what is sometimes called a mosaic structure. Lattice distortions of the second kind may well occur in polymers as a result of their chain structure. For example in a highly drawn crystalline fibre the regions between neighbouring crystallites may contain molecules that are highly parallel even though they are not arranged in perfect crystal register. Thus they will possess a certain short range order that would permit each of the chain atoms to be associated with a particular crystal lattice point from which it is displaced. Thus although within such a region there may be some short range order an extended region of this kind would not possess long range

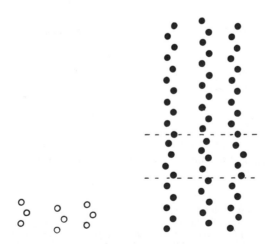

FIGURE 8.8 Schematic representation of paracrystallinity. The repeat motif for the perfect crystal is shown by the open circles and can be used to generate the structures above and below the region limited by the broken lines. Ordering is less perfect within the region bounded by the broken lines, each atom being displaced from its lattice site by a different amount.

order and the intervention of such regions between crystallites, as shown in Fig. 8.8, prevents long range order within the assembly of connected undistorted crystallites. This concept of a structure with short range order that can be associated (locally) with a crystal lattice, yet in which the distortional displacements of the atoms are not on average zero about the lattice points (as required for the preservation of long range order) but are instead accumulative, has been given the name 'paracrystallinity'.*

The detailed mathematical analysis of the effect of lattice distortions of the second kind is beyond the scope of this book. An approximate result is that the intensities of diffraction peaks of several orders (p) are modulated by a factor $\exp\{-2\pi(\Delta d_2/d)^2 p^2\}$ where Δd_2 is the mean displacement from the perfect lattice spacing. The broadening, in terms of $((2\sin\theta)/\lambda)$, is given by

$$\beta_2 = \pi(\Delta d_2/d)^2 p^2/d$$

Small crystal size and defect broadening are independent of one another and if both are present they lead to a combined breadth given by

$$\beta_{tot}^2 = K^2/t^2 + \pi^2(\Delta d_2/d)^4 p^4/d^2$$

Separation of the small crystal size and defect contributions is possible if reflections of different orders are present. A plot of β_{tot}^2 versus p^4 should produce a straight line with intercept K^2/t^2 and slope $\pi^2(\Delta d_2/d)^4/d^2$.

This analysis does not provide information about translational disorder and for a more complete treatment it is necessary to refer to the original works of Hosemann. It should be noted that although this approach may help in the development of an overall view of the structure and defect content of polymer crystals, it is unlikely that routine replication of this analysis will provide information essential to the solution of practical problems.

8.3.4 Conformation determination and radial distribution analysis

The treatment of diffraction outlined in section 8.3.1 relates to crystals. Now consider instead the scattering obtained from an array of N atoms which are not necessarily arranged on a regular crystal lattice. Thus using equation (8.3) for the total scattered amplitude, the intensity can be written

---

*Hosemann, who introduced the idea of paracrystalline order in polymers, claims that his studies show that all polymer crystals should be described as paracrystals, and that perfect crystal regions do not exist. If this is true, the structure depicted in Fig. 8.8 cannot occur. Even if this were so Fig. 8.8 does illustrate paracrystalline structure and emphasizes the relationship between the perfect crystal and the paracrystal. Hosemann's results are clearly the consequence of the highly defective state of (most) polymer crystals (exceptions may be those grown by solid state polymerization, a type not studied by Hosemann) and his view must be acknowledged. Thus when the term 'crystal' is used elsewhere in this book it should perhaps more correctly be replaced by a description such as 'a paracrystalline region with a low degree of paracrystalline disorder'.

$$I = |\psi|^2 = \sum_{n=1}^{N} \{f_n(\theta)\}^2 + \sum_{\substack{n \ j \\ n \neq j}} f_n(\theta) f_j(\theta) \cos 2\pi \mathbf{K}' \cdot \mathbf{r}_{nj} \qquad (8.11)$$

where $\mathbf{r}_{nj}$ is the vector joining the nth and jth atoms. In an isotropic sample the vectors $\mathbf{r}_{nj}$ are randomly oriented as long as a large enough sample is taken; this is true whether the sample is non-crystalline or polycrystalline. For this case the spherical average of $\cos 2\pi \mathbf{K}' \cdot \mathbf{r}_{nj}$ can be evaluated exactly and equation (8.11) becomes

$$I = \sum_{n=1}^{N} \{f_n(\theta)\}^2 + \sum_{\substack{n \ j \\ n \neq j}} f_n(\theta) f_j(\theta) \frac{\sin 2\pi K' r_{nj}}{2\pi K' r_{nj}} \qquad (8.12)$$

One way to proceed is to calculate the scattered intensity from equation (8.12) for particular model structures and to compare the computed results with the experimental scattering distribution. An alternative approach that can be used without any previous knowledge of the structure is to obtain the 'radial distribution function'.

To illustrate how this is conducted we simplify the analysis by considering the case in which the scattering is from one type of atom only. This is sometimes exactly true (as in a carbon sample) and may give a rough guide to the scattering from a hydrocarbon if only the carbon atoms are considered. Thus equation (8.12) becomes

$$I = Nf^2 + f^2 \sum_{\substack{n \ j \\ n \neq j}} \frac{\sin 2\pi K' r}{2\pi K' r} = Nf^2 \left\{ 1 + \int_0^{\infty} 4\pi r^2 P(r) \frac{\sin 2\pi K' r}{2\pi K' r} dr \right\} \quad (8.13)$$

where $P(r)$ is the probability that a particular atom is located a distance r within dr from the chosen origin, and is known as the radial distribution function. To obtain $P(r)$ from the experimental scattering distribution (I) a Fourier transformation is performed:

$$4\pi r P(r) = \left(\frac{2}{\pi}\right)^{1/2} \int_0^{\infty} q I'(q) \sin qr \, dq$$

where $q = 2\pi K' = 4\pi \sin \theta / \lambda$ and $I' = (I - Nf^2)/Nf^2$

The problem with this method when applied to polymers is that it is difficult to discriminate between inter- and intra-molecular spacings. Thus the procedure mentioned before in which the scattering intensity is compared to that calculated for model structures possessing reasonable conformations has been preferred by Windle, Mitchell and Lovell. To further assist the separation of the inter- and intra-molecular effects they prepared oriented samples as well as isotropic samples. The analysis required further modification when applied to the (uniaxially) oriented samples, one approach being to develop a 'cylindrical distribution function'. Further details of the analysis of scattering from non-crystalline polymers based on this approach can be found in the original papers [1–3].

8.3.5 *Small angle X-ray diffraction (SAXD)*

X-rays are scattered primarily by the electrons within the solid sample and diffraction effects correlate primarily with the atomic arrangement, for the electron density follows this closely. There can, however, be electron density variations on a much larger scale as well and these also give diffraction effects. This occurs if there is a variation in the density with which electron density correlates. If this variation is periodic it gives rise to diffraction maxima in exactly the same way as a periodic atomic lattice does. Generally, however, the periodicity is not as regular as a crystal lattice and the diffraction effects are less precise and are more difficult to interpret quantitatively.

With polymers several kinds of structures are encountered that possess density variations and it is worth listing them first, for this will help in the subsequent discussion of the application of the technique.

Semicrystalline polymers often possess a lamellar structure. The crystalline lamellae are generally much denser than the non-crystalline regions and hence possess much higher electron density. The lamellae often have a uniform thickness, though in some cases a bi-modal distribution of thicknesses occurs with thicker lamellae growing first, followed at a later stage by thinner lamellae within the regions between the earlier ones. The lamellae are often in the range 10–30 nm thick, but both thinner and thicker examples have been observed. If the lamellae are stacked they constitute a periodic structure in which the so-called 'long period' includes the lamellar thickness plus the inter-lamellar spacing. Also in semicrystalline polymers, if the lamellae are organized into spherulites, the interspherulitic region will generally have a low density and may lead to a spatial density variation that corresponds to the spherulite diameter (generally in the range $1-1000\,\mu m$).

Block copolymers often phase-separate to form lamellae or rods or spheres of one phase in a continuous matrix of the other. The two phases usually have different (electron) densities.

Voided structures, of which the most appropriate polymer example for SAXD study is probably the crazed structure, have a very large density variation, the voids having practically zero density.

From this list it is possible to estimate the range of scattering angles of interest using Bragg's law as an approximation. That is, if the periodicity of the structure is D, then the scattering angle β is given by

$$2D \sin (\beta/2) = \lambda \qquad\qquad (8.8a)$$

Thus for a periodicity of $D = 20$ nm (of the order of the anticipated long period in lamellar structures) we have for CuK_α radiation, $\beta = 0.44° = 0.0077$ rad, confirming that we can replace equation (8.8a) with the small angle approximation, $D\beta = \lambda$.

For many applications the simple application of Bragg's law to SAXD is adequate but the single spacing so obtained may not be easy to relate to the morphology of the sample. The scattering curve depends not only on the size and spacing of the scattering elements but also on their shape, the size distribution and orientation distribution. The scattering can be calculated for model structures, but to start with the measured scattering curve and derive the structure is virtually impossible, though it is sometimes possible to deduce that the system consists of rods rather than spheres etc. Orientation of the periodic structures is revealed by arcing of the diffraction maxima, just as in wide angle diffraction, (see section 8.5.3) and useful information can be deduced from this without sophisticated data analysis. Introduction to the more detailed analysis of SAXD can be found in standard texts [4, 5].

8.4 Practical aspects

There are several basic arrangements in use for conducting X-ray diffraction studies and some of them have several variants. The method chosen in any particular case is dictated primarily by the structural characteristics of the sample and by the information sought. The major structural types that are of interest are single crystals, polycrystalline aggregates with the crystals randomly oriented ('powder sample') and polycrystalline aggregates in which the crystals show preferred orientation. Polymer single crystals large enough for X-ray study are rare, though polydiacetylene crystals polymerized in the solid state from crystals grown from monomer have provided samples suitable for analysis by this technique. Powder samples are common, forming when crystallizing thermoplastics are cooled from the melt; a spherulitic structure is of this kind. Preferred orientation is often found in processed polymers if the fabrication procedure involves molecular flow followed by rapid cooling or cooling under stress. An important subgroup is that showing fibre orientation in which each crystal is oriented such that a particular crystal direction is parallel or nearly parallel to a single direction (the draw axis or the fibre axis). In this case the crystals have only this one common direction and are randomly oriented about it. In polymer fibres this structure is enhanced by drawing. Operations such as blow moulding or rolling may cause the development of a texture in which more than one crystal direction has a preferred orientation.

On examination of the Laue conditions it is evident that when a mono-energetic parallel beam of X-rays impinges on a single crystal, diffraction can be obtained only for certain discrete crystal orientations. If the subject is a single crystal it is therefore necessary to arrange to change the crystal orientation in a systematic way that permits a range of conditions to be investigated and a number of methods have evolved in which the crystal is rotated or oscillated during exposure to the X-ray beam, sometimes with an accompanying synchronous movement of the recording system (e.g. photographic film or counter). These methods require the prior identification of a prominent crystal axis. This is often possible from an inspection of the crystal morphology. Rotation and oscillation methods can also be used with samples having fibre orientation.

When a sample consists of an aggregate of a large number of randomly oriented crystals the Laue conditions will always be fulfilled by a fraction of the crystals. All possible diffraction reflections will be made and the consequence is that cones of diffracted X-rays are produced with cone angles $2\theta_1$, $2\theta_2$, $2\theta_3$ etc. where θ_1, θ_2, θ_3 etc. are the Bragg angles corresponding to the different sets of lattice planes that have non-zero structure factors (Fig. 8.9).

X-ray studies of polymers generally use standard equipment operated in standard ways, and it is not necessary to describe them in detail here. The following outline concentrates on principles and draws attention to those aspects of most importance in polymer studies.

8.4.1 *Rotation and oscillation methods*

In rotation and oscillation methods a single crystal sample is mounted with a prominent symmetry axis vertical and bombarded with a narrow horizontal beam of X-rays. The sample is rotated slowly about the vertical causing

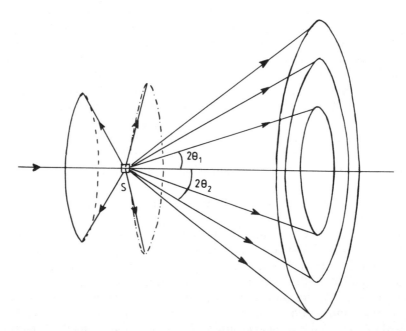

FIGURE 8.9 Diffraction from a powder sample located at S in the path of a monoenergetic parallel X-ray beam.

the reciprocal lattice to rotate in synchronization. Reciprocal lattice points therefore intersect the Ewald sphere (which remains stationary) at various instants and strong diffracted beams are produced at these times. The diffracted beams can be recorded on photographic film that can be mounted flat with the film normal parallel to the incident X-ray beam or around a cylinder with its axis coincident with the axis of rotation of the sample. The cylindrical mounting has the advantage that reflections with large Bragg angles are conveniently recorded. The positions of the reflections on the photographic film enable the determination of the layer plane spacings and gives some information on the symmetry of the crystal. Unambiguous identification of each reflection with a particular set of lattice planes ('indexing' the diffraction pattern) is not possible, however, because it is not possible to determine what rotational position the crystal occupied when any particular reflection was recorded and overlap of reflections occurs. In the oscillation method this problem is overcome by restricting the angle of rotation to about 5–15° so that overlap is unlikely. Several photographs recorded at different azimuthal settings are required to provide adequate information to index the crystal. Another approach designed to permit unambiguous indexing is to move the film while rotating the crystal, and a number of geometrical arrangements and procedures are based on this idea (section 8.4.2).

Although the rotation and oscillation methods are designed for single crystal analysis, the same basic geometry can be used for fibre studies. The fibre axis is vertical and if the crystals contained within the sample are randomly oriented about the fibre axis and present in sufficient number it is not necessary to rotate the sample since representative crystals will be providing every possible reflection at any azimuthal setting. It is preferable, however, to rotate for this improves the averaging process compensating for any accidental preferred direction (in addition to the fibre axis) when the number of crystals sampled by the X-ray beam is finite.

8.4.2 *Moving film methods*

In the rotation method the reflections obtained with a single crystal do not reveal its symmetry unambiguously even when photographs are taken about several rotation axes. This is because it is not known at which azimuthal angle any particular reflection was obtained. In order to overcome this problem a number of techniques have been designed in which the recording film is moved in synchronization with the rotation of the crystal.

In the Weissenberg method [6, 7, 9] the crystal sample is rotated about an axis perpendicular to the X-ray beam, and the cylindrical film, which is co-axial with the rotation axis, is moved parallel to the axis in synchronization with the rotation of the crystal. The sample is placed inside a cylindrical screen with a hole to admit the incident X-ray beam and a single equatorial slit. The slit selects a particular layer line cone, but all other diffracted X-rays are intercepted and do not reach the recording film.

In the 'precession' methods [6, 7, 9] developed by De Jong and Bowman and by Buerger the crystal and film movements are more complex than for the Weissenberg method. Detailed descriptions of these methods and the interpretation of the diffraction patterns obtained cannot be attempted here (see, however, [6, 7]). They are essentially for single crystal analysis and are rarely used for polymer structure analysis, though they have some limited use in studies of well aligned materials. An example of the application of the precession method to a well aligned fibre is given in [4] and an example of the Weissenberg method applied to a doubly oriented sample of nylon-6 is provided in [5].

8.4.3 Powder photographs

Samples containing a large number of randomly oriented crystals of the same kind ('powder' samples) provide convenient subjects for X-ray study. If such a sample is bombarded by a parallel beam of X-rays a fraction of the crystals will be so oriented to fulfil the Bragg condition for each family of lattice planes. Thus for each family of lattice planes having a non-zero structure factor there will be a collection of diffracted X-rays lying on the surface of a cone which has the incident X-ray direction as axis and a cone semi-angle of 2θ (Fig. 8.9). If a flat photographic film is placed with its normal parallel to the axis a series of concentric rings will be recorded. Each ring consists of a series of spots and becomes continuous if there is a sufficiently large number of crystals in the sample. It is evident that some reflections will not be recorded even when using

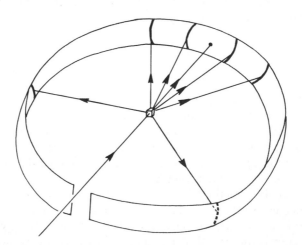

FIGURE 8.10 Debye-Scherrer camera for powder diffraction.

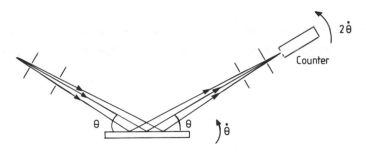

FIGURE 8.11 Basic layout of a diffractometer. Note that there is a focusing effect that reduces the effect of any incident beam divergence.

a large area of flat film and/or a short distance between the sample and the film. To overcome this problem a Debye-Scherrer camera can be employed (Fig. 8.10). With this arrangement a cylindrical strip of film with the sample at its centre intercepts part of each diffracted cone including those for which 2θ is greater than $\pi/2$. The Bragg angles and d-spacings can be measured simply from the processed film. The sequence of d-spacings may permit the identification of the crystal space lattice.

8.4.4 Diffractometer methods

As an alternative to methods based on recording X-ray intensities using photographic film an X-ray counter can be used in an instrument known as a diffractometer. The most common arrangement used is shown schematically in Fig. 8.11. The sample is normally a powder sample and is carefully mounted so that its surface is parallel to the bed of the sample mount assembly. A slightly divergent X-ray beam impinges on the sample which is rotated slowly. At any given position a multiplicity of Bragg reflections are excited with a crystalline powder sample but only those which pass into the counter are recorded. The counter is placed in a such a position that diffraction is recorded only when the lattice planes are parallel to the sample surface. This means that the counter must be mounted on a circular track so that its axis is always directed towards the sample and moves in synchronization with the sample when it is rotated. To fulfil this condition the axis of the counter must rotate through twice the angle through which the sample turns. Thus diffraction maxima will be observed at those positions for which the planes lying parallel to the surface fulfil Bragg's Law, i.e. $2d_{hkl} \sin \theta_{hkl} = \lambda$. Since λ is known and $2\theta_{hkl}$ can be read directly from the counter setting, d_{hkl} can be determined. It is normal practice to have the sample and counter rotated very slowly at a constant rate using a common motor and appropriate gearing. The counter has a very fast response and gives a signal proportional to the X-ray intensity. The signal is amplified and displayed on a meter, and is normally fed also to a chart recorder to provide a continuous trace of intensity versus 2θ. The peak positions can be determined quite accurately, but if the intensities are required it is essential that the incident X-ray intensity is constant over the period of time that data is collected (from several minutes to several hours). This is because the information about each region of the scattering distribution (i.e. each different Bragg reflection) is gathered sequentially instead of simultaneously as in the powder photograph method. Thus the X-ray tube and the associated circuitry must be very well stabilized.

Diffractometers have been designed for single crystal analysis, the most popular modern instruments being four circle models. The crystal can be rotated about three axes while the counter tracks along the fourth circle to maintain the 2θ condition with respect to the incident beam. The latest instruments offer sophisticated computer control and data processing facilities. As with the film methods for single crystal analysis, the reader is referred to standard texts and manufacturers' literature for details of the layout and operation of the instruments, and analysis of the data. Useful background in this topic can be found in [7] and [5].

8.4.5 Small angle X-ray diffraction

Since such small angles are involved (see section 8.3.5) it is clearly essential to have a well-collimated incident X-ray beam and to be able to measure scattered intensities very close to the unscattered beam. This means operating under a vacuum to avoid gas scattering, which occurs primarily at small angles. The elimination of dust is essential for the same reason.

For an introduction to some of the experimental arrangements that have been developed see [4] and [5]. Focusing crystals are sometimes used and although interpretation of diffraction using a single X-ray beam is most straightforward, slit sources are sometimes used giving higher intensities but a modified intensity distribution. Both film and diffractometer methods have been developed. The intensity changes very rapidly at low angles, with a range of 10^5 or more possible over the region of interest. Thus if the photographic method is used several different exposures may be required.

A fairly recent development is the use of a synchrotron source which provides a very intense X-ray beam*. The diffracted X-rays are measured using an array of linear counters (1024 pixels on the Stanford synchrotron), which may be placed a large distance from the sample (e.g. 5 m) to achieve the required spatial separation of the low angle beams. The source is sufficiently intense to permit simultaneous measurement in all channels in less than a second. This means that dynamic experiments can be conducted, with

---

*In a synchrotron electrons or protons are injected from a linear accelerator into a doughnut-shaped evacuated tube where they are trapped into circular orbits by a magnetic field. X-rays are emitted as a consequence of the (radial) acceleration of these charged particles in circular motion.

structural information being obtained from small angle scattering from samples during periods when the structure changes rapidly, for example during crystallization or during deformation under an applied load. An alternative arrangement, giving 0.3 s time resolution, has been used at Cornell University, and includes a transmission fluorescent screen placed in the path of the low angle diffracted X-rays. This is coupled with fibre optics to an image intensifier on to which a video camera is focused sending the signal to both a display monitor and a storage tape.

8.5 Applications to polymer characterization

8.5.1 Crystal structure determination

Polymer crystal structures are determined using the standard methods of X-ray analysis, though the task is often more difficult than for most other classes of crystal. There are several reasons for this.

1. Large perfect crystals cannot be grown from most polymers, thus eliminating a number of modern diffractometric techniques.
2. Intensities fall off rapidly with 2θ because the weak bonding in certain directions leads to a large temperature-vibration coefficient. Analysis must then be based on fewer reflections, reducing precision.
3. The exact location within the crystal lattice of pendant atoms and groups along the molecular axis may have only a secondary effect on the diffracted intensities, reducing the sensitivity.

Thus the approach is usually to attempt to devise model structures, and compute the diffraction that would occur from them for comparison with that observed. Once a fairly good fit is obtained further refinement, such as determination of the setting angle of the planar zig-zag in the case of a crystal formed from molecules in the all-*trans* conformation, or the exact location of pendant atoms, can be investigated by examining further test structures based closely on the original one. This approach is not unique to polymer crystal analysis, but is relatively more important than with many other materials because of the difficulties listed above and, on a more positive note, because polymers often lend themselves more readily to this method. This is because the molecular structure is often known and from a knowledge of the likely bond angles and a consideration of steric hindrance of side groups a reasonable guess at the structure of the crystal can often be made.

The complete determination of a crystal structure is a rather specialized activity and is required only when a new crystallizing polymer is synthesized though existing structures may be re-determined if better samples or more advanced equipment or methods become available. Thus although the determination of crystal structure is of importance the number of such determinations is limited and we will not dwell unduly on this topic. Refinement of the structure requires an understanding of X-ray diffraction and the use of sophisticated procedures that are beyond the scope of this book. Some familiarity is of value, however, because it is of direct relevance to the much more commonly encountered situation in which crystal identification is required. By this we mean that the diffraction obtained from an unknown polymer is used to identify it. Thus we can separate two quite distinct kinds of study. In one the objective is to determine an unknown crystal structure, often from scratch; in this case it is permissible to prepare the best possible sample, using careful crystal growth and/or annealing procedures. In the second the task is to analyse a specimen 'as it comes'. This often means extracting a sample from a fabricated polymer in which some orientation is present, and may be used to advantage, but no attempt can be made to produce the maximum amount of orientation. Sometimes very little or no orientation will be present, or the sample may contain foreign material, such as reinforcing filler, and this may even be crystalline.

Although other methods would usually be employed instead of (or as well as) X-ray diffraction, some circumstances may arise in which it gives valuable information, and some examples are cited below:

1. In polymer blends phase separation may occur and if one or both of the phases crystallize this may be indicated by X-ray diffraction. For example, polyethylene and polypropylene do not co-crystallize and diffraction peaks from both polyethylene and polypropylene crystals can be expected from a blend if phase separation occurs, as happens in most compositions.
2. If a polymer is polymorphic, X-ray diffraction may be used to show which type or types are present. This may relate to the crystallization conditions, including processing conditions in a fabricated polymer, or to a post-crystallization modification. For example, orthorhombic polyethylene may be transformed to the monoclinic modification by deformation and this may be detected by X-ray diffraction.

In examples such as these the presence of the crystals can be confirmed from the layer plane spacings obtained from the X-ray diffraction experiment making reference to a list of values obtained from the known crystal structure(s). Layer plane spacings may alternatively be tabulated in the literature, the most comprehensive source being that published by the American Society for Testing Materials (ASTM X-ray Powder Data Diffraction file). The ASTM index also gives the relative X-ray diffraction intensities for the several observed reflections. Although this is normally reliable for identifying systematic absences (forbidden by structure factor) it should be remembered that the sample used in the 'reference' determination may have been in a different form to the subject under study, and differences in sample orientation or diffraction geometry may lead to a different ranking of intensities.

Examples of complete crystal structure analyses by X-ray diffraction can be found in [4] and [5].

8.5.2 *Measurement of strain*

Wide angle X-ray diffraction (WAXD) is sometimes used for the accurate measurement of a selected lattice spacing in a crystalline polymer in order to determine the strain in the crystal in the corresponding direction. Extremely accurate measurements are essential and great care must be exercised to achieve the best results, for in studies of this kind it is usually required that small fractions of one per cent strain are followed with precision. This method can be used as the basis of the determination of the elastic modulus of polymer crystals within a semicrystalline sample. The strain is measured as a function of the applied stress for a series of loadings. Unfortunately calculation of the modulus is not trivial, for the crystalline and non-crystalline components have different stiffness and cannot deform identically. This is why a direct measurement of the macroscopic stress versus strain characteristic does not provide data from which the crystal modulus can be derived. We can define two cases: (1) *uniform strain*, in which both the crystalline and non-crystalline components deform equally, which would result in the stiffer (crystalline) component bearing a greater proportion of the stress than its volume fraction; and (2) *uniform stress*, in which the softer component is deformed to a greater extent than the stiffer one. Although it might be expected that a real material will show an intermediate behaviour, there is evidence that this may not always be so, thus complicating the analysis still further. Thus the application of this method for modulus measurement is almost exclusively confined to well-aligned samples, and even then must be used with caution.

WAXD has also been used to measure the shrinkage stress in thermosetting resins. This was achieved by adding small inorganic particles at chosen locations during the casting of the resin, then measuring their lattice parameters after the resin had cured. The lattice strain was hence determined and from a knowledge of the modulus of these 'tracer' crystals, the stress exerted on the crystals by the shrunken resin could be estimated. This was taken to be equal to the shrinkage stress, at that location, assuming that the presence of the tracer crystals did not disturb the internal stress field.

8.5.3 *Preferred orientation analysis*

In earlier sections we have referred to polycrystalline samples in which the crystal orientation is random or has a strong (uniaxial) orientation with a particular crystal axis lying closely parallel to a given direction for all crystals. In many instances polycrystalline samples have an intermediate degree of preferred orientation. For example, the crystal axis that contains the molecule axis in a polymer crystal will tend to orient parallel to the machine direction in a forming process such as fibre drawing or extrusion, but only a fraction of the crystals will be exactly aligned to this direction. In a well-ordered sample the majority of the rest will lie within a fairly narrow cone of directions about the most favoured axis, with perhaps 99% lying within a cone half angle of say 5°; a less well ordered sample may have say 90% lying within a cone half angle of 10°. Describing orientation in these terms is inadequate, however, for there are many different ways of satisfying such a description. For example, in the hypothetical case cited above in which 90% of the crystals are oriented so that the chosen axis lies within a cone of half angle 10°, two quite different ways in which this could arise are (1) 80% within 1° and 10% between 1° and 10°, with a preference for the low end of this range, or (2) none within 5° and 90% between 5° and 10° with a preference for the higher end of this range. Thus to obtain a complete description of the degree of preferred orientation in a uniaxial sample, the population density* of crystals must be obtained as a continuous function of misorientation angle.

In the case of polycrystalline sheets, orientation is often produced in more than one direction, particularly if the material is subjected to stretching in two orthogonal directions during forming. In this case preferred orientation (or 'texture') requires specification with reference to more than one axis. The information is most conveniently presented in the form of pole figures, which can be derived from X-ray diffraction data. This requires a great deal of measurement, though for some purposes a useful qualitative impression of the amount of preferred orientation can be obtained much more easily.

The properties of polymeric materials depend in a very sensitive way on the orientation distribution – much more so than other polycrystalline materials. This is because of the extreme anisotropy of polymer crystals. For example, the Young's modulus of a polyethylene crystal is 50–100 times greater in the molecule chain axis direction (crystal c-axis) than transverse to it, and a large part of this anisotropy is carried forward to a highly aligned macroscopic polycrystalline sample. Hence the interest in developing methods for assessing orientation distributions.

To begin with, consider the rotation pattern obtained with a single crystal in the form of a fibre. It consists of spots arranged on layer lines, and if a polycrystalline sample consisting of crystals with perfect fibre orientation were to be examined an identical pattern would be obtained, even without rotation. If, however, the polycrystalline sample contains crystals with imperfect fibre

FIGURE 8.12 (a) Schematic diagram of part of an arced diffraction pattern from an oriented polycrystalline specimen; (b) rotation pattern from polypropylene oriented by drawing at room temperature; (c) rotation pattern from polyethylene oriented by drawing at 75°C, draw ratio ~ 25:1, giving highly oriented material with high stiffness (as described in Sandilands, G.J. and White, J.R. (1977) *J. Mater. Sci.*, **12**, 1496). (X-ray patterns courtesy of D P Thompson.) In (b) and (c) the draw direction is vertical.

*It is frequently found that there is a variation in size of crystals in a polycrystalline sample, and it may happen that the size distribution differs in different directions. Thus the amount of material present with a particular orientation may not be proportional to the number of crystals in that orientation. Most measures of orientation, including those derived from X-ray diffraction data, depend on the amount of material rather than the number of crystals, but this is not a disadvantage for this is more likely to correlate with the macroscopic physical and engineering properties of the material.

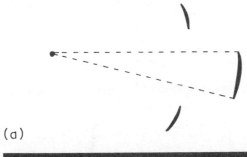

(a)

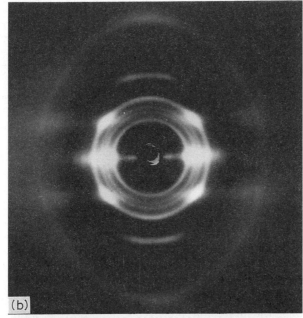

(b)

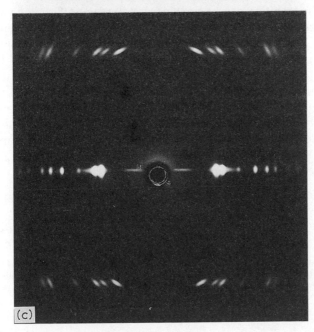

(c)

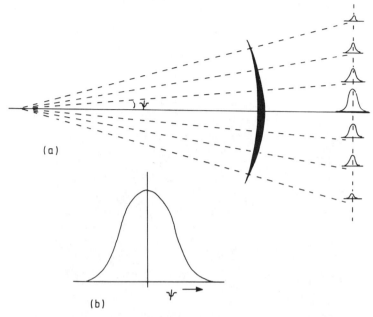

FIGURE 8.13 (a) Schematic diagram of a single arced diffraction maximum. The broken lines show positions at which the radial intensity distribution is sampled, and the corresponding profiles are shown on the right hand side; (b) azimuthal intensity distribution.

orientation, then the spots become arced (Fig. 8.12). The degree of misorientation is indicated by the angle subtended by the arc, and this can be used as a crude measure of the degree of preferred orientation. A more complete appraisal of the orientation can be made using the same diffraction pattern by measuring the intensity of the arc at selected intervals, as shown in Fig. 8.13(a). The intensity can then be plotted versus the azimuthal position (Fig. 8.13(b)). The intensity required is the 'integrated intensity' which is proportional to the area beneath the intensity profile, after having subtracted the background intensity*. The orientation distribution will determine the shape of the plot represented by Fig. 8.13(b), which therefore contains more information than a single measurement of the angle subtended by the arc. For a true fibre orientation this may be sufficient, but when higher orders of texture development are present the most useful way of representing the preferred orientation is through the preparation of pole figures.

(a) POLE FIGURES

Pole figures are based on the method of stereographic projection, and this requires brief introduction. Consider a single crystal placed at the centre of a

---

*The intensity profile can be obtained from a photograph using an optical instrument known as a microdensitometer or a microphotometer. The characteristics of the photographic film must be known and the exposure should be in the 'linear region' so that the optical density on the film can be taken to be proportional to the X-ray intensity.

sphere, with a prominent crystallographic axis vertical (i.e. coincident with the polar axis of the sphere). Construct normals to each family of lattice planes through the centre of the sphere, and extrapolate them to intersect with the surface of the sphere. Some examples are shown in Fig. 8.14 in which the multi-faceted crystal has lattice planes parallel to each of the faces (as well as in many other directions). A stereographic projection can now be constructed by joining each of the points of intersection with the sphere surface to the south pole of the sphere. The pole for a particular family of planes is the point of intersection of the corresponding line with the equatorial plane. If a second crystal, lying in a different direction, is now taken, the pole for this family of planes lies at a different position on the equatorial plane. If several crystals are taken the pole positions for a chosen family of planes will thus be distributed in some way over the equatorial plane. This constitute a 'pole figure' and if there is preferred orientation present there will be a corresponding concentration of poles. If there is a very strong preferred orientation the poles will lie very close together. The concentration of poles for a chosen set of lattice planes in a polycrystalline sample can be represented by contour plots, joining positions of equal concentration of poles within the pole figure. A different pole figure is required for each set of planes. Thus, the information that they portray is reasonably straightforward, though clear thinking is required to fully appreciate the significance of a set of pole figures representing the orientation ate the significance of a set of pole figures representing the orientation distributions of several chosen families of planes. An example is given in Fig. 8.15 which shows results obtained by Desper and Stein [8] for a lightly crosslinked polyethylene sheet that was heated to 160°C, stretched uniaxially 100% then cooled to 104°C and allowed to crystallize.

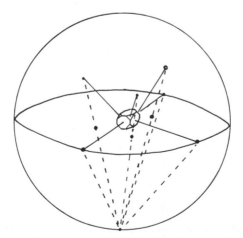

FIGURE 8.14 Stereographic projection. The crystal is placed at the centre of the sphere and normals constructed to each family of lattice planes. The points at which the normals intersect the sphere are joined to the south pole, and the points at which these lines (shown dashed) intersect with the equatorial plane form the stereographic projection.

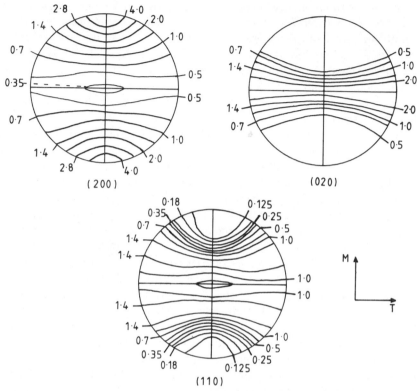

FIGURE 8.15 Pole figures obtained from a lightly crosslinked polyethylene sheet, stretched 100% while molten then cooled in the stretched state. The (200) poles show a strong preference for the machine direction, whereas the (020) and (110) poles are cylindrically distributed about M. Note that quite different distributions were obtained for 200% extension (After Desper, C.R. and Stein, R.S. (1966) *J. Appl. Phys.* **37**, 3990.)

The data required to produce pole figures can be obtained either by diffractometer methods or from photographs. In both cases the sample must be presented to the incident X-ray beam in a multitude of different orientations. This must be done in a systematic way, and equipment is available to do this automatically with some diffractometers. A detailed description of pole figure preparation and a worked example can be found in [4] and the construction of pole figures using photographic instead of diffractometric data collection is described in [9].

8.5.4 *Measurement of crystallinity*

The crystallinity of a semicrystalline polymer can be determined by analysing the intensity of X-ray scattering. The intensity is most conveniently measured by means of a diffractometer, and an example of a diffractometer trace

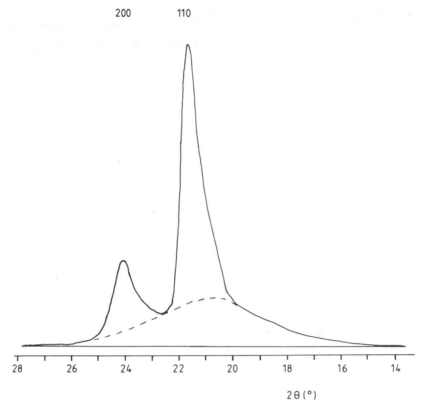

200   110

28   26   24   22   20   18   16   14

2θ (°)

FIGURE 8.16 Diffractometer trace from low density polyethylene. The dashed line shows the estimated position of the amorphous scattering intensity.

obtained with a sample of low density polyethylene is shown in Fig. 8.16. Prominent diffraction peaks are present and can be identified as corresponding to the (110) and (200) reflections from the orthorhombic polyethylene crystals. In addition it is evident that there is significant scattering at positions remote from the crystal peaks that cannot be attributed to them. This is because there is scattering from all of the material present, including the amorphous phase. It can be assumed that the amplitude scattered by an atom is the same whether it resides within a crystal or within the amorphous phase. When the contributions scattered from atoms within the crystal are summed the scattered energy becomes concentrated into a narrow angular range at the Bragg position by highly developed constructive interference, whereas destructive interference ensures that there is no scattered energy from the crystal in other directions. With the amorphous phase, scattering is found in all directions, though a very diffuse peak is normally observed. This is because there are present favoured interatomic bond lengths, giving rise to weak constructive interference in particular directions.

Because the crystalline peaks are much sharper than the amorphous diffraction it is often possible to separate them. If this can be achieved then the integrated intensity of the crystal diffraction peaks divided by the integrated intensity of the amorphous diffraction can be taken to be equal to the ratio of crystalline material to amorphous material. Separation of the crystalline and amorphous diffraction is usually not easy. In the case of polyethylene, the maximum of the amorphous diffraction lies beneath one of the crystal diffraction peaks and the height and the exact shape are therefore not easy to determine. If the exact shape of the amorphous scattering curve is known it is possible to apply a fairly rigorous procedure to achieve separation. This is done by adjusting the amplitude of the amorphous scattering curve until it matches the amplitude of the experimental curve at (all) positions remote from the crystal peaks. An example of this is given in [4]. Unfortunately in most cases it is not easy to obtain the exact shape of the amorphous scattering curve because it is almost impossible to prepare a suitable, totally amorphous sample on which to make the measurement. Taking again the example of polyethylene, if a melt is prepared in the form of a thin film then cooled rapidly to cryogenic temperature an amorphous sample may be obtained, though it is questionable whether its structure could be assumed to be the same as that in the amorphous phase of the material when conventionally cooled to form a semicrystalline solid. Furthermore the X-ray experiment would have to be conducted at cryogenic temperature, otherwise crystallization would commence. Similar objections attach to attempts to obtain a scattering curve for the amorphous phase by making measurements on a melt.

Thus the line sketched on to Fig. 8.16 to represent the amorphous scattering has been constructed with reference to the following criteria: (1) it must coincide with the total scattering curve at points remote from the crystal peaks; (2) the peak should occur at approximately $2\theta = 20°$ (for CuK$_\alpha$ radiation, $\lambda = 0.1542$ nm); and (3) it should describe a smooth curve. A further construction line represents the 'background' which consists of components that do not contain information on crystallinity (incoherent scattering, thermal diffuse scattering).

If required, the amorphous curve could be subtracted from the total experimental curve to obtain the true shape of the crystal reflections. This is unnecessary, however, in the estimation of fractional crystallinity (f_c) for this requires only that the integrated intensities (which are proportional to the separated areas under the diffraction traces) are measured. If the total area attributable to crystalline diffraction is I_c (which will include the areas corresponding to both the (110) and (200) reflections in polyethylene, for example) and that under the amorphous curve is I_a, then

$$\frac{f_c}{1 - f_c} = \frac{I_c}{I_a} \quad \text{or} \quad f_c = \frac{I_c}{I_c + I_a}$$

Because this contains only the ratio of the intensities and is hence dimensionless, the units in which the areas are measured are irrelevant, and it is not necessary to know the factor required to convert area into X-ray intensity. A convenient way of obtaining this ratio is to cut out the relevant areas and weigh them: this method for measuring areas is quite accurate as long as the paper density mass (per unit area) does not vary.

In the analysis presented above no account has been taken of the polarization factor and the Lorentz factor, both of which describe variation in X-ray intensity with the angle of reflection. If the crystal and amorphous contributions were similarly distributed with respect to angle this would not matter (though it would then be impossible to separate them) but the crystal reflections are relatively sharp and will thus depend on rather discrete polarization and Lorentz factors, whereas the amorphous reflections are much more evenly spread over a wide angular range. Thus the overall effect of these factors may be different for the crystal and amorphous portions, and the simple approach outlined above may not be good enough.

In assessing the importance of these factors first consider the combined polarization factor and Lorentz factor for the Debye-Scherrer method:

$$(1 + \cos^2 2\theta)/\sin^2 \theta \cos \theta$$

$(1 + \cos^2 2\theta)/\cos \theta$ varies very slowly, and the error introduced by ignoring this factor is small, but $1/\sin^2 \theta$ varies significantly within the range of θ values of interest. It is therefore more rigorously correct to proceed as follows. First make the polarization correction, then convert the resulting intensity, $I(\theta)$, to $I(s)$, where $s = 2 \sin \theta/\lambda$, and then obtain the scattering curve in terms of $s^2 I(s)$. If the crystal component, $I_c(s)$, can be separated, as before, then the crystallinity is given by

$$f_c = \frac{\int s^2 I_c(s) \mathrm{d}s}{\int s^2 I(s) \mathrm{d}s} \tag{8.14}$$

This formula is a useful approximation, though to be of practical value integration limits within the observation range must be chosen. Normally this is not a problem, for with both the crystalline and the amorphous fractions $s^2 I(s)$ drops to insignificant levels at both low and high angles. The formula will underestimate the crystalline fraction if defects are present for these tend to contribute to the amorphous scattering curve. Thermal vibrations cause the same effect. Ruland [10] has developed a procedure for separating out these contributions and a description of this analysis can be found in [4].

Although it was noted above that for many polymers it is not possible to obtain a totally amorphous sample with molecular organization similar to that in the amorphous phase of the semicrystalline polymer for which f_c is required, there may be exceptions. In cases in which the onset of crystallization can be fairly closely controlled (e.g. by an experimentally established heat treatment or straining procedure) it may be possible to work with the amorphous scattering curves obtained from the semicrystalline sample so obtained and a totally amorphous sample given a treatment that falls just short of that required to cause crystallization. By a similar analysis to the one that led to equation (8.14), the amorphous fraction (f_a) is given by

$$f_a = \frac{\displaystyle\int_0^\infty s^2 I_a'(s)\, ds}{\displaystyle\int_0^\infty s^2 I_a(s)\, ds} = 1 - f_c \qquad (8.15)$$

where $I_a'(s)$ is the intensity in the amorphous scattering curve obtained with the semicrystalline material and I_a is the scattered intensity obtained with an equivalent sample that is totally amorphous. If the curves from both samples are identical in shape, as will be the case if the molecular organization in the amorphous regions of the semicrystalline sample is indeed identical to that in the totally amorphous sample, then the integrals can be replaced by the intensity at any chosen position (s_i)

$$f_a = \frac{I_a'(s_i)}{I_a(s_i)} \qquad (8.16)$$

This formula needs to be applied carefully to overcome two problems. Firstly, the thickness (t') of the semicrystalline sample may not be the same as that (t) of the totally amorphous sample, and secondly, the diffraction curves are obtained at different times so the incident X-ray intensity may not be the same. We will now show how both of these effects can be allowed for.

Consider the passage of the X-ray beam through an amorphous sample of thickness t (Fig. 8.17). The intensity at depth x is $I(x) = I_0 e^{-\mu x}$, where I_0 is the incident intensity and μ is the absorption coefficient. If we now consider

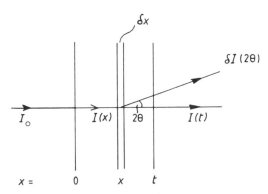

FIGURE 8.17 Monoenergetic parallel X-ray beam passing through a specimen, thickness t.

scattering taking place within a narrow slice of thickness δx at depth x, the intensity scattered into a chosen direction, 2θ, is proportional to the incident intensity at this position, $I(x)$, and to δx and is given by

$$[\delta I(2\theta)] = c(2\theta)I_0 e^{-\mu x}\delta x \tag{8.17}$$

Some of this will be lost by absorption in the remaining path through the sample, a distance $(t-x)/\cos 2\theta$. Thus the contribution to the scattering measured in the direction 2θ becomes

$$\delta I(2\theta) = c(2\theta)I_0 e^{-\mu x}\delta x \exp\left(-\mu(t-x)/\cos 2\theta\right) \tag{8.18}$$

where the square brackets have been omitted to distinguish it from the expression given in equation (8.17). The value of 2θ chosen should be such that the amorphous scattering is large, and should therefore be close to $20°$ if the sample is polyethylene and the radiation is CuK_α, the other condition being that it should not be close to a crystal diffraction maximum. It is convenient to replace $1/\cos 2\theta$ by $(1+\alpha)$, where $\alpha \ll 1$ ($\alpha \simeq 0.064$ when $2\theta = 20°$). Hence

$$I(2\theta) = \int dI(2\theta) = c(2\theta)I_0 \exp\left(-\mu t(1+\alpha)\right) \int_0^t \exp\left(\mu x\alpha\right) dx$$

$$= c(2\theta)I_0 \exp\left(-\mu t\right)\exp\left(-\mu t\alpha\right)\left\{\frac{\exp\left(\mu t\alpha\right)-1}{\mu\alpha}\right\} = c(2\theta)I_0 t \exp\left(-\mu t\right)$$

If the amorphous sample is now replaced by a semicrystalline sample the corresponding intensity becomes $I'(2\theta) = f_a c(2\theta)I_0' t' \exp\left(-\mu t'\right)$ where the crystalline intensity has been subtracted if present in this direction. Thus equation 8.16 should be replaced by

$$f_a = \frac{I'(2\theta)}{I_0't'e^{-\mu t'}}\frac{I_0 t e^{-\mu t}}{I(2\theta)} \tag{8.19}$$

Now the transmitted (undeflected) intensity obtained when the totally amorphous sample is in position is $I(t) = I_0 e^{-\mu t}$, and when the semicrystalline sample is in position is $I'(t') = I_0' e^{-\mu t'}$. Thus if both of these quantities are measured, equation (8.19) becomes

$$f_a = \frac{I'(2\theta)}{I'(t')}\frac{I(t)}{I(2\theta)}\frac{t}{t'}$$

and f_a is now expressed in terms of experimentally measured quantities.

One problem with this method as described is that the transmitted intensities $I(t)$ and $I'(t')$ are very much stronger than $I(2\theta)$ and $I'(2\theta)$ and it is unlikely that pairs such as $I(t)$ and $I(2\theta)$ could be measured satisfactorily on the same exposure. Alexander [4] describes a method in which this problem is overcome by the use of a 'reference' sample to attenuate the transmitted beam (but not the scattered X-rays), the same reference being used for both exposures so that the ratio of the two measurements remains unchanged as $I(t)/I'(t')$.

8.5.5 SAXD studies

The most common measurement derived from SAXD is the so called long spacing in crystalline polymers that possess a lamellar structure. This corresponds to the periodicity in a stack of lamellae and is thus equal to the lamellar thickness plus the thickness of the interlamellar (amorphous) region. SAXD has been used in many laboratory investigations of polymer morphology, often complementing other methods of measurement (see Chapter 14). Many laboratory studies have been conducted on solution grown crystals, precipitated on to a flat surface to provide a sample free from many of the complications that arise with crystals grown in a less controlled manner. They can be used to study the effect of heat treatment, for example, and by making SAXD measurements both during and subsequent to annealing polyethylene crystals it was found that some of the changes that take place are reversible. Reversible changes in long periods have also been observed in oriented low density polyethylene.

Many SAXD studies have been conducted on oriented polymers. The general shape (arcing) of the diffraction pattern gives an indication of the extent of crystal orientation. The SAXD peaks are broadened as a consequence of disorder and/or if there is a range of long period present. It is difficult to separate these two effects.

A synchrotron source has been used for SAXD examination of a single craze and has permitted comparison of the measured scattering with that predicted by calculation. The agreement was sufficiently good to suggest that a fairly reliable measurement of overall craze content of a sample could be derived from SAXD. Although the high intensity synchrotron source was needed to produce data from a single craze, this is not necessary for the latter application.

Synchroton SAXD studies have been used to study melting and crystallization of polyethylene and the effect of annealing. It has been shown that there are two mechanisms of lamellar thickening: melting and recrystallization take place over short time periods, followed by diffusion-controlled thickening over longer periods. Changes taking place in a few seconds have been followed. Some deformation studies have been performed using the same equipment, giving an insight into the deformation mechanisms taking place in real time. Further studies of this kind are anticipated and will undoubtedly increase markedly the understanding of the nature and sequence of deformation mechanisms in polymers under applied loads.

Another advantage of synchrotron dynamic structural studies is that they can be combined with techniques that are exclusively dynamic. For example, differential scanning calorimetry (DSC) provides information about changes occurring in a sample with changing temperature. SAXD with a synchrotron source can follow structural changes at rates compatible with DSC studies, and the two techniques have been combined in a study of crystallization of polyethylene.

8.6 Conclusions

X-ray diffraction is an extremely valuable technique for investigating polymer structure and morphology. For the most exacting tasks, such as the detailed determination of atomic positions within the crystal unit cell, a combination of good quality equipment used with care and according to strict procedures together with detailed analysis involving computation of results from trial structures is required. Such procedures are unlikely to be conducted routinely. Similarly a great deal of work is required to produce a full description of preferred orientation in a fabricated polymer. On the other hand estimates of crystallinity and an idea of orientation can be obtained quite readily by rather unsophisticated X-ray studies. Similarly, if an average value of the long spacing is adequate to characterize a sample it can be obtained fairly readily by SAXD, but to determine the distribution of long spacing is a much more difficult task, and the result will be subject to an unknown amount of error.

Synchrotron X-ray sources have opened the way to new research possibilities that can be expected to lead to a deeper fundamental understanding of crystallization and deformation mechanisms among other things, but such equipment will never be generally available.

References

1. Lovell, R. and Windle, A.H. (1981) *Polymer* **22**, 175
2. Mitchell, G.R. and Lovell, R. (1981) *Acta Cryst.* **A37**, 189.
3. Mitchell, G.R. and Windle, A.H. (1982) *Coll. Polym. Sci.* **260**, 754.
4. Alexander, L.E. (1969) *X-ray Diffraction Methods in Polymer Science*, Wiley, New York.
5. Kakudo, K. and Kasai, N. (1972) *X-ray Diffraction by Polymers*, Elsevier, Amsterdam.
6. Nuffield, E.W. (1966) *X-ray Diffraction Methods*, Wiley, New York.
7. Luger, P. (1980) *Modern X-ray Analysis on Single Crystals*, Walter de Gruyter, Berlin.
8. Desper, C.R. and Stein, R.S. (1966) *J. Appl. Phys.* **37**, 3990.
9. Henry, N.F.M., Lipson, H. and Wooster, W.A. (1960) *The Interpretation of X-ray Diffraction Photographs.* 2nd edn, Macmillan, London.
10. Ruland, W. (1961) *Acta Cryst.* **14**, 1180.

Further reading

1. Cullity, B.D. (1978) *Elements of X-ray Diffraction*, Addison-Wesley, Reading, Massachusetts.

Exercises

8.1 Using the diffractometer trace given in Fig. 8.16, estimate
   (i) the lengths *a* and *b* in the polyethylene unit cell;

(ii) the average crystal diameter (assuming distortion broadening is negligible);
(iii) the fractional crystallinity.
(You may find a magnified copy of Fig. 8.16 better to work with; this can be produced by photocopying.)

8.2 If a highly oriented sample of polyethylene loaded axially in an X-ray camera can support a maximum stress of $1\,GN/m^2$, calculate the maximum shift in the (002) peak that can be obtained using CuK_α radiation ($\lambda = 0.1542\,nm$). For polyethylene $c = 0.254\,nm$ and the Young's modulus in the chain axis direction can be taken as $285\,GN/m^2$.

8.3 Work out the structure factor rule for $hk0$ reflections in orthorhombic polyethylene. The molecule that passes through the centre of the unit cell is displaced a distance $c/4$ along the axis relative to those passing through the corners (see Fig. 8.5). Because of the symmetry, it is not necessary to know the setting angle, nor the angle between the C–H bonds.

8.4 Calculate the approximate location of the SAXD maximum obtained with CuK_α radiation for a mat of solution-grown polyethylene crystals in which the long period is 15 nm.

8.5 Make an analysis of the orientation in the samples used to produce the patterns shown in Fig. 8.12(b) and (c). What is required to make a more accurate quantitative assessment?

9

Transmission electron
microscopy and diffraction

9.1 Introduction

The transmission electron microscope (TEM) provides detailed structural information at levels down to atomic dimensions. Under favourable conditions the most capable instruments can resolve detail at the 0.1 nm level, but such high resolution examination is seldom possible with polymers. Nevertheless, it is possible to obtain information within the range 1—100 nm with varying degrees of difficulty. This is beyond the range of light microscopy and the TEM can provide information that can rarely be obtained by any other means. A further advantage of the TEM is that it can be rapidly adjusted to provide an electron diffraction pattern from a selected area, facilitating investigation of crystal structure and orientation, and enabling particular morphological features to be identified. The main disadvantage of the TEM is that it can be used only on thin samples, less than 1 μm thick, and preferably less than 100 nm thick. This is below the thickness of commercially important self supporting polymers (e.g. food packaging films) and less than that of most polymer coatings. Thus the TEM specimen requires special preparation. We can consider two groups of specimens. Firstly, there are specimens prepared in the laboratory directly in thin film form for the express purpose of conducting fundamental structural studies in the TEM. Secondly, if information is required about a polymer in its as-fabricated state (possibly after a period in service) a thin section must be cut before it can be introduced into the TEM.

There are two types of TEM, the conventional TEM (CTEM) and the scanning TEM (STEM). The construction of and image formation in the longer established CTEM will be described first and, because the image contrast mechanisms in the STEM are basically the same, the STEM will be dealt with briefly later, indicating its advantages and disadvantages for polymer study. Specimen preparation techniques and application of the TEM to polymer studies will be dealt with together. Other than the need to protect against mechanical damage and the inclusion of foreign material, there are no general rules concerning specimen preparation. Hence a representative selection of methods are described in the 'applications' section, along with some of the results that have been obtained.

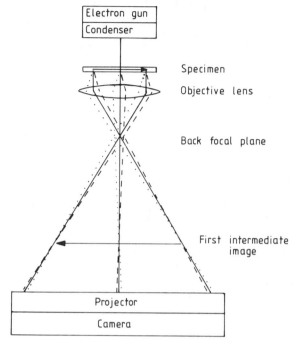

FIGURE 9.1 Layout of CTEM.

9.2 Layout of the TEM

There are many excellent books dealing with electron optics and the design and operation of the TEM, and it is not proposed to attempt a detailed description here. In the brief account that follows emphasis is placed on the region close to the objective lens because of its importance both in the formation of images and of electron diffraction.

The CTEM can be likened to the light microscope operated in transmission, consisting of a source (of electrons instead of light), a condenser system, a specimen stage, an objective lens and a projector system (Fig. 9.1). A major difference follows from the ready interaction of electrons with matter (including gas) for in the TEM the path from source to detector must be in vacuum so that the only origin of scattering is the specimen.

9.2.1 The electron source

The most common electron source consists of a hairpin tungsten filament heated with a low voltage source*. The filament is held at a large negative

---

*Lanthanum hexaboride and field emission sources can provide better coherence and higher beam intensity, but rarely offer any significant advantage in polymer studies.

potential and the electrons are accelerated towards an anode plate held at a small positive potential, forming a monoenergetic beam which passes through a central hole in the anode. The focusing action of the anode and a second electrode (the Wehnelt cap) produces a narrow crossover point below the anode. Beyond the anode the electron microscope column is held at earth potential so that the energy of the electrons when they pass through the specimen is determined by the filament potential. The potentials available in most TEMs are in the range 20—100 kV, though instruments providing 200 kV and even 400 kV are becoming more popular. Some specialized research microscopes offering up to 1 MV are in existence. In the first instance the purpose of the high voltage is to provide electrons with a small wavelength, for this allows the formation of high resolution images.

In light microscopy the ultimate resolution is diffraction-limited and is of the order of the wavelength, i.e. around 500 nm for visible light. Thus to form images of detail smaller than this, radiation of smaller wavelength is required and this led to the development of the TEM. Moving electrons have a wave nature, and after 100 kV acceleration the wavelength is approximately 0.0037 nm. This is below atomic dimensions, but resolution now becomes limited by the aberrations of the lenses and/or by the deficiencies of the detection system (e.g. the phosphor screen on which the image is displayed). Higher voltages produce shorter wavelengths and although this in itself is not of any benefit when resolution is determined by other factors, it permits the construction of lenses with slightly better image forming characteristics (see discussion in more advanced texts on 'contrast transfer function').

A second motive for building TEMs with higher voltages is that electron penetration increases, permitting the use of thicker specimens. Once the voltage is above 100 kV, relativistic effects become important, however, so that penetrating power does not increase in proportion to voltage, though the cost of the machine may increase more than in proportion to the voltage. With polymeric samples problems of radiation damage generally limit observation to an upper limit of between 100–150 kV, though some high resolution work has been conducted at 200 kV.

9.2.2 Lenses

The lenses consist of magnetic fields, which have a focusing effect on electrons, behaving in a manner similar to convex lenses in light microscopy. The lens fields are produced by electromagnets and focusing is achieved by altering their strength using the current control. This is much more convenient than moving lenses and/or the specimen parallel to the optic axis of the instrument as is required with the light microscope in which the lenses have fixed strength. In addition to the focusing action, the many lenses cause a rotation of the electrons about the axis which increases with the lens strength, and this has to be taken into account sometimes, for example when identifying crystallographic directions in images by reference to the corresponding electron diffraction pattern.

The condenser system normally consists of two lenses. The first one produces a demagnified image of the electron crossover (below the anode) and the second (weaker) lens projects this on to the specimen. The reason for this arrangement is primarily to provide separation between the condenser lens system and the objective lens. Both the first condenser lens and the objective lens are strong and therefore have short focal lengths (usually a few mm). The lens fields are not confined in the way that surfaces of a glass lens define its range of influence over light rays, and if the fields of two neighbouring lenses overlap they interfere so that if the condenser and objective lenses were to be located near to one another it would cause operating difficulties.

The objective lens is normally located immediately below the specimen, though in some cases an 'immersion lens' is provided, with the specimen located within the lens field. The immersion lens can be exploited to advantage to concentrate the electron probe further when the instrument is used in the scanning mode or for analytical work (see section 9.9). For conventional imaging it is convenient to regard the objective lens as equivalent to a convex glass lens in a light optical system, and a reasonable idea of its action can be gained from thin lens theory.

Finally comes the projection system which consists of at least two lenses. The intermediate lens forms a magnified image of the image formed by the objective lens, and the projector lens magnifies it still further, projecting it on to a phosphor screen that converts the electron energy into light for visual observation. Alternatively, the image can be recorded directly on to photographic film with the help of a camera located below the phosphor screen which can be tilted clear. Although the film plane is normally 10–30 mm below the phosphor screen there is no need to refocus because of the very large depth of focus. The intermediate lens plays a vital role in the formation of diffraction patterns in the CTEM (see section 9.4).

9.2.3 Translation and tilt coils

The modern CTEM is provided with several sets of electromagnetic coils which can be used to translate or tilt the electron beam. These facilities are used in the alignment of the microscope (sometimes in addition to a preliminary mechanical alignment) and may be required in dark field operation (section 9.5.1). The condenser translate coils cause the beam to shift, but it remains parallel to its original direction (Fig. 9.2(a)) permitting the illumination to be centred on the area under examination even at high magnification. The tilt coils cause the beam to tilt, and this is done in the manner shown in Fig. 9.2(b) using coils at different levels so that at a given position (e.g. the specimen plane) the tilt is achieved without translation and the area under observation remains illuminated. This is the way that tilted dark-field operation is set up in the CTEM.

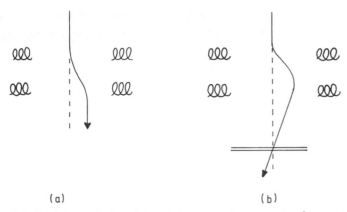

(a)   (b)

FIGURE 9.2 Combining electron beam deflections using two sets of magnetic coils to give (a) translation (but no tilt at the specimen) and (b) tilt (but no translation).

9.2.4 *The specimen stage*

The specimen stage must permit translation of the specimen to facilitate the selection of the area to be observed. To obtain structural information about crystalline specimens it is an advantage to be able to tilt the specimen so that the electron beam can be made to pass along different crystallographic directions, for this will change the electron diffraction pattern and the image contrast in an interpretable manner. Since the crystallographic orientation of a particular structural feature is not generally known before the specimen is inserted into the microscope it will have an arbitrary orientation within the microscope and for detailed work it is necessary to have either a double tilt facility (permitting tilt about any chosen axis to be achieved by the combination of tilts about two orthogonal axes) or tilt plus rotation (enabling rotation of the specimen until the chosen axis is aligned to the stage tilt axis). Some stages offer $\pm 60°$ tilt, but a more restricted range is often dictated by the restricted working space, especially when using a short focal length objective lens. The specimen stage is normally provided with an air lock chamber to facilitate insertion of the specimen into the microscope column without breaking the vacuum, the small air lock chamber being evacuated in less than a minute.

TEM specimens are normally placed on copper grids having square holes with edge lengths of the order of 50 μm for viewing through. The grids provide thermal and electrical conduction to the specimen stage, minimizing problems of overheating and charging.

9.3 Resolution limitation

The resolution of an optical system is limited ultimately by diffraction and the limiting value is given by $0.61\lambda/\alpha_0$ where λ is the wavelength of the radiation and α_0 is the half angular aperture. To produce image contrast an aperture is usually introduced below the objective lens and with a 100 kV CTEM this produces a value of α_0 of the order of 5×10^{-3} rad which corresponds to a limiting resolution of approximately 0.9 nm ($\lambda = 0.0037$ nm at 100 kV). At this level of resolution, aberrations of the electron optical system are important and the diffraction limit will be exceeded. The aberrations that are important are exactly equivalent to those in light optical systems and are chromatic aberration, astigmatism and spherical aberration.

Chromatic aberration can arise in two ways. Firstly, if the electron source is not monoenergetic the electron wave contains a spread of wavelengths. Secondly, if the magnetic lenses are subject to current fluctuations their strengths vary causing the focal length to change. Thus it is essential to have a well stabilized high voltage generator and well stabilized lens current circuits. These are not under the control of the operator, though in some laboratories refrigerated water at a constant regulated temperature is used to cool the lenses (rather than mains water) to improve stability. The critical time over which fluctuations must be minimized is the time to record a photograph, usually several seconds. For most applications a modern CTEM will not be limited primarily by chromatic aberration and it is highly unlikely to be a problem in polymer studies.

Astigmatism is said to occur if the focal length of a lens is not the same in all planes containing the optic axis. This is a common fault in the CTEM because the lens fields can be easily distorted by the presence of small foreign objects near to the lens. Contamination, which normally consists of condensed organic material, possibly crosslinked by the electron beam, often collects on the lens polepiece or the nearby aperture, causing astigmatism. Finger grease is a source of abundant contamination, but even with the most stringent precautions to keep the electron microscope column free from such material, traces of contaminant will cause the strength and direction of astigmatism to vary from day to day and from hour to hour. Thus both the condenser system and the objective lens of a CTEM are provided with 'stigmators', which are correction devices that enable the operator to make compensating adjustments. The basic design of stigmators is fairly straightforward and can be found in standard textbooks, while the procedures for adjusting them are described in the instrument manuals, along with suggestions for test specimens suitable for making the adjustments.

Spherical aberration is said to occur when the paraxial rays and those inclined to the optic axis are brought to a different focus. It is the aberration that attracts most attention in electron optics because it cannot be compensated for in the manner used in light optics, where a combination of convex and concave lenses is used. Spherical aberration can be minimized by restricting the angular range admitted, using an aperture. The use of a small aperture introduces diffraction limitation, however, and a compromise between the conflicting aperture size requirements may have to be reached ('optimum aperture'). The spherical aberration resolution limitation is proportional to a 'spherical aberration coefficient' (C_s) which is a property of the lens, and is approximately equal to its focal length. Thus it is advantageous to have a very strong lens $(C_s < 1 \text{ mm})$, though this will usually restrict the working space near to the lens making it difficult to provide a full range of specimen manipulation, such as double tilt.

Many CTEM studies of polymers are limited to low magnification observation because of their electron beam sensitivity (section 9.6). Consequently, resolution is often limited by the characteristics of the photographic emulsion. If the grain size of the emulsion is say $20 \, \mu\text{m}$ then if the microscope is being operated at $1000 \times$ magnification this is equivalent to 20 nm in object space, and this sets the spatial limitation on information contained in the recorded image. This is much worse than the electron optical limitation, and means that optimizing the aperture size, and even achievement of best focus, become irrelevant (see section 9.6).

9.3.1 Depth of field, depth of focus

The depth of field is the distance through which the focus of an optical instrument can be changed without causing noticeable deterioration in the image and is given by $2\delta/\alpha_0$ where δ is the limiting resolution. Because α_0 is normally very small in the TEM the depth of field is normally much larger than the thickness of the sample, which is then all in focus simultaneously (in contrast to a transparent optical section in which features at different levels require different settings of the focus). For example, when operating a TEM at the 5 nm resolution level with an aperture having $\alpha_0 = 5 \times 10^{-3}$ radians, the depth of field becomes $2 \, \mu\text{m}$. The corresponding condition in image space, which represents the distance through which the image observation recording plane can be shifted without causing impairment, is called the *depth of focus* and is obtained by multiplying the depth of field by M^2, where M is the magnification. Hence in the above example, if an appropriate (low) magnification of $5000 \times$ is used the depth of focus is 50 mm.

9.4 Electron diffraction

Before discussing image contrast it is necessary to deal with electron diffraction because the way in which images are formed with crystalline specimens and the interpretation of the contrast obtained depends on an understanding of diffraction. Furthermore, the electron diffraction pattern provides important complementary information, and is sometimes more important than the image itself.

Diffraction of a monoenergetic parallel beam of electrons by a crystal can be dealt with in exactly the same way as X-ray diffraction (Chapter 8). Thus

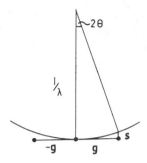

FIGURE 9.3 Ewald sphere construction showing the deviation parameters s for a set of lattice planes having diffraction vector **g**.

observable ('allowed') and 'forbidden' reflections are subject to the same structure factor rule. There are some important detailed differences that are the consequence of the very short wavelength of electron irradiation at the accelerating potentials employed in TEM, and of the use of very thin crystals required by the low penetration characteristics of electrons. Let us examine these features.

Firstly, if we consider Bragg's law ($2d \sin \theta = n\lambda$) and calculate from it the value of $\sin \theta$ for the first order reflection from the (110) planes in polyethylene (for which $d \simeq 0.41$ nm) when using 100 keV electrons ($\lambda \simeq 0.0037$ nm) the result is 0.0045. This corresponds to $\theta = 0.26°$. Thus the small angle approximation will almost always be valid when dealing with electron diffraction patterns (i.e. $2d\theta = n\lambda$), and it is noted that Bragg's law will only be satisfied by families of crystallographic planes that lie almost parallel to the electron beam.

Secondly, a consequence of using very thin crystals is that the Laue conditions are relaxed a little, and some constructive interference is still obtained when the crystal is rotated slightly from the Bragg position. This misorientation is normally represented by a misorientation vector (**s**) which describes the displacement of the reciprocal lattice point from the Ewald sphere (Fig. 9.3). The vector joining the origin of the reciprocal lattice to the reciprocal lattice point is **g**, and we can thus write $\mathbf{K'} = \mathbf{k} - \mathbf{k'} = \mathbf{g} + \mathbf{s}$. It can be shown that the diffracted intensity depends on s as $(\sin^2 \pi st)/(\pi s)^2$ where t is the crystal thickness parallel to **k**. The first minima of this function occur at $s = \pm 1/t$, and the intensity at higher values of $|s|$ is negligible. Thus one way of representing the range of the diffracting condition is to construct a 'spike' of length $2/t$ parallel to **k**, centred on each reciprocal lattice point. Diffraction occurs as long as the spike intersects the Ewald sphere, and because of the very large radius of the Ewald sphere ($= 1/\lambda$) it follows that many diffracted beams can be observed simultaneously (Fig. 9.4). Thus if the electron microscope is set for selected area diffraction, the pattern will consist of an array of spots. This

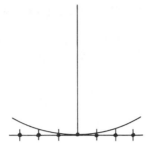

FIGURE 9.4 Ewald sphere construction for a thin crystal. Each reciprocal lattice point corresponds to a different set of lattice planes, and has a spike-shaped intensity function centred on it.

will represent a section through the reciprocal lattice, and if the subject is a single crystal, will reveal the symmetry of the particular zone axis along which the electron beam travels.

The deviation of a diffracted beam from the incident beam is 2θ. Thus if the electron diffraction pattern is allowed to fall on to a detection screen at a distance L from the specimen (Fig. 9.5) the separation (x) of the diffracted beam and the incident beam will be $x = L2\theta = L\lambda/d$. If lenses beneath the specimen are used to display the pattern as is usually the case, the effective value of L will not be equal to the separation of the specimen and the viewing screen, but can be found easily by using a calibration specimen which has known lattice spacings. Measurement of values of x for corresponding known d values permits calculation of the effective value of $L\lambda$, but the value may be subject to hysteresis effects if the lens excitation is changed, and appropriate precautions must be taken to preserve the accuracy of the calibration. If this is successfully

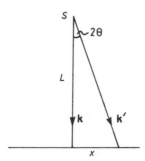

FIGURE 9.5 Geometry of diffraction with a specimen located at S, at a distance L from the observation plane. The undeviated transmitted beam and the diffracted beam are represented by wave vectors **k** and **k**$'$ respectively.

achieved then measurement of the values (x) of separation of diffraction spots from the origin in an 'unknown' pattern will enable calculation of the corresponding d values.

The procedure for displaying the diffraction pattern consists of reducing the excitation of the intermediate lens so that it focuses on the back focal plane of the objective lens, projecting the electron intensity at this plane on to the viewing screen. If, before the adjustment to the intermediate lens is made, an aperture is inserted at the position of the first intermediate image then the diffraction pattern will be obtained only from the area seen within the shadow of the aperture – 'selected area diffraction'. Care must be taken when adjusting the microscope to ensure that the lens excitations correctly locate the intermediate image at the plane of the selecting aperture, otherwise the diffraction pattern will not uniquely correspond to the area viewed in the image. The exact procedure differs from one microscope to another and is described in the operating manual.

9.4.1 *Interpretation of electron diffraction patterns*

The interpretation of electron diffraction patterns is usually more straightforward than of X-ray diffraction patterns. In the case of a single crystal the pattern represents a section through the reciprocal lattice plane that lies perpendicular to the incident electron beam. Thus the symmetry of this particular zone is revealed and the measurement of the layer plane spacings associated with it can be made. If the structure is known, the layer plane spacings enable each spot to be assigned to a particular family of planes and the pattern can be 'indexed' by choosing a self consistent set of indices for each spot, the orientation of the crystal can then be determined. If the structure is unknown it is unlikely that it can be deduced from a single pattern, but by tilting the crystal to obtain patterns corresponding to other zone axes, a complete structural analysis may be possible, but such a process is difficult with polymers because of radiation damage which restricts the time that a single area can be viewed. With a polycrystalline sample having randomly oriented crystals the pattern consists of a series of concentric rings and provides information identical to that derived from a powder photograph in X-ray diffraction. Sometimes a preferred orientation is present and this is revealed by arcing of the rings, indicating that particular crystal directions are present only within a restricted angular range.

A common application of electron diffraction is recognition of the presence of twinning or of polymorphism. Several studies have been conducted on polyethylene which, when deformed, can form twins and/or a monoclinic crystal form (recall that it normally crystallizes into the orthorhombic form). With twinning, diffraction spots are formed on the same radii but in different azimuthal positions. Polyethylene twins on {310} and {110} planes. If the twin plane is (310) then the structure on either side of the twin boundary is

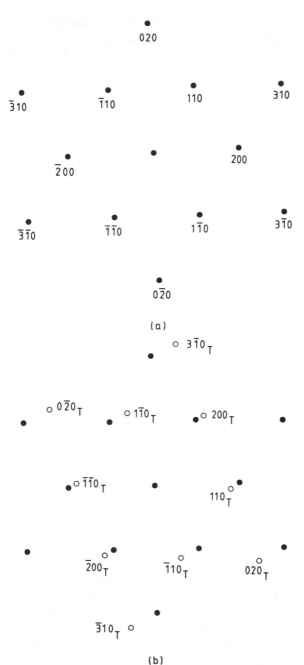

FIGURE 9.6 Sketch of diffraction patterns for orthorhombic polyethylene in the [001] orientation. (a) Indexed single crystal pattern; (b) pattern for a single crystal (as in (a)) plus a (310) twin: the twin spots are marked ○ and only these are indexed.

reflected in the (310) plane, and the diffraction pattern of the [001] zone (which contains the {310} planes) shows diffraction spots for both matrix and twin. The twin spots appear at positions obtained by reflecting the matrix spots in the (310) plane, or, equivalently, by rotating through $180°$ about an axis perpendicular to the (310) plane, i.e. the [310] direction in the reciprocal lattice plane (= diffraction pattern).

The single crystal pattern for orthorhombic polyethylene in the [001] orientation contains all $hk0$ reflections for which $(h + k)$ is even (structure factor rule). No reflections are observed in the positions that would index as 100 and 010 in the diffraction pattern, but the 200 and 020 reflections are present and form convenient reference axes. Since for polyethylene it happens that $a/b \sim 3/2$, then the lengths of the [200] and [020] vectors in the diffraction pattern are in the inverse ratio, $\sim 2/3$, making construction of a sketch of the diffraction pattern very easy on graph paper (Fig. 9.6(a)). If now the reflection or rotation operation described above is performed, new spots are generated, as shown in Fig. 9.6(b). This process can be repeated for twinning on the $(3\bar{1}0)$ or the {110} planes, and is left as an exercise for the reader.

More details of indexing diffraction patterns and their interpretation can be found in [1] and [2].

9.5 Image contrast

It is now necessary to examine how contrast arises in the TEM. If no contrast is produced, the high resolution capability becomes worthless. The most important type of contrast is diffraction contrast and this is the starting point in our discussion of image contrast.

9.5.1 Diffraction contrast imaging: use of the objective aperture

In Fig. 9.7 a crystalline specimen with a zone axis nearly parallel to the incident beam causes the formation of several diffracted beams, of which two are represented (by dashes and dots). All of the electrons diffracted at a particular angle (e.g. those represented by dashes) are united at a particular location in the back focal plane of the objective lens. The objective aperture is carried on a holder that allows it to be moved around in this plane. One possible setting is with the objective aperture centred around the transmitted electrons (Fig. 9.7(b)) so that all of the diffracted electrons are intercepted and only those electrons which have passed through the specimen without deflection reach the image. Contrast is produced in the image if different sites within the specimen transmit electrons with different efficiencies. For example, if a very thick region of the specimen is adjacent to a very thin region, most electrons passing through the thick part will be scattered and will fail to pass through the

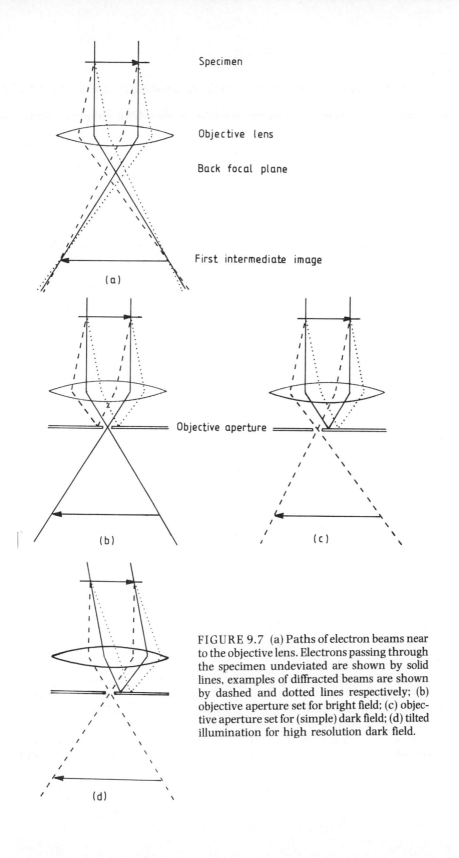

Specimen

Objective lens

Back focal plane

First intermediate image

(a)

Objective aperture

(b)

(c)

(d)

FIGURE 9.7 (a) Paths of electron beams near to the objective lens. Electrons passing through the specimen undeviated are shown by solid lines, examples of diffracted beams are shown by dashed and dotted lines respectively; (b) objective aperture set for bright field; (c) objective aperture set for (simple) dark field; (d) tilted illumination for high resolution dark field.

aperture causing the corresponding part of the image to be dark, whereas a high proportion of the electrons passing through the thin area will be undeviated and will reach the image making it bright. This is known as 'mass-thickness' contrast, because a similar effect is obtained if two regions have the same thickness but contain atoms with different atomic mass and transmit electrons with different efficiencies as a consequence. The most efficient transmitter of electrons is a hole which will appear brightest of all, and when the TEM has the objective aperture positioned in this way the instrument is said to be set for bright field observation.

If the specimen is crystalline then the aperture can be centred around one of the diffracted beams, as illustrated in Fig. 9.7(c). The result is a 'dark field' image formed from electrons that have suffered the chosen diffraction, so that only regions of the specimen having the required structure and orientation to cause such diffraction appear bright.

In order to set the instrument for diffraction contrast a selected area diffraction pattern is obtained first and the objective aperture is then brought into position and centred around the chosen beam, then the instrument is reset for imaging (by returning the intermediate lens excitation to the required value). If dark field imaging is obtained in this way the chosen electrons pass in a direction inclined to the objective axis and thus form an image which contains serious aberration. The alternative procedure is to use an electron beam tilt device (Fig. 9.2(b)) and to adjust the direction of the incident beam so that the chosen diffracted beam passes along the objective axis (Fig. 9.7(d)). The objective axis is 'marked' by setting the objective aperture in the 'bright field' position before operating the tilt circuit, then the tilt controls are used to steer the chosen beam into the aperture.

9.5.2 *Qualitative assessment of diffraction contrast*

Before attempting to introduce quantitative analysis of diffraction contrast, consider some simple examples for which the general image contrast features can be readily explained in a qualitative manner.

(a) POLYCRYSTALLINE SAMPLES

In Fig. 9.8 is shown a region in a polycrystalline sample in which some of the crystals show strong diffraction and others do not. If the microscope is set for dark field for a particular diffraction direction (e.g. the one represented by dashes) crystals diffracting into this direction (B in Fig. 9.8) will appear bright, all others will appear dark. In bright field, crystals diffracting strongly (e.g. B and C in Fig. 9.8) will appear darker than those which are not close to the Bragg position for any of their possible reflections (A, D, E in Fig. 9.8) because the intensity in the undeviated beam will be diminished when diffraction takes place.

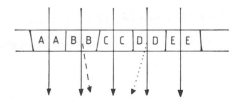

FIGURE 9.8 Schematic representation of a section through a polycrystalline film illuminated by a parallel beam of electrons.

(b) BENT CRYSTALS

Because electron microscope specimens are very thin they bend easily and it is easy to establish a situation like the one depicted in Fig. 9.9 in which in one part of the field of view the diffracting planes are at the Bragg position for a reflection (**g**) as represented by dashes, and in another part are at the Bragg position for reflection (− **g**) represented by dots. Thus, if the TEM is set for dark field using reflection **g**, a bright contour will be seen with its centre at A. If it is set for dark field using reflection (− **g**) a bright contour will be seen centred at C. In bright field there will be dark contours around A and C because of the substantial loss of electrons from the transmitted beam direction, and there will be a relatively bright region in between, centred at B.

(c) EDGE DISLOCATIONS

Consider a crystal containing an edge dislocation parallel to the specimen surface and suppose that the Bragg orientation is as shown in Fig. 9.10. Inspection of this diagram shows that in the regions of perfect crystal, remote from the dislocation, the planes are significantly misoriented from the Bragg position and thus produce only moderate diffraction. Just to the left of the dislocation the planes are rotated even further away from the Bragg position, causing the diffracted intensity to be still less, whereas just to the right of the dislocation the rotation caused by the presence of the dislocation is in the

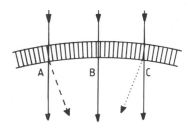

FIGURE 9.9 Bent crystal illuminated by a parallel electron beam. A set of lattice planes is shown oriented perpendicular to the crystal surface.

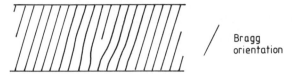

FIGURE 9.10 Thin crystal containing an edge dislocation lying parallel to the surface.

opposite sense and brings them much closer to the Bragg position. Thus diffraction is enhanced just to the right of the dislocation and the dark field image formed using the corresponding diffracted beam will show a bright line following the line of the dislocation and displaced slightly to the right of it, and a dark line parallel to it, displaced slightly to the left. The bright field image will be complementary to the dark field image to a first approximation. A similar approach can be used to predict the image contrast from a screw dislocation.

9.5.3 *The kinematical theory*

It is now appropriate to examine image contrast in a quantitative manner. The approach used is to calculate the intensity in a diffracted beam, which corresponds to the dark field image. To compute the bright field image the sum of all of the dark field intensities should be subtracted from the incident beam intensity. In practice there will often be only one strongly diffracting beam, and the bright field image will be roughly complementary to the dark field image obtained with this beam.

The simplest formulation is the kinematical approximation, which is good only for specimens with very low mass-thickness. Since polymers contain (mainly) light elements thin polymer samples often meet the requirements for this treatment. For this formulation to work the majority of electrons should pass through the specimen without being scattered, and a negligible number should suffer two (or more) scattering acts on their passage through the specimen.

The kinematical theory [1, 3] predicts that when a parallel electron beam passes through a crystal of thickness t, the amplitude scattered into the diffraction ($\mathbf{g}$) is given by

$$\phi_g = A \int_0^t \exp\left(-2\pi i s z\right) dz = A \frac{\sin \pi s t}{\pi s} \exp\left(-\pi i s t\right) \qquad (9.1)$$

where A depends on the incident beam strength and the elemental species causing the scattering. Thus the intensity in the dark field image formed with this diffracted beam is

$$I_g = |\phi_g|^2 = \frac{A^2 \sin^2 \pi s t}{(\pi s)^2} \qquad (9.2)$$

I_g depends on orientation and thickness and variations in intensity differ from one location to another according to the values of s and t. The maximum value of $I_g(= A^2 t^2)$ is approached as $s \to 0$, and equation (9.2) thus confirms that a bend contour should join regions of the sample for which $s \simeq 0$. I_g is also a function of t and equation (9.2) predicts that 'thickness fringes' will be obtained with a wedge shaped crystal, though polymer crystals are normally too thin for their formation. Thus the most important deduction from equation (9.2) is that if the deviation of the crystal from the Bragg position corresponds to a value of $s \geqslant 1/t$, then the intensity in the diffracted beam is negligible.

(a) CONTRAST FROM IMPERFECT CRYSTALS

When dealing with the effect of imperfections on the image it becomes imperative to consider the discrete nature of the scattering and to question the validity of the continuum approach used to derive equations (9.1) and (9.2). It is now necessary to introduce the 'column approximation' in which a notional column is constructed within the crystal. For a column constructed within a perfect crystal equation (9.2) provides an estimate of the intensity at the location of the column, but if a defect lies within the column or near to it, the atomic displacements caused by its presence have to be taken into account. This is achieved quite simply, as will be shown below, but first it is prudent to consider what size limitation should be placed on the column perpendicular to the z-direction. This depends on the Bragg angle and the thickness of the crystal, but it can be shown that even at less than 1 nm the interference caused by scattering in neighbouring columns will not seriously alter the intensity obtained. Thus the column approximation will serve more than adequately for all predictions of image contrast at the resolution limits set by the operating conditions used with most polymer crystal studies. Further discussion on the column approximation is found in more advanced texts on TEM.

If a defect is present with a strain field that can be represented within the column as $\mathbf{R}(z)$, then the integral in equation (9.1) becomes

$$\phi_g = A \int_0^t \exp\left\{-2\pi i \mathbf{g} \cdot \mathbf{R}(z)\right\} \exp\left(-2\pi i s z\right) \mathrm{d}z \qquad (9.3)$$

This will differ from one location to another as a result of changes in s and t as before, but particular attention should be paid to the area close to the defect. Note that if $\mathbf{g} \cdot \mathbf{R}(z) = 0$, the first term in the integrand becomes $\exp(0) = 1$, and the expression in equation (9.3) reduces to that in equation (9.1). For this condition the intensity at the defect is identical to that for a defect-free crystal, and the defect disappears. This is the basis of dislocation Burgers vector determination, as can be easily demonstrated for screw dislocations, for which the strain field ($\mathbf{R}$) is given by the equation $\mathbf{R} = \mathbf{b}\phi/2\pi$, where $\mathbf{b}$ is the Burgers vector and ϕ is the rotational displacement from a reference plane. Thus if the screw dislocation lies parallel to the plane of the foil then $2\pi\mathbf{g} \cdot \mathbf{R} = \mathbf{g} \cdot \mathbf{b}\phi$ which goes to zero when $\mathbf{g}$ is perpendicular to $\mathbf{b}$. Hence, by locating dark field images (and hence **g** vectors) for which the dislocation disappears (the direction of) the Burgers vector can be determined. With polymers this must be done retrospectively with images photographed at sub-visual intensities (see section 9.6.2), whereas with metals and other crystals that resist beam damage, it can be done actively while viewing the imaging screen.

9.5.4 *The dynamical theory*

The dynamical theories of electron diffraction attempt to take into account multiple scattering and the attenuation of the electron beam in the forward direction as it passes through the specimen. In the simplest dynamical formulation a 'two-beam' treatment is developed in which it is assumed that only one diffracted beam is strong so that the energy is shared between this beam and the forward beam, and energy interchange between these two directions is followed as the beam progresses through the crystal [1, 3]. As with the kinematical theory the 'column approximation' is invoked and the treatment of defects follows the same procedure, using the strain field around the defect to describe atomic displacements from the perfect lattice positions. More comprehensive 'many-beam' treatments take account of simultaneous diffraction into many different directions [1, 3]. The dynamical theories have found only limited application in the study of polymer crystals and they will not be discussed further here. It is sufficient to note that the kinematical theory gives a fairly reliable interpretation of most diffraction contrast features found in TEM images of polymer crystals, but that detailed quantitative contrast analysis may sometimes require results obtained by the more exact dynamical treatments.

9.5.5 *The effect of two beams reaching the image*

In the previous sections we have examined the intensity in a single diffracted beam. Consider now what happens when two beams reach the image. The electron beams approximate to plane waves and if they are inclined to one another, as is the case with two different diffracted beams or with a diffracted beam plus the transmitted beam, then it is expected that a periodic interference pattern will be formed when they merge. We will examine the characteristics of such a pattern.

(a) LATTICE IMAGING

Suppose a large objective aperture is inserted into the back focal plane of the objective lens and is positioned as shown in Fig. 9.11 so that the transmitted beam (O) plus one of the diffracted beams are admitted. Thus the image will contain contributions from both and the total wave function (ψ) can be represented by

$$\psi = \exp\left(2\pi i \mathbf{k} \cdot \mathbf{r}\right) + \phi_g \exp\left(2\pi i \mathbf{k}' \cdot \mathbf{r}\right) \tag{9.4}$$

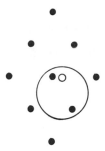

FIGURE 9.11 Schematic diffraction pattern showing position of objective aperture for two-beam lattice imaging. The transmitted beam is shown as ○.

where ϕ_g is the amplitude of the diffracted beam and where the transmitted beam is considered to have suffered only a small attenuation (kinematical approximation) and is given unit amplitude. It follows that the intensity can be written

$$I = |\psi|^2 = 1 + B^2 + 2B\cos(2\pi\mathbf{gr} + \varepsilon) = 1 + B^2 + 2B\cos\left(\frac{2\pi x}{d} + \varepsilon\right) \quad (9.5)$$

since $|g| = 1/d$ when d is the spacing of the crystal lattice planes that give rise to diffraction $\mathbf{g}$; B and ε are functions of s and t (cf equation (9.2)). This function has a periodicity in the x direction of d, and equation (9.5) predicts the appearance of a pattern of fringes parallel to the lattice planes and with a spacing equal to the interplanar spacing (with appropriate adjustment for the magnification of the image). Before such fringes can be observed, however, two conditions must be met. Firstly, the amplitude ($2B$) of the periodic component of equation (9.5) must be sufficient to provide visible contrast; this will depend on the thickness and orientation of the crystal. Secondly, the electron microscope must have adequate resolving power to show detail at the interplanar spacing level. The sensitivity to beam damage of most polymer crystals prevents routine operation at such resolution levels and precludes the use of this technique.

(b) MOIRÉ PATTERNS

Consider a specimen in which one crystal lies on top of another, as shown in Fig. 9.12(a). Let sets of planes $(h_1k_1l_1)$ in the upper crystal and $(h_2k_2l_2)$ in the lower crystal be close to the Bragg condition. In the particular case where the projections of both sets of lattice planes on the observation plane are parallel, the transmitted beam and both diffracted beams are coplanar and the corresponding diffraction pattern is as shown in Fig. 9.12(b) in which the transmitted beam (O) and the diffraction spots P_1 and P_2 from the upper and lower crystals respectively lie in a straight line. The diffracted beams will be

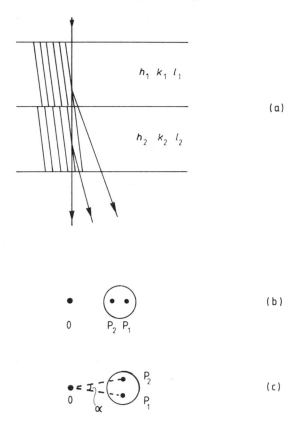

FIGURE 9.12 Formation of dark field moiré patterns: (a) two superposed crystals; (b) (part of) diffraction pattern and aperture position if the $(h_1k_1l_1)$ and $(h_2k_2l_2)$ planes are parallel and have different spacings; and (c) $(h_1k_1l_1)$ and $(h_2k_2l_2)$ planes have the same spacing but are misoriented by an angle α about the electron beam direction.

separated, as shown, if the spacings d_1 and d_2 of the $(h_1k_1l_1)$ and $(h_2k_2l_2)$ planes are different. The diffracted beams will hence be plane waves inclined to one another by an amount proportional to $(OP_1 - OP_2)$, i.e. proportional to $|\Delta \mathbf{g}| = |\mathbf{g}_1 - \mathbf{g}_2|$ where $\mathbf{g}_1$ and $\mathbf{g}_2$ are the corresponding diffraction vectors. Thus if an image is formed using these two beams by positioning the objective aperture as shown in Fig. 9.12(b), these two waves will interfere and produce a fringe pattern with periodicity $1/|\Delta \mathbf{g}|$, in analogy to the result in the previous section.

Now

$$|\Delta \mathbf{g}| = \frac{1}{d_1} - \frac{1}{d_2} = \frac{d_2 - d_1}{d_1 d_2}$$

Hence the periodicity of the fringe pattern is $d_1 d_2/(d_2 - d_1)$. When formed in this way the fringe pattern is called a 'parallel moiré pattern'. It is instructive to construct optical gratings on transparent paper; if two such gratings with slightly different spacings are superimposed and made parallel, a moiré fringe pattern will be visible. If d_2 and d_1 differ by 10% so that $d_2 = 1.1\ d_1$ then $d_1d_2/(d_2 - d_1) = 11\ d_1$. Thus the spacing of the moiré pattern is much greater than that of the individual gratings. This magnification allows resolution of the moiré pattern even if the individual grating patterns cannot be resolved, and TEM moiré patterns are very common even with specimens with which lattice images are difficult or impossible to obtain. If the lattice of one of the crystals is disrupted by the presence of a defect (most commonly a dislocation) a magnified impression is obtained ('moiré magnification'). Thus features such as dislocations can be imaged even when the resolution limit would suggest they could not be seen. The visibility will disappear when $\mathbf{g} \cdot \mathbf{R}$ is zero, just as with conventional diffraction contrast.

If the two crystals have a relative misorientation (α) about the incident beam direction then P_1 and P_2 will lie on different radius vectors. Consider first the case of two similar crystals with $(h_1k_1l_1)$ and $(h_2k_2l_2)$ belonging to the same family so that $OP_1 = OP_2$, leading to the diffraction pattern shown in Fig. 9.12(c). Now $\Delta\mathbf{g}$ is directed along P_1P_2 and has length $2g_1 \sin \alpha/2$, and if a dark field image is formed using P_1 and P_2 a moiré pattern is formed with periodicity $1/(2g_1 \sin \alpha/2) = d_1/(2 \sin \alpha/2)$, inclined perpendicular to $\mathbf{P_1P_2}$. If planes $(h_1k_1l_1)$ and $(h_2k_2l_2)$ are misoriented about the beam direction and have different spacings the moiré pattern has spacing $d_1d_2/\{d_1^2 + d_2^2 - 2d_1d_2 \cos \alpha\}^{1/2}$.

Up to now the discussion has referred to dark field moiré patterns. They are formed by interference between two singly diffracted beams which, if the crystals are both at similar misorientations from the Bragg position, are likely to be of similar amplitude, favouring good contrast. It is possible to obtain moiré patterns in bright field as well, as a consequence of double diffraction. In Fig. 9.13(a) part of the beam diffracted in the first crystal suffers a second diffraction in the lower crystal. This produces a double diffraction spot at $P_{1,2}$ in the diffraction pattern shown in Fig. 9.13(b) (for the 'parallel' configuration) or Fig. 9.13(c) (rotation). If the objective aperture is set for bright field and includes $P_{1,2}$ as well as the transmitted beam (O) then these two inclined beams will interfere, producing a fringe pattern with the same periodicity as the dark field pattern produced by $P_1 + P_2$. If the specimens are thin and are composed of elements with low atomic numbers then the transmitted beam will be much stronger than the doubly diffracted beam and poor contrast will be obtained. Thus with polymers much more useful work has been conducted on dark field moiré patterns than on bright field moiré patterns which have inferior contrast often below the threshold of visibility.

9.5.6 *Phase contrast electron microscopy*

When a small objective aperture is used and the image is formed from the transmitted beam only or from a single diffracted beam the image contrast can

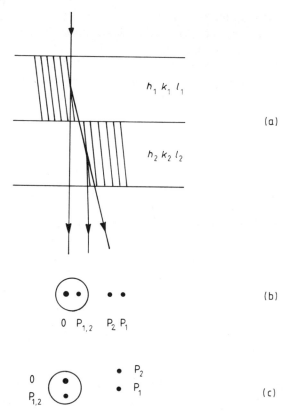

FIGURE 9.13 Formation of bright field moiré patterns. (a) Two superposed crystals; (b) (part of) diffraction pattern and aperture position for parallel moiré pattern; (c) diffraction pattern and aperture position for rotation moiré pattern.

be worked out from the diffracted amplitudes, as outlined in sections 9.5.3 and 9.5.4. In section 9.5.5 it was shown that the image intensity depends on the phase difference between the two beams when the aperture is large enough to admit both. To examine the phase dependence more carefully, consider the general case where a large objective aperture is used and admits the transmitted beam and several diffracted beams. The total wave function becomes

$$\psi = T + \sum_j D_j \exp(i\chi_j) \tag{9.6}$$

where T is the transmitted amplitude and D_j is the amplitude of the jth diffracted beam, for which the phase term is $\exp(i\chi_j)$. The phase angle has the form

$$\chi = \gamma + \frac{2\pi}{\lambda}\frac{\Delta f \beta^2}{2} - \frac{2\pi}{\lambda}\frac{C_s \beta^4}{4} + \chi_{ab} \tag{9.7}$$

where γ is the phase change on scattering, and will usually be close to $\pi/2$, Δf is the amount of defocus, β is the scattering angle, C_s is the spherical aberration coefficient and χ_{ab} includes aberrations other than spherical aberration, i.e. astigmatism, chromatic aberration, plus any contribution from mechanical vibration. If χ_{ab} is assumed negligible in a good machine, then

$$\psi = T + \sum_j D_j \exp(i(\pi/2 + \chi_j')) = T + i\sum_j D_j \exp(i\chi_j') \tag{9.8}$$

$$\chi' = \frac{2\pi}{\lambda}\frac{\Delta f \beta^2}{2} - \frac{2\pi}{\lambda}\frac{C_s \beta^4}{4} \tag{9.9}$$

If only one diffracted beam is strong, or alternatively if the image is formed from many beams all from the same family of planes so that β is the same for each, then ψ can be written

$$\psi = T + iD\exp(i\chi_j') \tag{9.10}$$

Hence the intensity is given by

$$|\psi|^2 = T^2 + D^2 - 2TD\sin\chi'$$

The background intensity in regions free from phase effects is $(T^2 + D^2)$ and the maximum contrast is therefore $4DT/(T^2 + D^2)$ and occurs when $\sin\chi' = \pm 1$. Since χ' is a function of both Δf and β, the phase contrast changes with defocus, and is different for features for which the prominent scattering is at different angles β. This is true for 'small angle scattering' by features with characteristic size or periodicity as well as for Bragg-diffracted scattering. Further discussion is found in more advanced texts on TEM under the heading of 'contrast transfer function', but an important result can be immediately deduced from equation (9.9). Maximum contrast is obtained when $\chi' = \pi/2$, and if the second term on the right hand side of equation (9.9) is ignored for the time being, this gives

$$\Delta f = \lambda/2\beta^2 \tag{9.11}$$

Thus if the effect of spherical aberration is ignored this expression gives the defocus required to image with greatest contrast features for which the favoured scattering angle is β.

Now in the case of lattice imaging β is equal to 2θ, where θ is the Bragg angle, and using Bragg's law ($2d\theta = \lambda$) we find that the optimal value for Δf becomes $d^2/2\lambda$. Hence, for 100 keV electrons and a lattice spacing of 0.4 nm the optimal defocus is ~ 22 nm. It is easily confirmed that this result is not changed very much by taking into account spherical aberration. The significance is that it shows that defocus must be adjusted to optimize contrast when working under phase contrast conditions. This is important, not only in lattice resolution imaging but also with other periodic structures with much larger periodicity which diffract at much smaller angles and require a correspondingly greater defocus. There are some important examples of polymers with this type of morphology.

9.6 Operating conditions and procedures for beam-sensitive materials

Before attempting to perform TEM studies on polymers, it is recommended that the reader gains some familiarity with the standard operating procedures for CTEM, including the formation of selected area diffraction patterns, and bright field and dark field images, using less beam-sensitive specimens.

With beam-sensitive materials observation at conventional levels of illumination is impossible. For example, when polyethylene crystals are viewed at levels of illumination normally chosen for stable specimens, diffraction contrast rapidly disappears and the diffraction spots become diffuse. Even at low magnifications (say 3000 ×) for which the illumination must be spread over a wide area, producing a relatively low beam intensity, diffraction contrast diminishes to almost negligible levels within a few seconds, yet polyethylene is by no means the most beam sensitive polymer. Thus special operating procedures are required.

9.6.1 Damage processes and accelerating potential

In order to establish the best operating conditions it is helpful to consider the possible damage processes. Polymers have low melting points and poor thermal conductivity so that melting under the heating action of the electron beam is possible. It has been established that this effect can easily be prevented, yet rapid damage as described above still occurs. It is therefore necessary to invoke a radiation damage mechanism. The effect of electron bombardment may differ from one polymer to another; in some polymers chain scission may occur and in others crosslinking. There is some evidence that in chain folded crystals damage occurs preferentially at the fold surface. Whatever the mechanism, accumulation of point defects eventually destroys the true crystal order, though in some cases it appears that some co-operative rearrangement of molecules may take place in addition to the localized formation of point defects.

At this point it is pertinent to consider whether there is an optimal choice for the accelerating potential. If the accelerating potential is increased the electrons are less likely to interact with the specimen and therefore produce less damage. However, if the electrons do not interact with the specimen, they contain no contrast information so no benefit is gained here by increasing the potential. The critical factor is the ratio of electrons scattered inelastically (meaning energy is transferred to the specimen, as must happen when damage occurs) to those scattered elastically. This ratio decreases slowly as the accelerating potential increases, but even though this argues for using a high voltage there may be an opposing requirement set by the detection system where, in the case of a photographic emulsion, the higher transmission of higher energy electrons means that the photographic grains may require a longer time to achieve the required exposure level. Hence there is a fairly broad range of suitable potentials, and 100 kV is a popular choice. It is expected that knock-on collisions, in which carbon atoms are removed from their bound state within the polymer, may occur on direct bombardment by electrons within the range 150 KeV to 200 KeV, and it is advisable not to work in this range (or higher) unless it can be established that this does not represent a serious problem for the polymer under examination.

9.6.2 Beam current

Much more critical than the voltage is the beam intensity, which must be as low as possible at the specimen. Low intensities can be achieved by using a low beam current and by adjustment of the condenser lenses. A low beam current will help to extend the filament life. A small condenser aperture is selected to minimize the source strength, and the beam is spread using the condenser lenses. The disadvantage of these conditions is that the image on the phosphor screen is too faint to see. An image can be recorded on a photographic plate or film, using a long time exposure. It is common to use a very fast emulsion (which may have the disadvantage of being a little more grainy than a conventional electron image emulsion) but even then exposures of about 60 s are common even at low magnification. Some quantitative aspects of this problem are considered in the following section.

9.6.3 Choice of magnification

Let us assume that damage accumulates linearly with time at a particular intensity level. For a particular value of intensity (I_0) suppose a corresponding time (t_0) just causes unacceptable damage (sometimes related to a chosen fractional fall in the intensity in a diffraction spot). Therefore $I_0 t_0$ defines the lifetime at the specimen. If the image has linear magnification M and if it is assumed that a fraction (η) of the electrons reaches the image, then the intensity in the image will be $\eta I_0 / M^2$. Suppose that at this intensity the emulsion requires an exposure of t_i. Thus if $t_i < t_0$ an image can be recorded before unacceptable damage has been caused.

If the microscope is set for dark field then the value of η will be determined by the reflection used and by the misorientation (s) for that reflection. It will generally be true that $\eta \ll 1$ for a dark field image. If for the (hkl) reflection η has the value η_{hkl} then the corresponding time required to expose the recording emulsion is $t_i(hkl)(\propto M^2/I_0\eta_{hkl})$. If it is desired to record images using several diffracted beams then the total time required will be

$$\sum t_i(hkl) \propto \frac{M^2}{I_0} \sum \frac{1}{\eta_{hkl}} \tag{9.12}$$

where the summations are taken over all the chosen combinations of *hkl*. Clearly, it must be arranged that $\sum t_i(hkl) \leqslant t_0$, if all images are to be successfully recorded. The only adjustment available is the magnification (M) and low magnification operation is chosen to improve the chance of fulfilling this condition. An added advantage of low magnification imaging is the corresponding large field of view, which is especially valuable when working at sub-visual levels as this increases the chance of including features of interest in the area recorded. The chief disadvantage of low magnification recording is that the resolution becomes limited by the characteristics of the detection system and is worse than the electron optical resolving power. Finally, inspection of expression (9.12) shows that if dark field images are required from n diffracted beams of roughly equal intensity then the maximum operating magnification must be $M_1/\sqrt{n}$ if M_1 is the maximum magnification at which a single image can be recorded.

9.6.4 *Focusing and other adjustments*

From the previous paragraphs it is evident that polymer studies in the TEM should be conducted at very low beam intensity and at low magnification. At the lowest beam intensities the image on the phosphor screen is not bright enough for visual observation. The following procedure is therefore recommended:

1. Set the condenser to give sufficient brightness to focus the image, choose a suitable feature and focus the objective;
2. Spread the beam to give low intensity;
3. Move the specimen so that a fresh area remote from that damaged during focusing is brought into the centre and photograph it immediately.

Selection of the area to be photographed must be as rapid as possible. At the lowest beam intensities, as required by the most sensitive specimens, the image intensity may be too low for visual observation and it may be uncertain that an area containing the required features is actually in view until the photograph has been processed. Recall that the field of view is large at low magnification, increasing the chance that an area selected 'blindly' contains interesting features. The large depth of field ensures that the specimen remains in focus when shifted unless it has pronounced undulations or is tilted. Finally, it is easy to confirm that if images are required in both bright field and dark field there is no need to refocus between each condition. It can be deduced that the difference in phase produced by spherical aberration between the transmitted beam and a diffracted beam at an angle β is equivalent to a change in focus (Δf) given by $\Delta f \beta^2/2 = C_s \beta^4/4$.

Taking once again as a typical example the value of $\beta(= 2\theta)$ corresponding to the 110 diffracted beam in polyethylene when using 100 keV electrons, and setting $C_s = 3$ mm we find $\Delta f = 122$ nm, well within the defocus limitation given by the depth of field. Diffracted beams from the same family of planes all have the same defocus value, and the defocus difference between dark field beams from planes with different interplanar spacing will also be negligible. This is true when using the shifted aperture procedure; there should be no defocus discrepancy when using the tilted beam procedure.

When recording a series of images from a single area the beam current should be switched off when observation or recording is not in progress, for instance when the camera film mechanism is being operated.

9.6.5 *Photographic emulsion and resolution*

The resolution of a photographic emulsion is limited by its grain structure. Thus if the granularity of the emulsion prevents recognition in the image of structural features less than 20 μm (a typical value) this limits the resolving power of the TEM to 20/M μm referred to object space where M is the linear magnification. In general, faster emulsions have larger grain sizes so that the benefit gained by shorter exposure times is reduced by the need to operate at a higher magnification to achieve a particular target resolution.

9.7 Polymer studies in TEM

9.7.1 *Solution-grown crystals*

A substantial body of knowledge about polymer crystallization and growth habit has accumulated from TEM studies of crystals grown from solution. They are in many ways ideal subjects for TEM observation since they are very thin and uniform, though they suffer from beam sensitivity.

(a) PREPARATION OF SOLUTION-GROWN CRYSTALS

The polymer most studied using this technique is linear ('high-density') polyethylene (HDPE). There are several solvents for HDPE but a popular one is xylene in which it dissolves at about 120°C. If a dilute solution is made (say 0.1% by weight HDPE) then cooled slowly, the polymer crystallizes in the form of thin platelets. Crystallization commences when the temperature falls to below 100°C. Once a crystal forms it continues to grow (unless the temperature is increased again) until the polymer is all used up. If the solution is allowed to cool continually crystals nucleate over an extended period of time so that when crystallization is complete there is a range of different crystal sizes. For scientific investigation of crystal growth the following procedure is normally used:

1. Dissolve HDPE in xylene at elevated temperature (say 120°C). (The concentration may have to be higher than the final target concentration for crystallization if dilution is applied at stage 4.)
2. Cool to a temperature at which crystallization will proceed (say 75°C) and allow nucleation to continue for 15–60 min;

3. Heat the solution slowly (4–10°C/h) to a temperature (T_s) at which the crystals begin to redissolve (> 100°C). Crystal nuclei remain after this process as long as neither the temperature nor the residence time is too great.

4. Cool the solution rapidly to the crystallization temperature (T_c) (usually 65–90°C). To ensure that T_c is reached in minimal time the solution at T_s is decanted into solvent already controlled at T_c. The nuclei formed during the earlier steps are present in sufficient number to ensure that subsequent spontaneous nucleation is almost negligible by comparison and that crystallization proceeds predominantly via these pre-existing nuclei. Hence when crystallization is complete the vast majority of crystals present will have begun to grow simultaneously at the beginning of the final stage of the process. Therefore they are identical in size and have grown under known controlled conditions. This method of preparation is known as 'self seeding' and can be used to investigate the effect on crystallization rate and crystal habit of such factors as crystallization temperature, solution concentration and molecular weight.

(b) GENERAL APPEARANCE OF SOLUTION-GROWN CRYSTALS

The crystals formed in this way are usually too small and/or too fragile to span the holes in a TEM specimen support grid and have to be placed on a support film before mounting on the grid. The support film normally chosen is evaporated carbon which is easy to prepare, is sufficiently continuous and strong to be self-supporting at thicknesses of 5–20 nm, is quite transparent to electrons at this thickness, and, being amorphous, shows no diffraction contrast so that it is hardly visible in TEM images of polymer crystals supported on it.

A satisfactory method of specimen preparation is to float an evaporated carbon film on to water, pick it up on a TEM grid and allow the water to evaporate, and then to place a drop of a suspension of solution-grown crystals in the original solvent on to the grid. The xylene evaporates rapidly and the crystals deposit on the carbon support film. The specimen is now ready for insertion into the TEM, but for some purposes one further preparation step is employed, namely 'shadowing'. This technique involves placing the specimen (on the grid) into an evaporation unit and evaporating a small quantity of a metal (usually gold or a gold-palladium alloy) at an angle (Fig. 9.14). A thin layer of metal deposits over the polymer crystal and the support film, but regions in the shadow of the crystal, as on the left hand side of Fig. 9.14, remain uncoated whereas heavy deposits form at steps facing towards the evaporation source, as on the right hand side of Fig. 9.14. Shadowing greatly assists visibility in the bright field electron image, and the permanent metal deposit persists long after the crystallinity is destroyed by the electron beam so that morphological (rather than crystallographic) information can continue to be gathered. The shadow length can be measured and if the evaporation angle

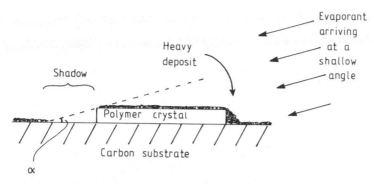

FIGURE 9.14 Shadowing.

(α) is known, the height of the crystal can be estimated (= shadow length $\times \tan \alpha$). Although the presence of a very small quantity of shadowing material does not seriously interfere with the diffraction pattern, it is normally preferable to omit the shadowing step if the information sought is mainly to be obtained from electron diffraction and diffraction contrast imaging.

Polyethylene crystals prepared in this way are usually diamond shaped (Fig. 9.15) with a height of 10–15 nm. Electron diffraction shows that the crystals are oriented with the molecular axis parallel to this short dimension, for the pattern corresponds to the (001) section of the orthorhombic structure. This discovery led to the various models for chain folding in polymer crystals. Discussion of this topic is beyond the scope of this book, but it should be emphasized how important TEM studies have been in the development of the subject.

Other observations made in TEM studies of HDPE solution-grown crystals were the presence of pleats in some cases and ridges in others. These suggested that the crystals had not grown as flat lozenges. From the evidence provided in the TEM together with some careful experiments in which the crystals still in suspension were studied in the light microscope, it was shown that the crystals grow as hollow pyramids. The molecular axis is parallel to the pyramid axis and when the crystal deposits on to a stiff substrate, slip parallel to this axis permits flattening of the crystal.

(c) DIFFRACTION CONTRAST STUDIES WITH SOLUTION-GROWN CRYSTALS

When hollow pyramid polyethylene crystals collapse on to a flat substrate as well as slip in the molecular axis direction some tilt may occur (Fig. 9.16). If the indexing scheme shown in Fig. 9.16(c) is adopted, then the sense of tilt of the $(1\bar{1}0)$ planes in sector B (along OQ) is opposite to the sense of tilt of the same family of planes in sector D, along PO. The (110) planes will be much less tilted in sectors B and D, than the $(1\bar{1}0)$ planes, however, and will remain more closely parallel to the original pyramid axis which coincides with the normal

FIGURE 9.15 Bright field image of shadowed multilayered solution-grown polyethylene crystals. A rather heavy shadowing deposit has been used for clarity.

to the support film on to which the crystal flattens. Conversely the (110) planes will tilt in sectors A and C, while the ($1\bar{1}0$) planes in these sectors will remain parallel to the normal to the flattened crystal plane. It is expected that this sectorization will be evidenced in dark field imaging. Taking the example of the structure depicted in Fig. 9.16, if the ($1\bar{1}0$) reflection is used then sectors A and C will be close to the diffraction condition in the untilted condition with the electron beam normal to the substrate plane. (Recall that the ($1\bar{1}0$) planes remain almost parallel to this direction in sectors A and C.) On the other hand

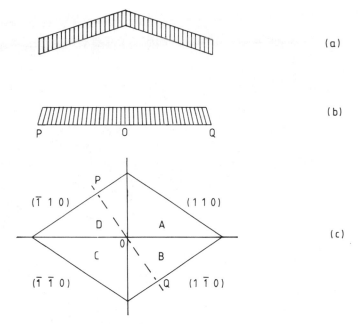

FIGURE 9.16 Collapse of a pyramidal crystal: (a) section before flattening; (b) section after flattening; (c) plan view.

to make sector B diffract into the $(1\bar{1}0)$ beam the specimen must be tilted (i.e. OQ must rotate clockwise (Fig. 9.16(b)) and to make sector D diffract a tilt in the opposite sense is required. Studies of this kind are difficult because of the limited lifetime of crystals in the beam. Examples can be found in the literature, however, but it has also been shown that there are other structural forms and that the growth morphology differs with different crystallization conditions.

Solution-grown polymer crystal preparations frequently contain multiple layered stacks of crystals. This is especially so if the self seeding procedure is not carefully followed. Bilayered crystals are common and often provide interesting diffraction contrast. The crystals tend to be aligned almost parallel to one another, indicating that this is a growth feature rather than a case of accidental superposition on drying down. If there is a small miosorientation a moiré pattern is formed and may reveal dislocations by moiré magnification. In some cases, primarily with crystals grown from low molecular weight polymer (e.g. HDPE with $\bar{M}_n < 10\,000$, especially < 4000) the image of a bilayer with a small misorientation consists of a regular network of wavy lines instead of straight parallel fringes (Fig. 9.17). This is indicative of a twist boundary consisting of a network of screw dislocations. In other circumstances parallel moiré patterns are found. When the sector boundaries of the two superimposed crystals are not coincident a spacing change in the pattern is observed on crossing the boundary in one of them.

FIGURE 9.17 Simple dark field image of a multilayered stack of solution grown polyethylene crystals, showing moiré patterns and, near the bottom left-hand corner, a screw dislocation network.

It has been argued that if chain folding is regular and occurs only on the growth plane in a particular sector (e.g. the (110) in polyethylene) then the spacing between these planes will be slightly different to that of the crystallographically equivalent planes that do not contain folds (i.e. the ($1\bar{1}0$) planes in the (110) sector in our example) as a consequence of the different fold surface conformations.* Thus the moiré pattern spacing change has been interpreted to indicate a difference in spacing between, for example, (110) planes in a (110) sector and (110) planes in a ($1\bar{1}0$) sector, and therefore as confirmation of regular chain folding. Interpretation is by no means straight-forward because of the possibility of the crystal lattice being strained or tilted during the flattening process, for both types of distortion would produce changes in moiré pattern spacing. More detail on the analysis of moiré patterns at sector boundaries can be found in [4].

(d) DEFORMATION STUDIES OF SOLUTION-GROWN CRYSTALS

If a drop of solvent containing solution-grown polyethylene crystals in suspension is placed on a deformable substrate, such as a polyester film (e.g. 'Mylar' or 'Melinex') and the crystals are allowed to dry down on to it they are found to adhere sufficiently well that they deform in an affine manner with the substrate if it is subsequently deformed in tension. The following procedure has been developed to enable crystals deformed on such a substrate to be transferred to a thin film TEM specimen which has the principal stress axis direction 'marked' on it.

1. Deposit the crystals as described above, apply a tensile strain and clamp the substrate in the strained configuration (Fig. 9.18(i));
2. Transfer the strained substrate plus crystals to an evaporation plant and deposit a small amount of shadowing material at a shallow angle, making sure that the projection of the shadowing direction on to the specimen plane coincides with the (known) strain axis. Although subsequent inspection of the shadows thrown by a large number of crystals should enable the shadowing direction (and hence the strain axis) to be identified, this is made much easier if a small number of micron-sized polystyrene spheres (which can be purchased in suspension in water from electron microscope laboratory suppliers) is distributed on to the surface. The shadows thrown by the spheres mark accurately the shadowing direction and hence the deformation axis (Fig. 9.18 (ii));
3. Pour a concentrated solution of poly(acrylic acid) in water over the crystals and allow to dry (Fig. 9.18 (iii));

---

*If there is a difference in spacing between the (110) and ($1\bar{1}0$) planes the crystal is not strictly orthorhombic. The distortion is very small, however, and is normally ignored.

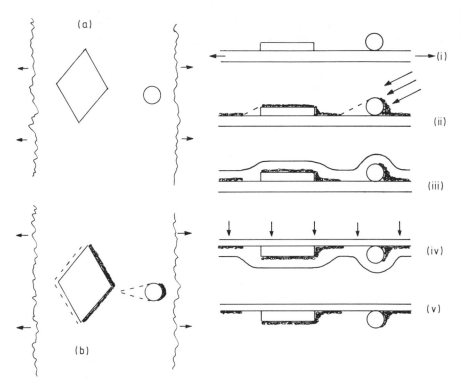

FIGURE 9.18 (i) Deformation of substrate and flattened crystals, with a polystyrene sphere nearby; (ii) shadowing; (iii) application of polyacrylic acid film; (iv) carbon evaporation on to inverted specimen after removal of substrate; (v) poly(acrylic acid) dissolved away. (a) Plan view of (i); (b) plan view of (v); the dashed lines show the extent of the shadow, and the direction of the shadow from the sphere indicates the strain axis.

4. Peel away the polyester substrate, invert the specimen and evaporate carbon on to it (crystal side up) in a normal direction (Fig. 9.18 (iv));
5. Dissolve away the poly(acrylic acid) on a water surface (Fig. 9.18 (v)). The specimen is now ready to be broken into pieces and mounted on to TEM grids. It is at this stage that the memory of the original deformation direction would be lost, and why marking it by shadowing is important.

As long as the amount of shadowing material used is not too great, electron diffraction patterns from the deformed crystals can be obtained. In studies of polyethylene using this technique it has been found that deformation can lead to twinning and/or transformation from the orthorhombic into the monoclinic phase. Relating the deformation direction to the crystal axes has enabled identification of the range of stress directions over which each deformation mechanism operates.

If the substrate for deformation is replaced by one to which the crystal adheres only partially so that it does not deform affinely with it then at sufficiently large strains the crystals fracture. It has been observed that cracks running parallel to the {110} growth faces are clean in the corresponding sector, whereas cracks inclined to the growth face, including those parallel to the (110) planes in a (1$\bar{1}$0) sector are bridged by fibrils. It has been suggested that this indicates that chain folding occurs predominantly on the growth face so that separation along these faces does not demand breaking of molecular chains.

9.7.2 *Solution-cast films*

Thin uniform polymer films can be cast on to flat surfaces from solution. If the solution is sufficiently dilute and if good spreading is achieved the film may be thin enough for electron transmission. The film can be stretched before inserting into the electron microscope, permitting study of deformation mechanisms, or crystallization in the presence of flow-induced molecular orientation.

(a) CRYSTALLIZATION OF POLYISOPRENES

Cis-polyisoprene (natural rubber) crystallizes at sub-ambient temperatures. To make a film for TEM study a solution (say 1–2%) is made in a suitable solvent (e.g. benzene) and cast on to deionized water at room temperature. The solvent is immiscible with water and the solution spreads out to form a thin film. The solvent evaporates, leaving behind a thin film of *cis*-polyisoprene. Pieces of the film are next picked up on TEM grids and transferred to a low temperature cabinet for crystallization. The crystals remelt on warming the sample to room temperature and for direct observation in the TEM a low temperature specimen holder is required, together with a suitable low temperature transfer module permitting transfer of the specimen from the crystallization cabinet to the microscope without the temperature rising to the melting temperature. Although such equipment has been developed and is used quite extensively in microbiology and biomedical laboratories, an alternative method of approach is most commonly used with polymers, namely staining. The purpose of the stain here is twofold: firstly the stain marks certain morphological features selectively, enhancing their contrast, and secondly it 'fixes' the structure because the stain remains in position long after melting (on heating) and/or radiation damage (during TEM observation) has destroyed the original structure.

In the case of *cis*-polyisoprene, osmium tetroxide (OsO_4) is a suitable stain. It reacts selectively with the amorphous phase, attacking the double bonds and building up deposits containing heavy osmium atoms. Presumably OsO_4 can diffuse relatively easily through the loosely packed molecules in the amorphous phase, but cannot penetrate the crystals. OsO_4 is provided in the vapour

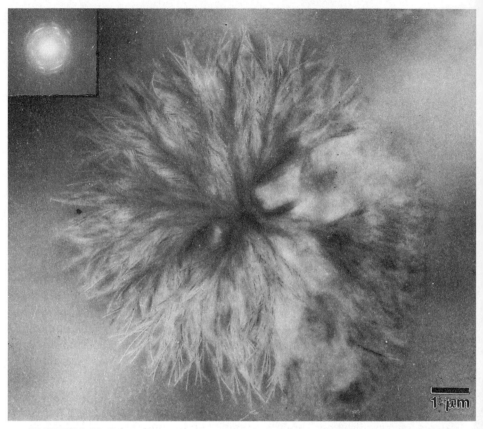

FIGURE 9.19 Spherulite grown in a *trans*-1, 4-polyisoprene film crystallized for 41 min at 48°C. Inset is an electron diffraction pattern taken before crystallinity was destroyed. (Courtesy of C.K.L. Davies [6].)

phase using OsO_4 crystals which give adequate vapour pressure for this purpose even at sub-zero temperatures in the cold cabinets used for polyisoprene crystallization. The OsO_4 crystals are held in a side tube, sealed from the main tube in which the crystallizing samples are placed, for premature exposure to OsO_4 would inhibit further crystallization. When crystallization is complete, the side tube is opened into the main tube, permitting OsO_4 vapour to reach the sample. After staining is complete the sample is removed and allowed to warm to room temperature. The crystals melt, but the regions in which they were located remain unstained, surrounded by the stained phase that was amorphous even at the low temperature. Consequently, the (formerly) crystalline regions appear bright in the TEM bright field image, surrounded by the dark (stained) phase. In the example shown in Fig. 9.19 the subject is *trans*-1, 4-polyisoprene, from which

(a)

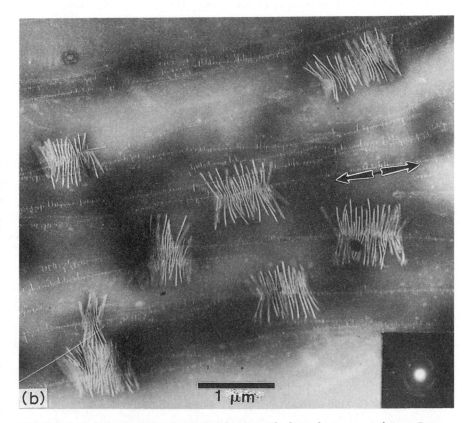

(b)

FIGURE 9.20 (a) Schematic of modified dividers. The limit device is not shown. Better control is afforded by using draftsman's dividers; (b) row-nucleated crystals in a *trans*-1,4-polyisoprene film strained 150% and crystallized for 25 min at 53°C. The arrow shows the direction of strain. Inset is an electron diffraction pattern, taken before crystallinity was destroyed. (Courtesy of C.K.L. Davies [6].)

samples can be prepared in a similar manner to that described above for *cis*-polyisoprene, but with the added convenience that it melts above room temperature. Even after staining, which can be used to terminate crystallization after a chosen time, crystals remain and their structure can be investigated using transmission electron diffraction as long as very low beam exposure conditions are used. Figure 9.19 shows a fairly well developed spherulite. Further examples of staining are discussed later in this chapter.

A simple modification of the above procedure permits the study of crystallization in films containing oriented molecules. After the solvent has evaporated from the cast film part of it is stretched then held in the stretched state for the rest of the preparation. A convenient way of doing this is to use a pair of draftsman's dividers modified by replacing the points by cylinders aligned parallel to the axis of rotation (Fig. 9.20(a)). It is a simple task to provide restrainers that limit the separation of the cylinders to a pre-selected minimum and maximum so that if the cylinders in the minimum position are gently lowered on to the surface of the film (supported on water) and then opened to the maximum position, the film will suffer a pre-selected extension between the cylinders. Figure 9.20(b) shows row nucleated structures produced in *trans*-1, 4-polyisoprene using 150% extension, and stained after straining and crystallization at 53°C.

(b) DRAWING OF SEMICRYSTALLINE POLYMERS

A useful laboratory technique for preparing samples of semicrystalline polymers with a fibre orientation is to draw a solution cast film from a heated substrate. This produces TEM specimens with special characteristics that have provided useful information about polymer structure and morphology. It is not yet clear how relevant this is to bulk polymers, but many of the key parameters in such processes as fibre drawing are imitated in this technique (temperature, cooling rate, draw ratio etc.) and it is expected that strong analogies should occur.

The method is to cast the polymer from dilute solution as before, but this

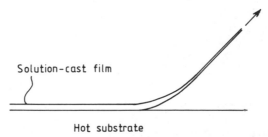

Solution-cast film

Hot substrate

FIGURE 9.21 Drawing a thermoplastic after casting from solution on to a hot substrate.

FIGURE 9.22 Electron diffraction from a drawn polyethylene film. Drawn polyethylene films imaged here were provided by D.C. Yang (see also Yang, D.C. and Thomas, E.L. (1985), *J. Mater. Sci.*, **19**, 1098).

time on to a heated substrate. Most studies reported to date have used either orthophosphoric acid or glass. Because of the elevated temperature, normally close to the crystal melting temperature (e.g. about 125°C for linear polyethylene), the solvent evaporates quickly. A rod or some other suitable implement is touched on to the surface and drawn away, pulling with it film which draws as it detaches from the substrate (Fig. 9.21). If the conditions are properly adjusted the film can be drawn off continuously, giving a narrow (a few mm) very long tape. A further annealing process can be added prior to TEM inspection. A carbon film is normally evaporated on to the specimen to improve stability in the electron beam as a result of improved thermal and charge conductivity.

Films produced in this manner are often thin enough for direct TEM inspection. Electron diffraction shows that the molecular axis is predominantly parallel to the draw direction, lying in the plane of the film. Specimens

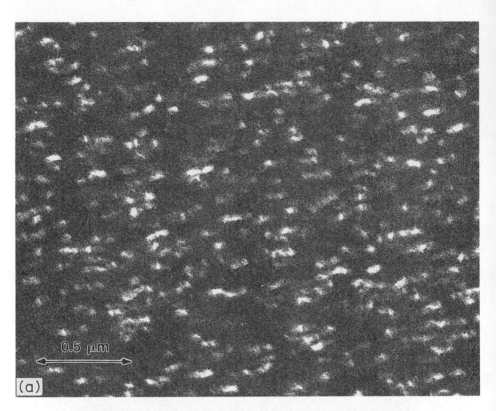

(a)

0.5 μm

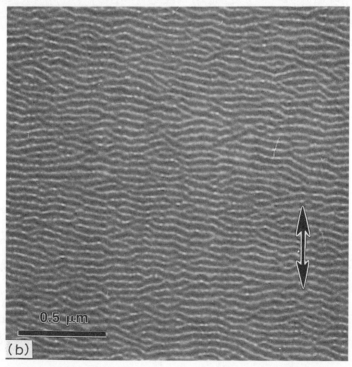

(b)

0.5 μm

230

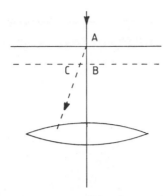

FIGURE 9.24 Formation of ghost images. If the objective lens is de-focused, so that the image plane is conjugate with plane BC, the image of a point A in the specimen formed by the diffracted beam through C will be displaced from the image formed by the transmitted beam through B by an amount equal to BC times the image magnification.

prepared from crystalline polymers show a pronounced fibre orientation (Fig. 9.22). The electron image shows that the film contains small crystals, arranged in lamellae oriented perpendicular to the fibre axis. Although evidence for this can be found in the form of diffraction contrast in both bright field and dark field conditions, a more clear representation of the structure can be obtained using phase contrast (Fig. 9.23). To do this the objective aperture is withdrawn and the objective lens defocused to provide maximum contrast for the lamellar period. For example if the period is 50 nm (equal to the lamellar thickness plus the thickness of the interlamellar region), the defocus required (equation (9.11)) is about 0.34 mm (very large in electron microscopy practice). Phase contrast persists even when inspection of the diffraction pattern confirms that crystallinity has been destroyed in the electron beam. Although the reason for this has not been established, it could be the consequence of continued superior order or greater density in the regions that were formerly crystalline, or of the crystalline regions being thicker than the non-crystalline regions.

An interesting observation can be made in phase contrast imaging if crystallinity has not been destroyed. If the objective aperture is large enough to admit diffracted beams as well as the transmitted beam then defocus causes the diffracted beam image of a particular crystal to be shifted relative to that formed by the transmitted beam (Fig. 9.24). Consider a crystal oriented close to a Bragg position. It will diffract more electrons than its neighbours, assumed less close to a Bragg position, and the transmitted beam will be less intense than in neighbouring regions. In bright field this crystal will appear darker

FIGURE 9.23 Images of drawn polyethylene film: (a) dark field: only crystals in the appropriate Bragg orientation appear bright; (b) phase contrast. The approximate draw direction for both is shown by the arrow in (b).

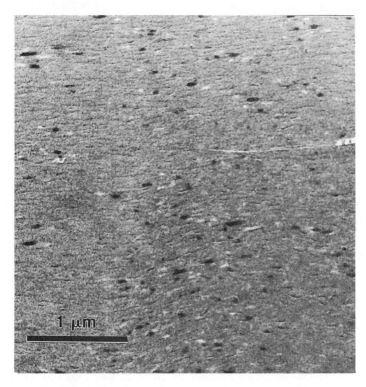

FIGURE 9.25 Ghost images in drawn polyethylene film viewed in defocus bright field with no objective aperture present.

than the surroundings. If no objective aperture is present the electrons 'lost' to the diffracted beam will contribute to the image and if it is focused almost no contrast will be found. On defocusing, the diffracted electrons will shift (in the direction of the diffraction vector) by $\Delta f 2\theta$ and the crystal will appear darker than its surroundings again, and will have associated with it, displaced in a particular direction, a 'ghost' image corresponding to the diffracted beam (Fig. 9.25). The exact orientation of the particular crystal can therefore be deduced.

In dark field only those crystals diffracting into the objective aperture appear bright. Generally only a small fraction satisfies this condition for a particular film orientation and aperture position, and the lamellar texture may not be revealed by the dark field image alone (Fig. 9.23).

9.7.3 Crazes in thin films

The examination of the detailed structure of crazes is beyond the resolution capability of the light microscope and their morphology has been investigated using the TEM. Some model studies have been performed on thin cast films which were mounted on a TEM grid before straining to form crazes (copper TEM grids deform easily). This technique has been particularly successful with polystyrene, which is fairly resistant to electron beam damage.

Although crazes in cast films are convenient for laboratory investigation it is important to know whether they are similar to those formed in the bulk. This has been shown to be so by examination of sections cut from crazed bulk polymer samples. Improved contrast was achieved in some studies by impregnating the interconnecting voids with sulphur or sulphur + iodine which acts as a stain, and probably also improves the sectioning behaviour. Sectioning is discussed further in section 9.7.4.

9.7.4 Sections

In order to obtain samples that are truly representative of the bulk, thin sections must be cut using an 'ultramicrotome'. This is similar in concept to the 'microtome' used for thin light optical sections, described in Chapter 11, but is designed to produce sections thin enough for electron transmission. It is sometimes necessary to cool the sample and the knife to obtain satisfactory sections from polymers. Much development of this technique has been undertaken for biological and medical applications (sectioning of cells etc.) but it remains a most exacting experimental technique for polymer studies. The following list of applications is arranged according to classes of material, and includes mention of techniques for sectioning other than by microtomy, and of the complementary technique, staining.

(a) SEMICRYSTALLINE BULK POLYMERS

Detailed investigation of the internal structure of polymer spherulites probably represents the most coveted goal of polymer electron microscopists. Despite the enormous incentive to develop specimen preparation methods there are still no satisfactory routine sectioning procedures for semicrystalline polymers for direct observation. Sectioning of polymeric material is always difficult, but with semicrystalline polymers the difference in stiffness between the crystalline and non-crystalline phases probably makes matters worse, as does the presence of favoured fracture paths, dictated by the morphology and not by the cutting geometry. Sections are therefore frequently of poor quality and often contain distortion that results from the cutting process and which further hinders accurate interpretation of the images.

Considerable success has been achieved with the help of a staining technique 'chlorosulphonation', in which the stain is applied before sectioning and helps to improve the mechanical properties required for satisfactory microtoming as well as improving contrast. Again this technique was developed for polyethylene and has found only limited application with other polymers. The main steps in the procedure are as follows:

1. Cut a small sample from the bulk polymer (< 3 mm thick) and immerse in

chlorosulphonic acid for several hours at 60°C (or much longer at room temperature). The sample blackens;

2. Remove and rinse using sulphuric acid then water;
3. Embed in epoxy resin and cut sections using the ultramicrotome;
4. View in the TEM. Chlorine and sulphur atoms attach to the lamellar surfaces causing the lamellae to appear much brighter than the surrounding region when viewed end on or side on in bright field. Enhanced contrast can be obtained by adding a further stain by immersing in aqueous uranyl acetate (0.7% for 3 h) between steps 3 and 4. Examples can be found in [4] and [5].

(b) FIBRES

Fibres are generally much easier to section than bulk polymers as long as the section required is parallel to the fibre axis. Polymer fibres can usually be split parallel to the axis, and some polymers even form microfibrils thin enough for electron transmission without sectioning in some circumstances (e.g. ultra-drawn polyethylene subjected to further mechanical deformation). In order to cut a section using an ultramicrotome, the fibre must be mounted in a resin. Careful positioning is essential to ensure the chosen section is cut. The block containing the fibres is then mounted in the ultramicrotome and sectioned (possibly at cryogenic temperature). Sections are mounted on TEM grids and a thin layer of evaporated carbon can be deposited to reduce charging and thermal instability problems. Sections inclined to the fibre axis often show that deformation has been caused by the sectioning procedure and any information obtained from them must be treated with caution.

If the fibre fibrillates easily then it may not be necessary to use an ultramicrotome to produce thinning parallel to the fibre axis. The simplest alternative technique is to place the fibre between two pieces of adhesive tape and strip them apart. Part of the fibre will be left on each piece of tape. The process is repeated until sufficient thinning has been achieved. By using the same pieces of tape repeatedly a multitude of fragments can be positioned side by side, giving a specimen with a fairly dense concentration of microfibrils over the whole grid square rather than one single fibre width. After the thinning process is complete a carbon film is evaporated on to the tape (fibrils upwards). Finally, the carbon film plus fibril fragments are separated from the tape by immersion in a solvent for the tape adhesive, and mounted on TEM grids. The disadvantage of this procedure is that the original positions of the microfibrils within the fibre are unknown. A hybrid preparation technique may be possible wherein the fibre is mounted in resin, partially sectioned in an ultramicrotome, then fragments stripped from the part remaining in the resin block by adhesive tape or a similar stripping technique. An alternative stripping agent to adhesive tape is poly(acrylic acid) which is applied as a concentrated solution in water and allowed to dry before peeling away.

Specimens produced from fibres formed from polymers with stiff backbones containing ring structures have been found to be less sensitive to irradiation damage than most other polymers. As far as the authors are aware, the first example of a lattice image obtained with a polymer was obtained with such a specimen, made from poly(*p*-phenylene terephthalamide) ('Kevlar').

(c) BLOCK COPOLYMERS AND OTHER TWO-PHASE POLYMERS

Block copolymers often display phase separation and form very regular structures in which one of the phases appears as spheres or rods or lamellae dispersed within the other (continuous) phase. TEM studies of sections have shown the size and periodicity of these structures to be extremely regular. Phase contrast conditions are often valuable for examining such specimens.

Rubber modified 'high impact' polystyrene has been the subject of several TEM investigations. The size and dispersion of the rubbery phase is of interest, but additionally specimens which have been subjected to various kinds of mechanical tests have been examined in an attempt to determine the deformation mechanisms. Crazes in the continuous polystyrene phase have been found nucleated by the easily-deformed rubbery phase.

9.7.5 *Other applications*

(a) SOLID STATE POLYMERIZED FILMS

Considerable interest has been attracted by the family of polymers based on polydiacetylene because of their unusual electronic and optical properties. They can be made by solid state polymerization of crystals of monomer, and if the original crystal is grown in thin film form the polymerized product is suitable for TEM study. These specimens are more resistant to electron beam damage than the average crystalline polymer and have provided examples of lattice images. Their superior TEM lifetime has enabled more detailed study of defects using diffraction contrast than is possible with many other polymers.

(b) ETCHING AND REPLICATION

Replication, in which the surface undulations of the sample are replicated in a thin film which is then detached and transferred to the TEM (often after shadowing) was once a popular technique: details may be found in [5]. It has now been largely superseded by the scanning electron microscope (SEM) for which specimen preparation is much simpler. The reason for including reference to replication here is that it has been used in conjunction with a surface etching technique to reveal significant information in a series of notable micrographs produced by Bassett and co-workers [4]. It is admitted that the same information could be obtained using SEM, but the replica is stable in the electron beam and can thus be examined in the TEM for long

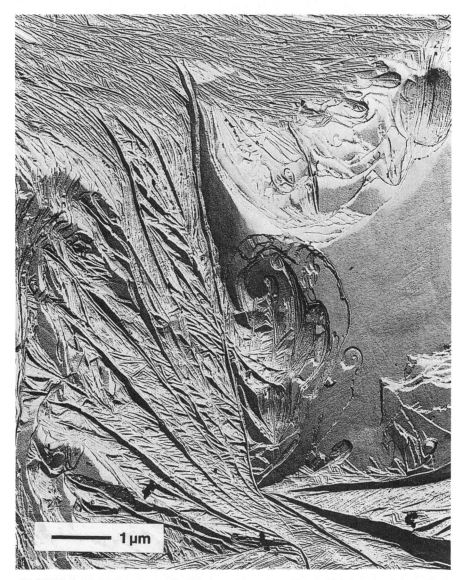

FIGURE 9.26 TEM image of a section through a blend of linear and branched polyethylene after permanganic etching. This technique has revealed in great detail the lamellar structure. Several thick dominant lamellae are seen in the centre, which crystallized isothermally. Most are edge on, but on the right hand side is a region in which the lamellar fold surfaces are parallel to the section surface. Near the top, crystallization took place after subsequent quenching, and the dominant lamellae are thinner than those in the central region. In all regions still thinner lamellae are seen to have formed at a later stage between the dominant lamellae.

(Courtesy of D.C. Bassett and A.S. Vaughan. More details of the permanganic etching technique and its application are to be found in Bassett, D.C. (ed.) (1988), *Developments in Crystalline Polymers*, 2nd edn, Elsevier Applied Science Publishers, Barking.)

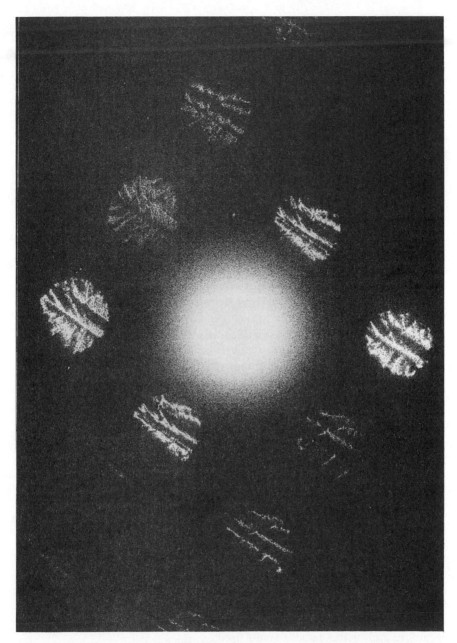

FIGURE 9.27 Multiple dark field image from a poly(butylene terephthalate) crystal grown in a film of a block copolymer containing 73% by weight poly(butylene terephthalate) and 27% by weight poly(tetramethylene ether glycol). This specimen was provided by R.M. Briber (see also Briber, R.M. and Thomas, E.L. (1985), *Polymer*, **26**, 8).

periods at convenient levels of magnification and brightness. Resolution is limited by the replica itself, but the faithfulness with which it follows the sample surface is likely to be comparable with the performance offered by the best SEMs (and better than the average).

More important than the replication procedure is the etching technique used in this study. Etching procedures, in which material is removed selectively at the surface of a sample to reveal morphological details, are introduced in connection with other techniques in Chapters 10 and 11. Many attempts have been made to reveal the lamellar arrangement in spherulites in semicrystalline polymers using organic solvents or acid etchants. Early attempts at acid etching were often too destructive, but significant success has been obtained with potassium permanganate dissolved in sulphuric acid (< 7%). As with other etchants the interlamellar material is attacked preferentially, but this treatment does not appear to penetrate far into the material and does not leave the specimen embrittled. Figure 9.26 illustrates the kind of result that can be achieved using this method.

9.8 Multiple dark field imaging

A method that may occasionally be of value in polymer studies is multiple dark field imaging. In this method the CTEM is set for selected area diffraction then the intermediate lens is defocused. The effect is to spread each of the diffracted beams (and the transmitted beam) into a disc that contains a low magnific- ation image of the chosen region (Fig 9.27). Each diffraction spot contains only electrons that have suffered the appropriate deviation and therefore represents a dark field image. The magnification of the images in the discs is very low, normally less than 1000 × , and resolution is limited by the photographic film, as described for conventional imaging at low magnifications. The magnific- ation is limited by the need to keep the discs separate and high magnifications require a large defocus, causing a loss of resolution in the electron optical image. Alternatively, by adjusting both objective and intermediate lenses a focused image can be produced, but when this is done the image in each different diffracted beam comes from a different region of the specimen, which will often be contrary to requirement. The advantage of this technique is that several dark field images can be recorded simultaneously and for a short time- intensity level since the magnification is low. This is valuable when using beam-sensitive materials, though it can only provide information at low resolution levels because of the limitations produced by low magnification recording and/or the defocus levels used.

9.9 Scanning transmission electron microscopy (STEM)

In STEM the lenses above the specimen are used to focus the electron beam into a very fine probe at the specimen. The probe is scanned across the specimen under the command of scanning coils in the same way as in the conventional scanning electron microscope (Chapter 10). The interaction of the electron beam with the specimen produces various signals, including secondary electrons and X-rays, as explained in section 10.2.1, and these can be used to produce images or elemental analysis in the same way as in the SEM, given the appropriate detectors and counters. For STEM, however, an electron detector is placed below the specimen and the intensity is measured continuously as the beam moves across the specimen. If an objective aperture is placed on the axis of the instrument the undeflected electrons are admitted so that those regions that scatter strongly provide a weak signal, which is conventionally presented as dark in the image on the cathode ray display whereas those regions that scatter weakly give a strong signal and appear bright in the image. (See section 10.2.2 for more discussion on the formation of scanning images.) Thus the contrast should be the same as that for bright field CTEM. For dark field the aperture is used to select a diffracted beam just as in CTEM, though the choice of whether the incident beam is tilted or the aperture is shifted will depend on details of the electron optical system.

Thus to a first approximation images obtained in CTEM and in STEM should be identical, though the two electron optical systems are not strictly compatible in their most popular configurations and this leads to detailed differences in the images [6]. To see why this should be, note that the focused probe in STEM is conical with a cone half angle typically 5×10^{-3} radians, whereas the angular spread of the illumination in CTEM is very much smaller. Thus in STEM a range of incident directions is present and the Bragg position is not precisely defined; strong Bragg diffraction is obtained for a range of crystal orientations. In CTEM the Bragg position is discrete and a small departure from the exact Bragg position has a large effect on image intensity, as predicted by image computations (which usually relate to parallel illumination). In polymer applications the most significant consequence of this to date has been that when viewing polycrystalline films in dark field, the number of crystals that are in a reflecting position is much greater in STEM than in CTEM because of the range of incident directions present in STEM.

If the beam is held stationary and the objective aperture is removed a diffraction pattern is formed in the same way as in CTEM. Because the incident beam is conical the Bragg diffracted beams are too, and the diffraction pattern consists of discs, not spots. With polymers the intense focused beam causes very rapid destruction of the tiny area sampled under these conditions. Most modern microscopes offer the facility of providing a more parallel probe over a wider area (say a micron in diameter, compared to a STEM probe size of typically a nanometre). This is a more useful sampling size and allows detailed structural variations to be investigated.

During STEM imaging the beam moves rapidly from one position to the next so that destruction of the specimen does not necessarily take place even though the intensity is very high within the probe. Damage rates at a given image magnification are comparable to those in CTEM. STEM does have one distinct

240 *Transmission electron microscopy and diffraction*

advantage over CTEM with respect to radiation damage, and that follows from the fact that the area exposed to the beam is very closely controlled. Thus an area can be selected and imaged at high magnification while focusing is performed, then the magnification can be reduced to leave the damaged region in the centre of an area that has otherwise remained unexposed and which can be photographed immediately without any further refocusing. This is often more convenient than the procedure described in section 9.6.4 for CTEM in which the damaged area is not confined to that chosen for viewing.

References

1. Hirsch, P.B., Howie, A., Nicholson, R.B. *et al.* (1977) *Electron Microscopy of Thin Crystals*, 2nd edn, Kreiger Pub Co, Huntington, New York.
2. Andrews, K.W., Dyson, D.J. and Keown, S.R. (1971) *Interpretation of Electron Diffraction Patterns*, 2nd edn, Adam Hilger, London.
3. Thomas, G. and Goringe, M.J. (1979) *Transmission Electron Microscopy of Materials*, Wiley-Interscience, New York.
4. Bassett, D.C. (1981) *Principles of Polymer Morphology*, Cambridge University Press.
5. Sawyer, L.C. and Grubb, D.T. (1987) *Applied Polymer Microscopy*, Chapman and Hall, London.
6. Loretto, M.H. (1984) *Electron Beam Analysis of Materials*, Chapman and Hall, London.
7. Davies, C.K.L. and Ong Eng Long (1977) *J. Mater. Sci.* **12**, 2165.

Further reading

1. Thomas, E.L. (1984) in *Structure of Crystalline Polymers* (ed I.H. Hall), Elsevier Applied Science Publishers, London, chapter 3.

Exercises

9.1 Make three copies of Fig. 9.6(a), showing the diffraction pattern for orthorhombic polyethylene in a [001] orientation. On one of them mark the twin spots for twinning on $(3\bar{1}0)$ and on the others perform a similar operation for twinning on (110) and $(1\bar{1}0)$ respectively.

9.2 Draw an optical grating by carefully ruling bold lines with thickness roughly equal to the spacing between them on to a transparent film. A periodicity of 2–3 mm is suitable. Construct a second one with the same spacing and a third one with a larger periodicity (say 10% greater). If you have access to photographic facilities a single master can be used to produce copies at appropriate magnifications on to film.

Superimpose the gratings with the same spacings and observe the effect of different relative rotations. Superimpose the gratings with different spacing and make them parallel to observe the 'parallel moiré pattern'.

Observe the effect of misorienting these gratings. Draw an impression of an edge dislocation in a lattice having the same spacing (at positions distant from the dislocation) as one of those already constructed. The strain associated with the dislocation should diminish as the distance from the dislocation gets larger. Obtain moiré patterns with this grating plus (in turn) one of the defect-free gratings.

9.3 Consider a flat polyethylene crystal, 15 nm thick, with its *c*-axis parallel to the electron beam. If the microscope is operated at 100 kV ($\lambda \sim 0.0037$ nm) calculate the deviation parameter (s) for 110 reflections and 200 reflections. What fraction of the maximum diffraction intensity will be obtained in the 110 beams in this position? What is the minimum tilt required to be applied to the specimen to make one of the 110 reflections disappear? (For polyethylene take $a = 0.742$ nm and $b = 0.494$ nm.)

9.4 In section 9.7.2(b) it was noted that a defocus of 0.34 mm is required to obtain maximum phase contrast with a specimen having a periodicity of 50 nm. What is the displacement of the ghost image corresponding to a {110} reflection in drawn polyethylene when observed under this condition?

10

Scanning electron microscopy

10.1 Introduction

Scanning electron microscopes (SEMs) first became widely available in the late 1960s. They are quite sophisticated instruments but are relatively easy to operate and the information they provide comes in the form of magnified images and is normally easy to interpret. Many readers will be familiar with SEM pictures of the eye of a fly, or of a whole flea, and although the biological world often provides subjects for spectacular pictures of popular appeal the technique is no less important in materials studies. The SEM is widely used in studies of polymers, but there are limitations which are caused by specimen charging that occurs in the SEM when the subject is non conductive, and by structural damage caused by the high energy electron beam when it impinges on the specimen. In this chapter we introduce briefly the principles of SEM imaging, then consider the particular problems presented by polymers and the way these difficulties are countered. In addition we will describe ancillary equipment that enables the identification of the elemental constituents of the area of the specimen under examination ('microanalysis').

10.2 Design and operation of the SEM

10.2.1 Interaction of electrons with solid surfaces

In section 10.2.2 we describe how the SEM is built around the interaction between a beam of electrons (accelerated to 1–50 keV) and a solid surface on to which it impinges. The following types of interaction may occur:

1. Some electrons are 'backscattered' as a consequence of the electrostatic attraction between the negatively charged free electron in the incident 'primary' beam and the positively charged nucleus within the specimen ('Rutherford scattering'; Fig. 10.1(a));
2. Some primary beam electrons interact directly with electrons within the atoms of the specimen, knocking them free ('secondary electrons'; Fig. 10.1(b));

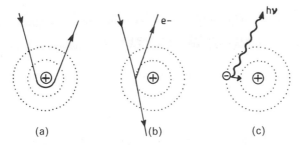

FIGURE 10.1 Interactions of a high velocity electron with an atom: (a) Rutherford scattering in which the negatively charged electron is deviated, often through a large angle, by the field of the positively charged atomic nucleus; (b) the incident electron collides with a bound electron, knocking it out of its orbit to become a 'secondary electron', e^-; (c) a characteristic X-ray photon, $h\nu$, is produced by a higher energy electron falling into the state vacated at (b).

3. After a secondary electron has been removed from an inner shell, an electron from a less tightly bound state falls into the inner shell, with the emission of a photon ('$h\nu$') which is often in the X-ray range of the electromagnetic spectrum, (Fig. 10.1(c)).

Rutherford scattering is an elastic process and the backscattered electrons change direction without loss of kinetic energy. If they are turned through a sufficiently large angle they travel back towards the surface and may escape. They may then be detected and measured by apparatus arranged adjacent to the specimen.

The secondary electrons have a much lower energy (usually 0–200 eV) but a proportion of those which travel towards the surface escape and these too may be detected and measured. The process by which a secondary electron is produced is an 'inelastic' interaction in which energy is lost by the primary electron. The primary electron continues with slightly diminished energy and may make many similar collisions before its energy falls below the threshold for interactions of this kind. Thus the number of secondary electrons escaping from the surface may exceed the number of primary electrons. The number of secondary electrons produced per primary electron (the 'secondary yield') depends on the atomic species in the specimen and on the angle between the primary beam and the surface. This can be exploited to form image contrast (see section 10.2.3). In a similar way the elemental composition and inclination of the surface determine the fraction of primary electrons that is backscattered and their angular distribution.

Each X-ray photon has a wavelength (and hence energy) characteristic of the electron transition made between the two atomic states and is specific to the element in which it is produced. Thus by measuring the wavelength or the energy of the photon, the element can be identified.

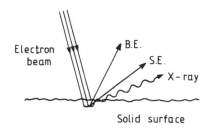

FIGURE 10.2 Schematic beam-specimen interaction showing backscattered electrons, BE (cf. Fig. 10.1(a)); secondary electrons, SE (cf. Fig. 10.1(b)) and X-rays (cf. Fig. 10.1(c)).

The interactions described above are not confined to the surface atoms. The high energy electron beam can penetrate a significant distance (of the order of $1\,\mu m$, depending on the electron energy and on the atomic species in the target). A backscattered electron may suffer further scattering acts before reaching the surface and may not escape, and this must be taken into consideration when assessing the probability of obtaining backscattering from any particular depth within the specimen. Similarly secondary electrons may be produced at quite large depths within the specimen (for primary electrons may suffer repeated inelastic collisions before falling below the threshold energy for further interaction) but the probability of escape diminishes as the depth of production increases. X-rays may also originate at some distance into the interior of the specimen; those which travel towards the surface have a high probability of escape.

The most common (and most widely exploited) interactions are summarized in Fig. 10.2. There are other ways in which the electron beam interacts with the specimen and which can be utilized to produce information about the surface but these are not of importance in polymer studies.

10.2.2 Layout of the instrument

The first requirement of the SEM is a monoenergetic beam of electrons. The designs of electron guns used in SEMs are similar to those used in TEMs, with the simple heated tungsten 'V' filament the most popular source. For the best resolution the electron beam must be as narrow as possible, requiring a lanthanum hexaboride source or a field emission source, but these are more expensive and often less convenient to operate. For the majority of polymer applications resolution is limited as a consequence of electron beam damage and/or by the presence of a conductive coating layer, so that it is very rare that the provision of a better source would enhance the information available and the tungsten 'V' filament is quite adequate.

The electron beam is accelerated, as in the TEM, by holding the filament at a

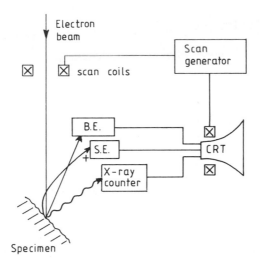

FIGURE 10.3 Layout of the SEM. BE = backscattered electron detector; SE = secondary electron detector; CRT = cathode ray tube display.

large negative potential (usually between 1 kV and 50 kV) and holding the anode (and the specimen) at earth potential. The beam passes through a hole in the anode and is then focused on to the specimen using a lens or a system of lenses. As soon as the electron beam impinges on the specimen surface the interactions described in the previous section take place. Various detectors are arranged in the specimen chamber for the measurement of the several signals which are characteristic of the region of the specimen under bombardment. The electron beam can be moved over the surface of the specimen using a variable magnetic field provided by current-carrying 'scan coils' (see Fig. 10.3). When the beam moves to a different site the characteristic signal measured by any of the detectors may change, and this is exploited to form 'image contrast'. In the SEM the electron beam is made to move across the specimen surface in a regular TV-like raster, tracing out a rectangle line-by-line under the command of the scan coils which are fed by a 'scan generator'. The signal measured by any one of the detectors changes continuously in correspondence with the changing characteristics of the surface probed by the electron beam. The signal, suitably amplified, is used to control the brightness of the spot on a cathode ray tube (CRT) and the CRT scan is controlled by the same scan generator that controls the SEM beam position so that spatial correspondence between the specimen and the (CRT) image is maintained. Normally the signal from only one detector is displayed, but some instruments offer the facility to display more than one image simultaneously (on separate CRTs) or to mix the signals.

The detailed design of detectors is beyond the scope of this book, but a few points are worthy of mention.

Secondary electron detectors are provided with a small positive bias to attract slow moving secondary electrons emitted over a wide angular range. Line-of-sight backscattered electrons will also enter the detector and provide a small additional signal, but other backscattered electrons travel too fast to be pulled in by the small bias voltage. The signal is used to produce the 'secondary electron image' (SEI).

Backscattered electron detectors should have a large angular acceptance for they collect only line-of-sight reflected electrons. Backscattered electron detectors are solid state devices which exploit the high energy of the reflected electrons, and a small negative bias is usually applied to repel the secondaries and ensure that they do not contribute to the signal. The signal is used to produce the 'backscattered electron image' (BEI).

X-ray detectors are of two kinds:
Energy dispersive detectors are solid state devices which measure the energy of the X-rays. Each X-ray photon produces a voltage pulse proportional to the energy, hence identifying the element from which it originated. A pulse height analyser with rapid counting capability measures the rate for each different energy so that data for all elements are gathered simultaneously.

Wavelength dispersive detectors contain a focusing crystal plus a counter set so that it accepts only X-rays reflected from the focusing crystal in a mirror-like fashion. From a knowledge of the interplanar spacing of the focusing crystal, Bragg's law can be applied to determine the selected X-ray wavelength at any setting angle. The crystal and detector have to be rotated through a wide range of Bragg angles to complete an analysis and this is quite time consuming if the scan is made sufficiently slowly to ensure favourable counting statistics. The energy dispersive system is more popular because of the superior speed of analysis but the wavelength dispersive technique is capable of greater accuracy and can be used down to lower atomic numbers (to $Z = 4$, beryllium) more readily.

Finally, it should be noted that, as with the TEM, the electron path must be entirely within a vacuum and the electron gun, lenses scan coils and specimen chamber are encased in an evacuated 'column'. The movement of apertures and the manipulation of the specimen must be controlled via vacuum lead-throughs. To enable rapid specimen changeover an air lock chamber is usually provided to obviate the need to pump out the whole column. This also helps to minimize dust and other unwelcome extraneous material from entering the column and spoiling the optical performance.

10.2.3 *Image contrast*

(a) ATOMIC NUMBER CONTRAST

The yields of secondary electrons and of reflected backscattered electrons both increase with increasing atomic number. Therefore regions with different atomic compositions produce different signal magnitudes, though contrast from this source is not easily recognized unless the boundary between regions of different composition is sharp. The elemental composition information obtained using X-ray detectors is dealt with below.

(b) SURFACE TOPOGRAPHY CONTRAST

The yield of secondary electrons is especially sensitive to the local surface contours (Fig. 10.4). At site A the incident beam strikes a trough in the surface. A significant fraction of the secondary electrons which escape collide with the neighbouring surfaces and are collected by the specimen rather than travelling on to the collector. Consequently, the signal is low and this part of the specimen is displayed as a dark region in the image. Conversely, the chance of escape of a secondary electron produced when the beam impinges on the asperity at B is enhanced by the close proximity of the surface to the point of secondary production for a wide range of directions of travel and the signal will be large, producing a bright area in the image.

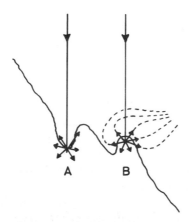

A   B

FIGURE 10.4 Effect of surface topography on the secondary electron signal. Secondary electrons emitted when the electron beam impinges on the cavity at A are mostly collected by the specimen and fail to reach the detector, registering a dark region on the image. Secondary electrons emitted when the beam impinges on the asperity at B escape easily and are drawn towards the positively charged detector (schematic trajectories shown for a detector located to the right of the diagram).

Both the probability of a backscattered electron escaping and being counted, and the angular distribution of backscattered electrons are also dependent on the local surface inclination. If two detectors are used the number of backscattered electrons collected by each will depend on the surface inclination, and the difference in the signals received by each will be characteristic of the surface topography and can be displayed. Alternatively, if the sum of the two signals is displayed instead it is representative of the elemental composition, so that atomic number contrast and surface topography contrast can be distinguished.

(c) X-RAY MAPS

If the SEM is equipped with an X-ray detector (either energy dispersive or wavelength dispersive) the signal characteristic of a particular element can be displayed on the imaging screen in synchronization with the scanning electron beam as in other imaging modes. In this case the signal consists of a short pulse of standard strength each time the X-ray count exceeds a set threshold and the image consists of a dot pattern in which the concentration of dots indicates the local abundance of the chosen element. Thus the distribution of different elements on a surface can be readily obtained, though it should be appreciated that the operating conditions are not the same as those required for accurate elemental composition analysis using the same detector–counter equipment.

10.2.4 Operational aids

Modern SEMs offer many facilities that aid the operation of the instrument and improve its performance and/or the appearance of the image. Some are discussed below.

(a) FOCUSING ON AN INCLINED PLANE

It is often advisable to tilt the specimen surface towards the electron collector to maximize the collection efficiency and so improve the signal to noise ratio and optimize image resolution. As a consequence the scanning electron beam will successively strike parts of the specimen at different heights and hence at different distances from the objective lens. Unless this is compensated for, the electron beam will be focused at one level only, and the image will show a gradual increase in defocus as the distance from this position increases, giving images which are sharp near the centre but which show considerable fuzziness near to the top and the bottom. In modern instruments this can be compensated for and the objective lens automatically changes strength according to the position of the electron probe on the surface (as determined by the scan coils).

(b) CONTRAST CONTROL

Sometimes the contrast within a region of interest lies within a range quite different from that within the general field of view which in normal circumstances dictates the instrument settings. For this reason the 'gamma' contrast facility is provided on many instruments and permits the contrast within a chosen interval to be enhanced while preserving the details of the rest of the field of view which can be presented without enhancement.

A common example where this facility is of value occurs when the features of interest lie within a hole. In this case signal collection is inefficient and the hole appears dark. Even though there may be large relative differences in signal from different features within the hole, they will be within a narrow range compared to the total contrast variation displayed by the whole field of view. Simply adjusting the general brightness or contrast controls to improve the viewing conditions for the features in the hole would cause the region surrounding the hole to become much too bright. Use of the gamma control permits the brightness at the dark end to be expanded without altering the upper limit to the brightness for the total field of view and this enables the details within the hole to be displayed with enhanced contrast.

(c) CHARGE NEUTRALIZER

This device consists of a low energy ion source to irradiate the specimen with a flux of ions to neutralize the charge which builds up on the surface when it is bombarded by electrons. The charge neutralizer is quite effective at low magnifications and can be valuable when used in conjunction with a deformation stage for studying polymer deformation processes in the SEM (see section 10.3.1(c)).

10.2.5 Resolution and depth of field

The resolution in a SEM depends on the size of the electron probe. The signal that is recorded at any instant comes from a region the dimensions of which are directly dependent on the probe size. The probe size is determined by the electron optics of the instrument, and the electron gun and the final focusing lens are of greatest importance. Spherical aberration and astigmatism limit the fineness to which the probe can be focused and can therefore control resolution just as in the CTEM. In practice the resolution is often dictated by what happens to the electrons after impinging on the specimen. As already noted, the electron beam penetrates a significant distance into the surface, and the electron–atom interactions that lead to the signals that are measured take place in a pear-shaped volume. The beam spreads out and information is collected from outside the boundary of the area of the beam projected into the specimen. For a well-focused beam this effect will produce the major limitation on resolution, and even a beam of 0.5 nm diameter or less is unlikely to provide a resolution of better than 2–3 nm. This problem is not encountered in STEM (or CTEM) because the lateral spreading of the probe (or beam) is negligible if the specimen is thin enough. Normal operation in SEM often involves a probe size of 5–10 nm diameter and a corresponding resolution limit that is made a little bigger by the beam spreading effect, though poorer resolution than this is common with unfavourable specimens.

The depth of field of a SEM is very large, just as with a TEM, because of the very small apertures used, again a feature that follows from the very small wavelength of fast electrons.

10.3 Observation of polymers in the SEM

There are two major problems in the observation of polymers in the SEM. Firstly, polymers are poor conductors of electricity and a charge rapidly builds up on the specimen when it is bombarded with an electron beam. If this happens in the SEM the electric field produced by the charge interacts with the incoming electron beam and with the electrons which produce the signal, distorting the image. A conductive coating must be applied to remedy this problem. Secondly, polymers are often damaged by the energetic impinging electrons, causing artefacts to appear. This restricts the operating conditions for the observation of certain polymers, requiring the use of low accelerating potentials, which may in turn limit resolution and reduce contrast. The ways in which these problems are tackled are described in the following sections.

10.3.1 Specimen preparation

(a) MOUNTING AND APPLICATION OF A CONDUCTIVE COATING

The SEM can accommodate relatively large specimens, typically 10 mm across. The specimen is attached by means of a fast-setting adhesive to a platform ('stub') which locates into the specimen stage inside the microscope by means of a spigot. With polymers it is important to choose an adhesive that does not attack the surface to be examined, remembering that a solvent vapour may do damage under some circumstances. The specimen can be removed from a bulk sample and trimmed down using relatively simple tools such as a small hacksaw or a scalpel. If information is required from one surface only, the specimen should be cut to a depth of 3–5 mm prior to mounting; this will enable the most favourable working distance (between the objective lens and the specimen) to be established with most SEMs. This is not very critical for most microscopes permit a generous height adjustment using the stage controls. If information is required from an extended region with several intersecting surfaces then a larger depth may be unavoidable.

After mounting the specimen on the stub a conductive coating is normally applied.* This is done either by evaporation or by sputtering. In most cases the coating material chosen is gold, a gold–palladium alloy, or aluminium. In evaporation the metal source is heated in a vacuum, usually using a tungsten filament which is resistively heated using a low voltage. The evaporant metal is normally in the form of a wire which first melts then remains attached to the filament by surface tension. Atoms evaporate singly or in small clusters and condense on any cold substrate in the vacuum chamber, including the specimen surface, which must be sited in the line of sight from the source. At the low pressures used in this method the atoms make hardly any collisions with residual gas atoms and hence very little scattering occurs between source and substrate. Thus to ensure an even coating on an undulating surface it is usual to provide a specimen holder that describes a planetary motion about the source, causing the specimen to orbit the source and spin about its own (spigot) axis during evaporation.

In sputtering the specimen is sited in an evacuated chamber near to a target made from the required coating material. A gas, usually argon, is admitted at a slow, controlled rate, ionized and accelerated through a high potential (1–10 kV) to impinge on the target with sufficient energy to knock out atoms. The target atoms then condense on to the specimen. In this case the pressure in the chamber is sufficiently high to cause coating vapour atoms to collide with gas atoms and change course repeatedly so that they impinge on the surface in all directions and there is no need to rotate the specimen to ensure an even coating. With polymeric specimens it may be necessary to limit the rate of vapour deposition or to provide some cooling to prevent the specimen from overheating and becoming damaged as a result.

Gold is a popular coating metal because it is easy to vapour deposit, and on bombardment with high energy electrons it gives a high secondary electron yield. The SEI signal will derive from interactions taking place within the coating layer so it is important that it follows the contours of the surface faithfully. When gold begins to deposit on a surface, nucleation of minute crystals is heterogeneous and islands develop. Eventually the islands impinge, but the film does not become continuous until the average thickness is approximately 6 nm, and at this stage the local thickness is very variable. Further deposition tends to smooth out the thickness variations and the favoured coating thickness is about 25 nm, giving reasonably good protection from charging, and reasonably faithful following of the surface contours. Continued deposition would lead to smaller features being obscured because of

---

*It has been demonstrated that by using very low accelerating potential and low magnifications, uncoated polymer specimens may sometimes be observed in the SEM, but the difficulties attached to this approach often outweigh the inconvenience of the extra coating step except when the presence of a foreign coating material must be specifically excluded.

the tendency of the evaporated atoms to attach to concave sites, so smoothing the surface. Gold–palladium coatings tend to have a smaller grain size and form an even coating at lower average thickness.

Carbon is an alternative coating material. It can be evaporated by the resistive heating of a pointed carbon rod, spring-loaded against a flat carbon surface, or can be sputtered. The advantage of a carbon film is that it becomes continuous at a very low average thickness (1–2 nm) and faithfully follows the contours of highly undulating surfaces. Its main disadvantage is that it has a poor secondary electron yield. Carbon coatings may be recommended when X-ray analysis is required because the carbon does not impede the escape of secondary X-rays. The high transparency of carbon to electrons means that it does not provide much protection for polymers which are susceptible to electron beam damage, unless very large thicknesses are used, then negating the advantage offered by the good coating property.

Finally it should be noted that it is not enough for the conductive coating to cover the surface to be examined but it must provide a continuous path down the sides of the specimen to the stub so that there is efficient conduction all the way from the area under the electron beam to the earthed specimen stage. This is often assisted by painting a conductive silver or graphite paste ('dag') on the side of the specimen and across the glue which attaches it to the stub.

10.3.2 *Other preparation procedures*

(a) SECTIONING AND POLISHING

For many purposes, for example if the subject of study is a fracture surface, the above preparation is all that is required. If, on the other hand, it is not surface details that are required but the underlying morphology then this has to be exposed by sectioning the sample. This may be done by deliberately fracturing the sample, possibly after first embrittling it by cooling in liquid nitrogen. This procedure is unlikely to provide a representative section because it will naturally follow the path of easiest fracture, but for this reason may be useful in revealing flaws or foreign particles. To obtain a representative section a specimen is prepared in a manner similar to that used in 'metallographic' studies. The section is first sawn (or may be cut flat using machine tool procedures such as milling) then mounted in an epoxy or acrylic cold-setting resin. It is then polished, first using abrasive papers of successively smaller grit size, then transferred to a polishing wheel on which abrasive particles (alumina is often used) are carried on a synthetic cloth. Water (or water plus a suitable detergent such as 'Teepol') is used liberally for lubrication and cooling during polishing. Note that the organic liquid extenders, (usually alcohols), used in standard metallographic procedures are aggressive towards many polymers, causing dissolution, swelling, etching or crazing, and should not be used unless it is established beforehand that they are benign.

During the polishing procedure the sample should be checked at regular intervals using a light microscope. At each stage the polishing marks left by the previous stage should be removed before proceeding to the next stage in which a finer polishing particle size is used. The process is complete when the most prominent surface scratches are those corresponding to the smallest particle size chosen, (e.g. 1 μm or, for the most exacting purposes, 0.05 μm). At this stage there will be no other topographic detail, but there will be light optical contrast if the material contains fillers (e.g. glass fibres, carbon black, pigments etc). Similarly, if such a specimen is examined in the SEM (after the application of a suitable coating) atomic number contrast may be obtained in the SEI or the BEI if fillers are present, and the elemental constituents of the fillers may be determined using the X-ray analytical technique.

(b) ETCHING

In many cases the morphological features of interest do not correspond to differences in composition and cannot be revealed by atomic number contrast or by X-ray microanalysis. It is sometimes possible to etch the surface of the sample (after polishing but before applying the conductive coating), and reveal the morphology by the consequent topographic contrast. If the sample is semicrystalline the morphology can often be revealed by etching; the non-crystalline parts are removed much more easily, leaving behind the crystalline component. A number of spherulitic polymers have been studied by this method.

Etching can be conducted in a solvent or a mixture of solvents and etching times can be as short as 30 s when conducted at elevated temperature. Polyethylene has been the subject of many studies and it has been found that in some circumstances suitable etching can be applied by inorganic liquids including fuming nitric acid, and the permanganic etch described previously in connection with TEM studies. Chromic acid has been used with polyethylene and several other polymers. Liquid etchants can also be used to produce contrast with multicomponent polymers by the selective removal of one component.

Etching may be obtained without the use of a liquid in some systems. The simplest case is that in which one component of a multicomponent polymer system can be made to evaporate by raising the temperature slightly in a vacuum. This technique does not have very wide application, but has been used to reveal spherulitic morphology in a sample of polypropylene, crystallized in the presence of a low molecular weight diluent which was subsequently removed in this way.

A method which promises more general application is ion beam etching. In this technique material is removed by the action of a high energy ion beam directed at the surface. Selective removal of material is required to produce topographical contrast and attempts have been made to interpret the details obtained using several polymers. However, in some carefully controlled studies it has been found that by changing the ion etching conditions the topography of the specimen can be made to change significantly and it is clear that while the underlying morphology of the polymer plays an important role in determining the surface features that develop, the etching conditions have a strong influence as well. Thus, although for low resolution applications (e.g. revealing the presence of spherulites) this method probably provides reliable information, it cannot yet be used with confidence for studies at higher resolution levels as required, for example, in examining the detailed structure of spherulites. Further research is required to develop this method to its full potential.

Finally, it has been reported that contrast appears or increases during electron beam bombardment in the SEM with many polymers. In the case of spherulitic polyethylene the lamellar appearance which was first revealed by etching prior to SEM observation was further enhanced in the SEM. In another study similar features developed during SEM examination of a specimen which had not been etched previously.

In conclusion, when dealing with specimens with which etching is required to reveal details of interest it is advisable to try several preparation procedures and to compare the results in order to establish which is the best method for the particular polymer and to see whether there is any indication of artefact production which is the consequence of the preparation procedure and not interpretable in terms of true specimen morphology.

10.3.3 *Use of special attachments*

The specimen chamber in the SEM is quite large, permitting the use of macroscopic specimens measuring several centimetres across in some cases. The available space is sufficient to accommodate ancillary equipment for conducting dynamic experiments in the SEM during viewing. The focal length of the objective lens is sufficiently large to allow a fairly generous working distance in the vertical direction and special attachments for deforming the specimen or indenting or scratching it inside the SEM have been built. The type of experiment that can be conducted is restricted by the special needs of the specimen environment, including the high vacuum requirement. Furthermore no electric or magnetic fields which would interfere with the electron beam or with the operation of the detectors can be permitted. If the dynamic experiment involves deformation at the surface then any coating present may be disrupted and it may be advantageous to install a charge neutralizer (see section 10.2.4(c)) and to use uncoated specimens. Low voltage conditions may also be required.

(a) DEFORMATION STAGE

The most common dynamical SEM experiments in polymer science and technology involve deformation. This often includes the study of fracture development. One of the simplest deformation test procedures is the uniaxial tensile test using a bar-shaped specimen gripped at both ends. For studies in the SEM it is an advantage if both grips move (at the same speed) so that the centre of the specimen remains stationary. If this is the part that is viewed it should stay in the field of view without major adjustment of the stage controls. If the grips travel in a horizontal plane there should be very little vertical movement of the viewed surface during deformation and the small variations that do occur normally come within the depth of field. In studies of this kind the large depth of field of the SEM may be even more advantageous than the resolution capability when comparing with the light microscope.

When dealing with miniature specimens of the type used for studies in the SEM, the measurement of the strain is not easily performed, especially if the strain field is non-uniform, as near to a crack. One method of overcoming this problem is to use a photolithographic procedure to deposit a regular grid of spots, $1\,\mu m$ in diameter and spaced about $10\,\mu m$ apart before deformation. The spots are visible in the SEI and can be used to determine the strain at any location and in any direction from photographs recorded during observation.

Even though low strain rates are normally employed the image appearance often changes too rapidly to record photographs at a slow scan rate in the conventional manner. Pictures can be recorded directly from the CRT operating at TV scan rate using fast film (e.g. ASA 400 and a shutter speed of $1/8$s), but the image has inferior quality at TV scan rates because of a poor signal to noise ratio. A further handicap with this technique is that the interval between successive photographs may be too long or difficult to measure and record accurately. More recently images have been recorded on to magnetic tape and successful video recordings of fracture processes in polymers have now been made in several laboratories.

10.3.4 *Artefacts: electron beam damage*

In all types of surface microscopy it is important to beware of artefacts. There are two kinds of artefact: firstly, foreign objects may become attached to the surface prior to or during specimen preparation; and secondly, modifications to the surface may occur during preparation for microscopy and/or during microscopical examination.

Artefacts are a special hazard with the SEM because the high resolution capability permits surfaces to be viewed at high magnifications at which the structure is often unfamiliar and artefacts may be difficult to recognize. Foreign objects are often easy to distinguish, though this may not be so if the subject has a rough, uneven surface. The presence of a conductive coating may further hamper the recognition of foreign objects. Sometimes it is of interest to examine and identify the foreign particle. For example, if the sample was taken from a component which has failed as a consequence of wear, foreign abrasive particles responsible for the wear may still be attached to the surface and may often be identified by their size and shape or by determining their elemental composition using the X-ray microanalytical attachment. Foreign objects which become attached to the surface during specimen preparation contain no useful information and should be eliminated by operating in a dust-free environment and using pure chemicals at all stages of specimen preparation. Tap water is a source of a host of organisms and other particles that are clearly resolvable in the SEM.

Artefacts that are the result of specimen preparation are quite common with polymers. This should be evident from the discussion of etching earlier on. When a surface is etched 'artefacts' are introduced deliberately and in a controlled manner, and reveal details of the structure which can be correctly interpreted, and in this case it is not usual to emphasize the fact that the detail is an 'artefact'. Artefacts can be introduced into some polymer specimens as a result of contact with aggressive liquids or vapours (e.g. organic solvents). It is therefore important that the possible consequence of all specimen preparation steps are carefully considered (recall the comment on the choice of glue).

It would be impossible to present a comprehensive list of artefacts or potential artefacts, but the above remarks should indicate the importance of applying a critical assessment of every detail seen in the SEM before attempting an interpretation; the idea that 'seeing is believing' can be an over-simplification.

All of the above remarks relate to artefacts produced at any time up to the introduction of the specimen into the microscope. With the SEM, however, the most serious artefacts often develop inside the instrument during observation. This is particularly true with polymeric specimens. With biological specimens introduction to the vacuum can cause dehydration and a consequent collapse of the structure, but this is not a problem with most synthetic polymers. If the polymer has been swollen by a solvent this may evaporate in the vacuum, disrupting the coating film. This problem can usually be avoided because the solvent will normally be removed during the coating operation (e.g. by holding the specimen in the vacuum for an extended period prior to deposition of the coating). Much more serious is the damage that is found to occur with many polymers on exposure to the electron beam. This damage can take many forms, and the rest of this section is devoted to a discussion of the more common ones, and remedies wherever these exist.

When viewing polymer specimens in the SEM it is frequently found that artefacts appear during observation of an area, sometimes immediately and sometimes after prolonged exposure to the electron beam. Sometimes the artefact takes the form of a highly damaged region with no recognizable structure and in this case the coating film becomes disrupted and serious charging occurs. In other cases the artefacts can take a characteristic form, such as elliptical blisters, or fissures, or a pattern of tiny cracks running in specific directions, and they often develop in a controllable manner. If this is so, the artefact may reveal some information about the specimen, though this has not yet been developed or exploited to any great extent.

The shape and size of the artefacts depend on the material and on the fabrication conditions, and also on the coating material and coating thickness. The time that it takes for the artefacts to develop depends in addition on the electron beam accelerating voltage and current. Some polymers are very resistant to artefact production (e.g. polystrene) whereas with others artefacts appear very readily (e.g. poly(oxymethylene)).

If a specimen is examined at high magnification, there is often a sharp change in contrast at the periphery of the irradiated area when subsequently viewed at low magnification. This phenomenon is not confined to polymers and other beam-damaging materials; it may be caused by contamination.* It is known as 'picture frame' contrast and with many polymers it appears so rapidly and with such clarity that it seems certain that it is not caused simply by contamination. In this case a possible cause would be removal of material from the irradiated region.

The processes that may contribute to artefact formation are as follows:

1. *Decomposition* of the polymer caused by electron irradiation. Volatile products may cause cracking of the coating layer as they try to escape. They may cause blistering when the coating layer is continuous if it becomes detached from the polymer and inflated by the volatiles.
2. *Temperature rise*. A significant temperature rise (tens of degrees) can occur when a surface is bombarded by an energetic electron beam, especially if the material is a poor thermal conductor. This may cause the production of volatiles, but another problem arises because the thermal expansion of the polymer is much greater than that of the metal coating layer, causing shear stresses to form at the interface, leading to detachment of the coating layer and blistering or the formation of tiny surface cracks (which may be confined to the coating layer).
3. *Chain scission*. If chain scission occurs this may lead to the loss of material within the irradiated area. It may be a step on the way to the production of volatiles. Polymers known to be prone to chain scission when irradiated by high energy electrons tend to form artefacts readily in the SEM. Polymers which crosslink on exposure to high energy electrons tend to be relatively stable in the SEM and to be free from artefacts except when operating at very high beam voltages and currents and at high magnification (giving high electron intensity at the specimen surface).

---

*Contamination in the SEM usually consists of a carbonaceous layer that becomes deposited on the specimen during electron beam bombardment. The source of the contaminant can be oil from the pumping system and finger grease, which may be introduced into the specimen chamber on the specimen or on the specimen stage if these are handled directly instead of using clean gloves and tools.

Finally, it has been found in studies of spherulitic polyethylene that contrast sometimes develops during SEM examination. This has been observed with specimens that had been previously subjected to a chemical etch. The contrast which developed during electron irradiation seemed to emphasize features already present rather than introduce new details. Contrast development has also been found with specimens not previously treated, and the features seen as a consequence seem to reveal genuine details of the spherulitic/lamellar morphology. In this case the description of 'artefact' may be regarded by some as a misnomer.

10.3.5 Operating conditions with polymer specimens

In section 10.2 no attempt was made to define the operating procedure, for details differ from one machine to another. The operator is best advised to follow the procedures laid down by the microscope manufacturer. There are, however, choices of operating conditions to be made by the operator, and now that the problems of specimen charging and artefact production have been discussed we are in a position to consider the action required when observing polymer specimens.

(a) BEAM VOLTAGE

Both charging and artefact production increase as the beam voltage is increased. With some materials it is found that observation at a particular magnification can be continued indefinitely as long as the voltage is kept below a certain level, but that above that value artefacts develop fairly quickly. Thus low voltage operation is usually advisable and is sometimes necessary to avoid artefact production. Resolution is not as good at low voltages, but for most purposes the ultimate resolution capability of the SEM is not required, so this is not usually a disadvantage. Many polymers can be observed using accelerating potentials in the range 5–10 kV, but it may be necessary to use 1–5 kV with more sensitive polymers. The form of the material can be important, e.g. fibres can be especially vulnerable because of inadequate heat dissipation.

(b) BEAM CURRENT

High beam currents increase artefact production and charging problems, but the current setting is usually largely determined by the requirements of the detection and display system. If the current is too small then the signal to noise ratio falls and image quality is lost. Fairly high currents are needed for backscattered electron imaging or X-ray microanalysis. In the latter case high voltages are required too. Thus for some polymers secondary electron imaging can be conducted perfectly adequately but damage occurs if conditions appropriate to one of the other techniques are set.

(c) MAGNIFICATION

The higher the magnification the greater the electron intensity on the viewed area: the same power falls on to a smaller area. Thus low magnification conditions offer the best chance of survival. On the other hand when operating at high magnification only a small region is bombarded and the neighbouring regions remain undamaged. Thus in some circumstances it is convenient to set a high magnification to focus the instrument, then to move rapidly to a neighbouring (undamaged) area and immediately take a photograph for which the only damage will be that produced during the picture recording period. Remember that with the SEM the magnification can be reduced without refocusing, and it is often advisable to record a sequence of pictures at different magnifications for details of interest often appear at different magnification levels. It is recommended that the operator works with a few chosen set magnifications because it is inconvenient to compare features using pictures of different areas taken at different magnifications.

10.4 Examples of SEM applications with polymers

Most scanning electron microscopy is conducted in the SEI mode, which offers the best resolution and most convenient operating procedures. This is especially true for polymer subjects because the problems of charging and artefact production are less serious with the operating conditions required for secondary electron imaging. In this section examples of the application of secondary electron imaging will be discussed first, then attention drawn to some circumstances in which other techniques can be used profitably. It should be noted that even when other techniques are applied the SEI will normally be used to search the specimen for a suitable area, and the SEI will be photographed to provide a record of the 'general appearance' of the chosen region for comparison with the specimen contrast features and information that are obtained using the other methods. It is advisable to record this image as soon as the area has been chosen, before artefacts have time to develop. It is important to return to the SEI after conducting the (prolonged) high current operation required for the other technique(s) to check whether this produced any artefacts.

10.4.1 *Secondary electron image*

The SEI is suited to most types of surface morphology examination. This can be best illustrated by discussing some common examples.

(a) FIBRE MORPHOLOGY

Much interest centres on the surface of polymer fibres because it controls most of their important properties. The fracture properties are governed largely by the size and distribution of surface flaws. The properties relating to textile applications are determined mainly by surface characteristics and the SEM can be used to examine surface finishes and the effects of drying, washing, wear etc. If the structure of the interior is also of interest it may be necessary to reveal it by peeling away the surface (which often has a quite distinct morphology, forming a 'skin') before mounting the specimen on a stub and coating it.

Polymer fibres are typically $10-20\ \mu m$ in diameter and often contain fibrils and microfibrils that are much finer still. As a consequence they are very easily damaged in the electron beam, even when coated with metal. It is difficult to dissipate beam-induced heat and the consequent temperature rise may cause expansion and hence movement during observation, adding a further difficulty that can usually be overcome with a combination of well-executed coating plus carefully chosen electron optical viewing conditions.

Fibres are formed by dry or wet spinning or by melt extrusion, all of which are normally followed by drawing to enhance orientation and hence improve mechanical properties such as stiffness, strength and dimensional stability. Similar structures can be produced directly from isotropic polymers by cold drawing or by solid state extrusion. By carefully controlling the conditions these techniques can be used to produce very high stiffnesses (for example values of Young's modulus in the fibre direction in excess of $100\ \mathrm{GN/m^2}$ have been reported for polyethylene) and these materials also develop strong fibrillation which is revealed in the SEM (Fig. 10.5).

(b) FRACTURE SURFACES

The study of fracture surfaces is important with all classes of materials for two reasons. Firstly, by examining specimens broken in the laboratory under carefully controlled conditions it may be possible to deduce the fundamental fracture mechanisms. Secondly, the microscopic examination of a field failure may indicate why the component broke. Such an assessment may often require reference to laboratory failures which provide a 'database', but the cause of failure is often revealed to be an adventitious flaw. For example with extruded polymer pipe a foreign particle is often found on the fracture surface, surrounded by markings that confirm that it had initiated fracture from its position embedded in the pipe wall. Identification of the foreign particle can often be achieved by X-ray microanalysis (see section 10.4.3). The large depth of field is often just as valuable as the high resolution capability of the SEM for fracture surface examination when the surface is rough.

In considering the features observed in the SEM inspection of polymer fracture surfaces it is convenient to begin with glassy polymers which at the scale of examination we are concerned with here can be considered to be homogeneous. Glassy polymers usually break by the formation of a craze through which a crack subsequently propagates. In some instances the craze

FIGURE 10.5 Fine fibres produced by drawing high density polyethylene at 75°C (Sandilands, G.J. and White, J.R. (1977), *J. Mater. Sci.* **12**, 1496).

may develop quite extensively before the crack begins to propagate whereas during fast fracture the crack front usually follows closely behind the craze front. Whether the development of the craze/crack lies at these extremes or somewhere in between, is determined by the material, the crack speed (and hence the rate of deformation which promotes the failure), the temperature and the presence of an aggressive environment. These factors also determine the craze thickness. If the craze and crack develop slowly, the appearance of the fracture surface is smooth and mirror-like under light-optical observation. A spongy, fibrous appearance is evident at the higher magnification available in the SEM, for a craze consists of a dense system of fibrils oriented parallel to the tensile axis and confined between two parallel planes lying perpendicular to the tensile axis. When the crack grows rapidly it no longer follows the mid-plane of the craze but jumps from side to side, leaving behind a fracture surface which has patches of craze, apparently equal in depth to the craze depth, separated by regions with negligible craze material where the crack has followed the craze boundary. Features interpreted as craze remnants have been seen on several thermoplastics and thermosets. They have been the subject of detailed discussion, and it is beyond the scope of this chapter to review this topic but an example is shown in Fig. 10.6.

Fatigue fracture, in which the material fails under repeated loading, can often be indicated on the fracture surface of glassy polymers by the presence of markings parallel to the fracture front. These are known as 'fatigue striations' if their spacing corresponds to the incremental crack growth rate, presumably

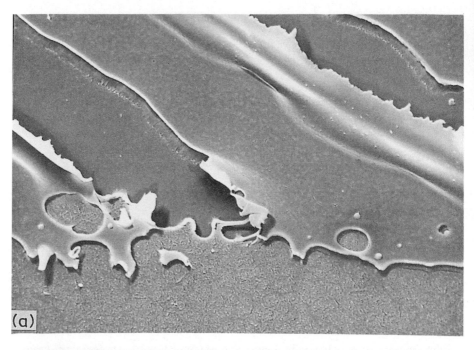

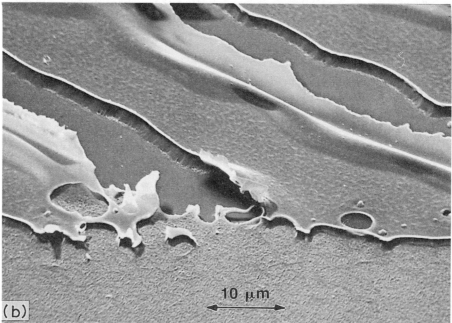

FIGURE 10.6 Polystyrene fracture surface showing craze remnants: (a) viewed almost normal to the surface; (b) tilted 45° (Courtesy of A.V. Iacopi).

indicating that the crack advances each load cycle. Examples have been observed in polycarbonate, poly(methyl methacrylate) and polysulphone. Alternatively 'discontinuous growth bands' are often found in which the spacing corresponds to many load cycles (typically 100) and indicates that the crack front remains stationary for several cycles then advances rapidly, possibly in a single cycle, then remains at the new position for several cycles before the process is repeated. Discontinuous growth bands have been observed in polystyrene, poly(methyl methacrylate), polysulphone and poly(vinyl chloride).

With semicrystalline polymers the presence of two distinct 'phases' (crystal and noncrystalline) has a significant influence on the fracture behaviour. Examples can be found in which the fracture path has apparently followed interspherulitic boundaries. On high magnification examination of the fracture surfaces from semicrystalline polymers the appearance is often closely reminiscent of the craze structure left on glassy polymer fracture surfaces.* Interspherulitic fracture is not a general phenomenon, and trans-spherulitic fracture has also been observed. It is often impossible to discern the underlying spherulitic morphology on a fracture surface when masked by the dense population of fibrils.

Fatigue fracture surfaces from semicrystalline polymers have revealed fatigue striations (polyethylene, nylon-6,6) and discontinuous growth bands (polyacetal, nylon-6,6) but there are often other markings parallel to the fracture front with a spacing much less than the incremental cyclic crack advance and independent of it. These 'microstriations' presumably correspond to structural features within the specimen. Such detail cannot be observed by light microscopy.

Finally, the SEM has been especially valuable in examining the fracture surfaces obtained with both glassy and semicrystalline polymers when the fracture has been caused by environmental stress cracking in which the polymer is in contact with an aggressive liquid, facilitating cracking at a stress much lower than that required in the absence of such an agent, and producing a quite different fracture surface morphology. It is worth noting that a pattern of crazes promoted on a polycarbonate surface by finger grease has been recorded using the SEI.

(c) FIBRE COMPOSITES

The mechanical properties of many polymers are greatly enhanced by incorporating fibrous fillers. The SEM can provide valuable information about

---

*This should not be surprising, for in many ways semicrystalline polymers appear to be very good candidates for craze formation since they have good cold drawing properties, and crazes consist of a special kind of cold drawn structure. We normally associate crazes with glassy polymers because they can be seen with the naked eye, whereas in semicrystalline polymers they cannot be seen by light optical methods. Furthermore their structure is less easily defined because the craze direction is influenced by the presence of the hard crystalline units and not determined exclusively by the stress system.

such materials, for example the locus of failure, the degree of adhesion between the polymer and the fibre, and fibre orientation. It is normal to observe a fractured surface, either the result of a mechanical test for which details of the fracture mechanism are sought, or broken deliberately to reveal the internal structure.

When a fibre composite breaks the crack is normally deflected repeatedly by the fibres, which tend to break at their weakest point even when this is some distance from the plane of fracture in the polymer matrix. Crack deflection provides a valuable increase in fracture path length and hence in fracture toughness, but a further contribution comes as a secondary consequence of this process. To separate the two parts of the component each broken fibre must be pulled out of the polymer 'tube' surrounding it, and even if the interfacial bonding has been disrupted by the passage of the crack there will still be frictional forces to overcome. The SEM enables measurement of the lengths of the fibres pulled out during this process, and a careful examination of the fibre surface may indicate at least in a qualitative way the degree of adhesion between the polymer and the matrix.

Short fibre reinforced thermoplastics that can be injection moulded are rapidly becoming popular for a wide range of commodities. The fibre orientation influences the properties yet distribution within the injection moulding is very complicated. Some information can be obtained by inspecting the fracture surfaces in the SEM, but here the large depth of field may sometimes be a disadvantage unless careful tilting sequences are conducted (i.e. several pictures recorded with the specimen tilted to make different angles with the electron beam because fibre orientation is not easily interpreted from a single image). An example is shown in Fig. 10.7.

(d) OTHER EXAMPLES

The SEM has provided important information in many other fields of polymer science. It is not intended to attempt a comprehensive review here, but brief mention of a selection of other examples will serve to illustrate the applications of secondary electron imaging.

(i) Polymer foams

Details of the cellular structure and measurement of the dimensions of the struts can be obtained using the SEM, but are unattainable with the light microscope (Fig. 10.8).

(ii) Weathering

Surface degradation of polymer articles used for outdoor applications can occur under the action of UV radiation and/or as a consequence of wind-borne particles. The SEM is ideal for examining the surface damage (Fig. 10.9).

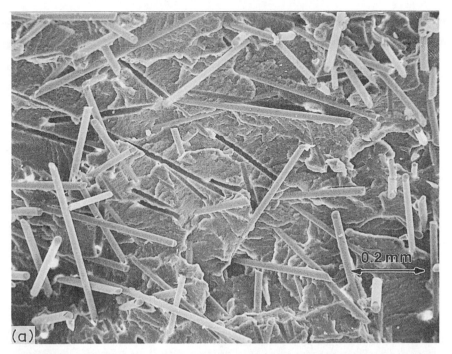

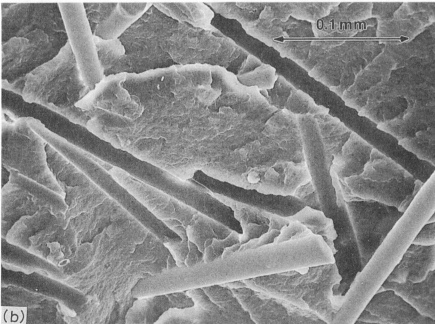

FIGURE 10.7 Fracture surface from injection moulded glass-fibre filled polypropylene: (a) a general view; (b) a higher magnification view showing some polymer-fibre adhesion on one fibre and several smooth fibre surfaces and channels indicating poor adhesion.

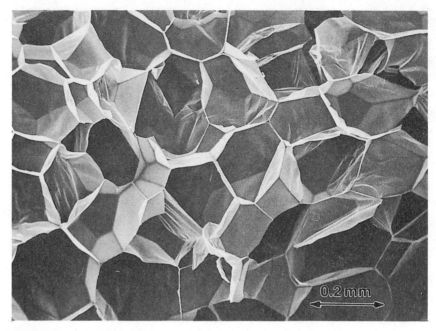

FIGURE 10.8 Expanded polystyrene foam.

(iii) Wear

Engineering polymers may be used as bearings or in other situations in which they rub against other materials, causing wear. The SEM again offers the best means of examining the surface damage.

10.4.2 Backscattered electron image

Backscattered electron images (BEIs) remain of very limited and specialized application in polymer studies. BEIs rely primarily on atomic number contrast, and polymeric materials usually have a fairly uniform elemental composition. Fillers may be seen with high contrast, though they are normally easily recognized in the SEI. Staining techniques can be used in some cases. Osmium tetroxide can be used to stain amorphous regions of a semicrystalline polymer or unsaturated regions (for example in a polystyrene/polybutadiene blend) and the heavy osmium-containing domains can be seen in the BEI. The domain structure of phase-separated blends can often be seen in the BEI if one component can be halogenated, permitting atomic number contrast.

BEI has also been used to monitor crack growth in polypropylene deformed inside the SEM. Images were obtained with a relatively poor vacuum (10^{-2} torr $= 1.3$ Pa) purposely set to obtain a satisfactory charge balance under BEI viewing conditions, allowing imaging without a coating film.

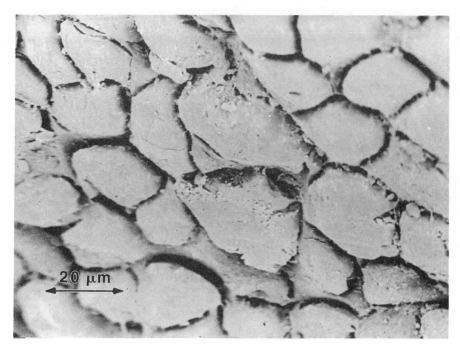

FIGURE 10.9 Polystyrene surface after 3 years weathering in Jeddah, Saudi Arabia. This area was originally a smooth injection moulded surface (More examples in Qayyum, M.M. and White, J.R. (1985) *J. Mater. Sci.* **20**, 2557).

10.4.3 X-ray microanalyser

X-ray microanalysis has likewise found only minor application in polymer science and technology. The characteristic X-rays of light elements cannot be measured with accuracy (or cannot be measured at all in some cases) with the detectors currently available in electron beam machines. Therefore, SEM microanalysers cannot be used for compositional analysis of polymers. They can be used to detect heavy atoms present as substituents or in additives or foreign particles, and this can be conveniently achieved using the mapping technique in which the relative abundance of a chosen element is displayed on the imaging screen. Some applications are discussed below to indicate the kind of problem that can be tackled with the help of the microanalytical facility.

(i) Identification of foreign particles

The microanalysis attachment can be used to identify crack-initiating flaws, and examination of fracture surfaces of failed extruded pipes has revealed a number of different species. Some were found to consist mainly of iron, presumably broken away from the processing machinery and carried along in

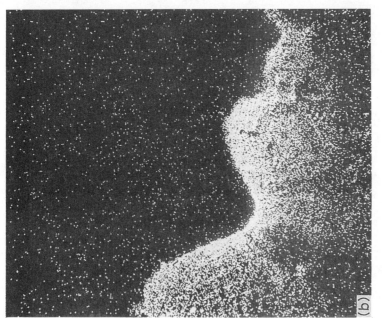

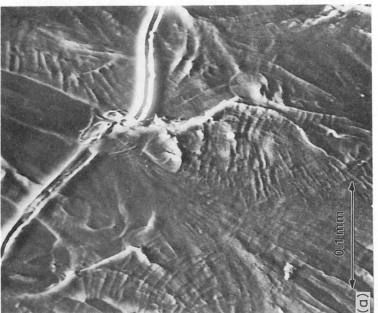

FIGURE 10.10 (a) Fatigue fracture surface from low density polyethylene. Note the surface striations in the bottom left hand corner. Note also a fissure in the top right hand quadrant; it was produced by radiation damage and the coating film became disrupted, leading to charging and the consequent white glare. (b) Al X-ray map of the same area showing the location of the fracture front at the time the aluminium coating was applied. The amount of Al deposited was very small, and required a long scanning time to generate this image; hence the noise level was high and accounts for the signal recorded in the upper part of the image, which should contain no aluminium (Singian, V.I., Teh, J.W. and White, J.R. (1976) *J. Mater. Sci.* **11**, 703).

the melt; some were found to be rich in calcium, presumed to be present in the form of calcium stearate, a processing aid; some were rich in titanium, possibly present as aggregates of titanium dioxide pigment.

(ii) Fracture of adhesive joints

When an adhesive joint fails the fracture path may pass within one or other of the joined materials or may pass within the adhesive or at one of the interfaces with the adhesive. The microanalysis attachment can be used to help the positive identification of material attached to the fracture surface. Similarly, the locus of the fracture path in rubber–metal seals has been identified in a system in which a primer and a top coat were present on the metal: the primer contained a chlorinated polymer plus titanium and zinc in sufficient concentration to be detected, whereas the top coat contained detectable levels of lead and bromine.

(iii) Fracture front marking

In the section on fractography it was noted that fatigue fracture surfaces often contain striations. For the purpose of post-fracture analysis it is important to know whether the striations are a reliable guide to successive crack front locations. With opaque samples this cannot be checked directly by observation during the fatigue experiment and to overcome this problem the following procedure was devised. A fatigue experiment on a notched low density polyethylene sample was interrupted, the sample was removed from the fatigue tester and placed in a jig that held the crack slightly open. The sample was then placed in a coating unit and a small amount of aluminium evaporated into the crack. The sample was then returned to the fatigue tester and the test continued until it broke. A layer of carbon was then evaporated on to the surface to provide conductivity and the sample was placed in the SEM. The amount of aluminium was too small to be seen in the SEI (Fig. 10.10(a)), but was clearly located by the X-ray mapping method (Fig. 10.10(b)). The fatigue striation pattern at the position of the aluminium front is uninterrupted, showing that the amount of aluminium used was sufficiently small to avoid serious modification of the stress at the crack front on resuming the fatigue test. Much more aluminium would have been required to be rendered visible in the SEI. The use of carbon as a final overall coating permitted penetration of the electron beam to the aluminium below, and permitted easy escape of the characteristic aluminium X-rays.

10.5 Conclusions

The SEM provides unique information about the surfaces of materials and is outstanding for topographic examination. Polymeric specimens must be prepared with care and viewing conditions must be chosen that avoid the problems of charging and beam damage, but for most purposes it remains an instrument that can be used routinely without much difficulty,

Further reading

1. White, J.R. and Thomas, E.L. (1984) Advances in SEM of Polymers. *Rubb. Chem. Tech.* **57**, 457.
2. Sawyer, L.C. and Grubb, D.T. (1987) *Applied Polymer Microscopy*, Chapman and Hall, London.

Exercise

10.1 Section a piece of polymer foam using a scalpel or a razor blade and examine the internal structure under a light microscope in reflected light. Compare what you see with the SEI in Fig. 10.8. If the foam is brittle enough break a piece and examine the fracture surface in the light microscope. Again compare with Fig. 10.8, but note also how inconvenient it is to observe such an undulating surface in a light microscope, requiring frequent refocusing.

11

Light optical techniques

11.1 Introduction

Many of the chapters in this book concentrate on a single type of instrument or a family of closely related instruments; this chapter includes several distinct techniques. Even then, others are omitted because it is considered more appropriate to include them elsewhere (e.g. the light scattering method for molecular weight determination, and spectroscopy in the visible range).

Although the equipment required to get the best results in some of the areas covered in this chapter is expensive, apparatus that permits execution of many of the techniques outlined here can be assembled from standard components or built quite cheaply, and the reader is encouraged to try them out whenever possible.

11.2 Refractive index

Many polymers are used as optical components and it is necessary to know their refractive index. There are several refractometers based on the identification of the critical angle for total internal reflection within a material at the boundary with a material of lower refractive index (Fig. 11.1). If the rays in Fig. 11.1 are reversed, it is evident that when rays in the lower right hand quadrant of the picture approach the interface at an incident angle greater than ϕ_c then refraction into the upper left hand quadrant cannot occur, and total internal reflection must take place. The Abbé refractometer is used for illustration because it is suitable for the examination of both solid and liquid films.

11.2.1 Abbé refractometer

In the version of the Abbé refractometer shown in Fig. 11.2 the lower prism is provided simply as a convenient way of mounting the specimen and illuminating it. The specimen, in the form of a film of refractive index n, is placed between the two prisms and illuminated with monochromatic light from below. The critical angle for extinction is determined at the exit face of the

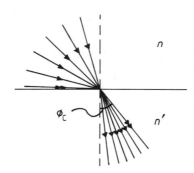

FIGURE 11.1 Refraction at a boundary between materials of different refractive index (n, n') where $n' > n$. The ray with the double arrowhead travels parallel to the interface in the upper material (angle of incidence $= \pi/2$) and passes into the lower material with angle of refraction ϕ_c.

upper crystal, and is indicated in terms of the angle of refraction (ϕ_c') in Fig. 11.2. If the prism has refractive index n' and the prism angle indicated in Fig. 11.2 is α, then

$$n = n' \sin \phi_c \quad n' \sin \phi = \sin \phi_c' \quad \text{and} \quad \phi_c = \alpha - \phi$$

By eliminating first ϕ_c, then ϕ, it can be shown that

$$n = \sin \alpha (n'^2 - \sin^2 \phi_c')^{1/2} - \cos \alpha \sin \phi_c' \tag{11.1}$$

11.3 Birefringence

If processing causes molecular orientation and/or moulding stresses to be present, the refractive index varies according to the direction of light propagation and the orientation of the plane of polarization. Birefringence is defined as the difference between the refractive index in two orthogonal

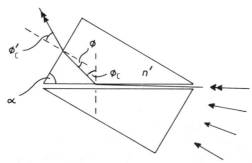

FIGURE 11.2 Abbé refractometer.

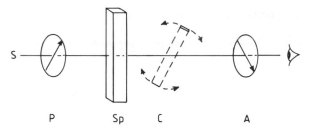

FIGURE 11.3 Apparatus for birefringence measurement: S, source; P, polarizer; Sp, specimen; C, compensator; A, analyser.

directions. It is normally easier and more accurate to measure it directly, as described below, than to measure the two relevant refractive indices separately.

11.3.1 Birefringence measurement

Figure 11.3 is a schematic representation of an apparatus for measuring birefringence. Plane polarized light travels in the x-direction from left to right through the specimen. The electric vector vibrates in the yz plane, and the polarizer is set such that the plane of polarization is at $45°$ to the vertical and a prominent axis of the specimen (e.g. the fibre axis of a drawn fibre) is set parallel to the vertical z-direction. If the amplitude of the electric vector is A then it can be replaced by components in the y- and z-directions, both with amplitude $A/\sqrt{2}$ (Fig. 11.4(a)). If the specimen is optically anisotropic these components propagate at different velocities. If the refractive index for light

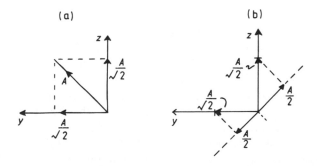

FIGURE 11.4 Vector amplitude diagrams: (a) decomposition of polarized light, amplitude A, into components parallel to the specimen 'unique' axis (z-axis) and perpendicular to it (y-axis); (b) recombination in direction of the analyser: the two resolved components will not be in phase if the specimen is birefringent.

with the plane of polarization in the xy plane is n_{xy} and that in the xz plane is n_{xz} then the components entering the specimen in phase will leave out of phase, having taken a different time to pass through. If the specimen measures Δx in the x-direction, the optical path difference for the two components is $\Delta x(n_{xz} - n_{xy})$ and is called the relative retardation. $(n_{xz} - n_{xy})$ can be written Δn_{zy}, representing the birefringence value measured with the z-axis of the specimen as the reference axis. The y-direction within the specimen must also be specified, except for cases in which only the fibre axis is unique, with isotropic properties in the xy plane. Thus, the phase difference between the two components is

$$2\pi\Delta x\Delta n_{zy}/\lambda$$

where λ is the wavelength, and their amplitudes can be written as

$$A/\sqrt{2} \quad \text{and} \quad (A/\sqrt{2})\cos(2\pi\Delta x\Delta n_{zy}/\lambda)$$

The analyser is set with its axis at $90°$ to that of the polarizer (Fig. 11.3) and when these components pass through it their amplitudes are reduced to

$$(A/\sqrt{2})\cos 45° \quad \text{and} \quad (A/\sqrt{2})\cos(2\pi\Delta x\Delta n_{zy}/\lambda)\cos 135°$$

respectively (Fig. 11.4(b)), giving a resultant on recombination of

$$A(1 - \cos 2\pi\Delta x\Delta n_{zy}/\lambda)/2$$

One technique, suitable for small values of Δn_{zy} (i.e. such that $2\pi\Delta x\Delta n_{zy}/\lambda < \pi$) is to measure the transmitted intensity and compare it with the incident intensity. Although this requires certain precautions associated with photometric measurements it does permit continuous monitoring during fairly rapidly changing conditions. An example of the application of this method is given in section 11.3.5(c).

In the analysis presented above, no reference has been made to the light source. In order to apply the photometric method it is necessary to use a monochromatic source (note the dependence on the wavelength, λ). If white light is used there is a mixture of wavelengths present, and for each wavelength the amplitude is altered by a different amount. Thus some colours become suppressed and on recombination the light is no longer white but has a colour characteristic of the relative retardation of the specimen. The colour sequence versus relative retardation is presented in the 'Michel–Levy colour chart' and if the colour obtained can be identified on the chart then the relative retardation can be read off directly. This method requires minimal equipment and extends the range of measurement to significantly higher values of $\Delta x\Delta n_{zy}$. Colour identification is subjective, however, and the upper limit of $\Delta x\Delta n_{zy}$ for which satisfactory observations can be made is below the levels sometimes encountered in polymer investigations. Finally, the colour gives only the magnitude of the relative retardation and does not indicate whether n_{xz} is larger or smaller than n_{xy}. If n_{xz} is larger, then the birefringence is taken to be

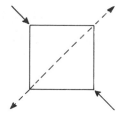

FIGURE 11.5 Production of variable birefringence using a block of a photoelastic rubber held between finger and thumb, applying a force in the direction shown by the solid arrows. The birefringence is proportional to the stress difference between this direction and the tensile principal stress axis perpendicular to it.

positive, and though the sign convention chosen is arbitrary, it is important to know the sign because it has important structural significance.

A convenient way of determining the sign of the birefringence utilizes a piece of photoelastic rubber held by hand between the specimen and the analyser. Application of pressure between finger and thumb causes the rubber to become birefringent and if it displays positive stress optical behaviour the tensile axis will give the direction of the plane of polarization for which the refractive index is greater (Fig. 11.5). If n_{xz} is larger than n_{xy}, and the rubber is held with the compressive axis in the z-direction then the relative retardation suffered in the specimen and in the rubber are in opposite senses. By gently increasing the pressure on the rubber the relative retardation can be made to cancel that produced by the specimen, indicated by the appearance of the black fringe, confirming the sign of the birefringence in the specimen. On the other hand, if n_{xy} is larger than n_{xz} increasing pressure on the rubber causes the appearance of colours progressively further along the Michel–Levy chart. If this is the case the birefringent rubber should be held with the compressive axis in the y-direction and a check made that the black fringe can be brought up, then confirming that the specimen has negative birefringence.

The apparatus in Fig. 11.3 is often modified by the provision of a compensator. This is a birefringent device that can be adjusted to compensate for the relative retardation imposed by the specimen in a similar way to that of the birefringent rubber described above, but capable of quantitative measurement. Suppose once again that the specimen has $n_{xz} > n_{xy}$, and that the compensator is inserted in such a way that its larger refractive index is obtained with light having the electric vector in the yx plane. Thus the relative retardation caused by the compensator is in the opposite sense to that caused by the specimen. The compensator is designed so that the amount of relative retardation is adjustable, and the position at which the relative retardation of the compensator is exactly equal in magnitude to that produced by the specimen is identified by setting for zero transmitted intensity. Several compensators are in use, but it will suffice to describe briefly the principle behind two of them by way of illustration.

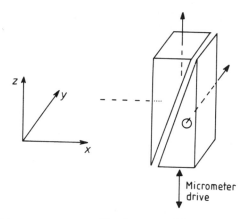

FIGURE 11.6 Babinet compensator. The direction of the optic axes of the two wedges are shown by the solid arrows. Light travelling in the x-direction is shown by the dashed line.

(a) THE BABINET COMPENSATOR

This consists of two birefringent quartz wedges cut with their optic axes perpendicular to one another, and mounted as shown in Fig. 11.6, one with the optic axis in the z-direction, the other with the optic axis in the y-direction. Thus when light propagating in the x-direction is plane polarized in the xz plane, the optical path length is much greater near to the top of the assembly shown in Fig. 11.6 than at the bottom. On the other hand, light having xy as its plane of polarization has an optical path length that is larger near the bottom than near to the top. Hence the relative retardation will vary along the length of the compensator and fringes are seen. When monochromatic light is used the fringes have a regular period and by measuring the displacement required to be applied to one of the wedges to cause the fringes to shift one fringe spacing, the compensator can be calibrated. When using white light, a central black fringe is seen, flanked on either side by coloured fringes, as given by the Michel–Levy chart. When a specimen is in the beam the fringe pattern is shifted, and the new position of the black fringe gives a measure of the relative retardation produced by the specimen.

(b) THE TILTING COMPENSATOR

This consists of a birefringent crystal plate that can be tilted about one of its principal optic axes. Thus different thickness as well as different refractive indices can be presented to the light beam, giving relative retardation (Fig. 11.7). The compensator is used with white light and is adjusted to bring the black fringe to the centre. The amount of tilt required to achieve this condition with the specimen in position is converted to relative retardation using calibration tables. The manufacturer's recommended routine should be

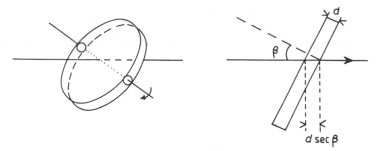

FIGURE 11.7 Tilting compensator. Light enters from the left and the thickness presented depends on the tilt angle, β.

carefully followed when making measurements using the compensator, and should include tilting successively in both senses to minimize errors. This type of compensator can be used for large relative retardations (up to 30 wavelengths).

11.3.2 General remarks on birefringent polymer specimens

Polymer articles rarely have uniform birefringence and when viewed in white light between crossed polars, transparent mouldings often display coloured fringes, the consequence of residual stress and/or molecular orientation. Even when fringes are not in evidence but a uniform colour is obtained, as may be the case with sheet or film, this does not necessarily indicate uniform birefringence. The total relative retardation can be determined by comparison with the Michel–Levy chart, or by using a compensator. If this is then divided by the sheet thickness the result is the average birefringence, but there may be a variation through the thickness of the sheet.

If there are variations in birefringence in the yz plane (i.e. fringes are seen when light propagates in the x-direction), measurements can be made at intervals, down to about 1 mm if necessary. If the variation in birefringence in the yz plane is rapid it may be preferable to make measurements using a polarizing microscope. Compensators are often designed for insertion into a microscope tube, and, provided the specimen is thin or can be illuminated with fairly parallel light, precise localized measurements can be made. When the variation in birefringence is through the thickness of the sheet, serial samples must be carefully sliced off for measurement, or layers must be removed and birefringence measurements made on the remainder at each stage (see section 11.3.5(b)).

A problem with polymer mouldings not encountered in investigations of crystals (in mineralogy, for example) is that the directions of the principal optic axes may vary from one position to another. It is therefore conventional to choose prominent axes, such as the machine direction in a fabrication process that may involve stretching or flow, to use as reference axes in birefringence measurements, without checking whether they coincide with the true principal axes. Such simplifications cannot be applied in more fundamental studies of the physics of birefringent polymers.

11.3.3 Sources of birefringence

(a) ORIENTATION BIREFRINGENCE

First recall that birefringence occurs when the velocity of light through the medium differs according to the plane of polarization. The velocity of propagation depends on the degree of interaction between the electromagnetic wave and the bonding electrons, a property expressed by the polarizability. The strength of the interaction depends on the angle between the electric vector and the bond axis and is greatest when they are parallel so that this condition gives the highest refractive index. Thus, for a polymer, molecular orientation will cause the material to be birefringent and the magnitude of the birefringence depends on the degree of molecular orientation and on the characteristics of the bonds; a double bond has a much greater polarizability than a single bond.

(b) STRESS BIREFRINGENCE

Polarizability changes when bonds are stretched or bond angles are deformed by the application of stress, causing a birefringent effect that is clearly connected with the principal stress axes. This type of birefringence is sometimes called deformation birefringence, but this can be misleading when dealing with polymers because orientation is often produced by deformation. Indeed, if a stress is maintained for an extended period, even a glassy polymer may suffer molecular conformational changes, causing a contribution to orientation birefringence. This is strictly different from stress birefringence, which should really refer only to reversible instantaneous changes that occur on loading or unloading. Note, however, that with rubbery materials the major change in birefringence on application of stress is caused by bond orientation and that this also develops almost instantaneously. For these materials the term 'stress birefringence' is often applied even though the behaviour is 'entropy elastic' rather than 'energy elastic'.

(c) FORM BIREFRINGENCE

Form birefringence may occur when the material contains two (or more) phases with different refractive indices, and is the result of distortion of the electromagnetic field at the interface. Thus parallel rods or lamellae of one phase within a matrix of different refractive index provide suitable morphologies and for the maximum effect their spacing should be about $\lambda/20$, where λ is the wavelength of the light. The two phases in block copolymers often separate in this manner, are often caused to be anisotropic by processing, e.g. extrusion, and the period is often of the required order.

11.3.4 Interpretation of birefringence measurements

The interpretation of birefringence measurements is best illustrated by reference to specific applications, some of which are discussed below.

In interpreting stress birefringence it is useful to evaluate a 'stress optical coefficient', which is the birefringence caused by the stress divided by the stress. For many cases of practical importance this can be taken to be a constant. The stress optical coefficient may be negative, as is the case with poly-(methyl methacrylate) (PMMA). Rheologists often work with flowing melts and may measure the stress sensitivity in this state and express the result as a stress optical coefficient. With some polymers this may differ in sign from that obtained with the same material in glassy form. For example polystyrene has a positive stress optical coefficient in solid form, but the birefringence in the melt becomes increasingly negative when the stress causing it to flow is increased. This is because molecular main chain orientation causes the phenyl groups to orientate with their planes tending to be perpendicular to the tensile stress axis. The overall polarizability of the polystyrene repeat unit is mainly determined by the phenyl group because of the π-bonding. Hence, the orientation birefringence is negative and the melt appears to have a negative stress optical coefficient. If the orientated melt is cooled rapidly, as might happen in a thin walled region of an injection moulding, the negative birefringence is preserved after solidification. The stress optical coefficient of the glass is still positive, however, and application of a tensile stress parallel to the reference axis (flow direction) reduces the magnitude of the birefringence.

When stress is applied to a rubbery material having a low modulus, the majority of the deformation is given by conformational changes. In the undeformed state the molecular segments are randomly directed and there is no birefringence. When conformational changes are produced by stress, preferred bond directions develop. This can be dealt with using a similar approach to that used in the theory of rubber elasticity. The molecule is replaced by an equivalent random chain where the statistical elements are assumed to possess two principal polarizabilities only, one parallel to the segment axis ($b_{\parallel}$) and the other perpendicular to it ($b_{\perp}$). These are added up along and perpendicular to the tensile stress axis, and if the rubber deforms in an affine manner the birefringence in the deformed system is

$$n_{zx} = \frac{2\pi}{45} \frac{(\bar{n}^2 + 2)^2}{\bar{n}} N_c \{(l/l_0)^2 - 1/(l/l_0)\} (b_{\parallel} - b_{\perp}) \qquad (11.2)$$

where $\bar{n}$ is the average refractive index of the undeformed network, N_c is the number of crosslinks per unit volume, and (l/l_0) is the extension ratio. If this is combined with another result from the theory of rubber elasticity:

$$\sigma = N_c kT\{(l/l_0)^2 - 1/(l/l_0)\} \tag{11.3}$$

then the 'orientational' stress optical coefficient can be derived as

$$C_{or} = \frac{2}{45} \frac{\pi}{kT} \frac{(\bar{n}^2 + 2)^2}{\bar{n}} (b_\parallel - b_\perp) \tag{11.4}$$

Thus it is predicted that the stress sensitivity of birefringence is independent of the cross-link density, a result confirmed by experiment.

11.3.5 *Applications of birefringence measurements*

(a) ORIENTATION ASSESSMENT

Consider the effect of a fabrication operation that causes a preferred molecular orientation to develop. In the case of fibre drawing the preferred orientation is parallel to the draw direction and the fibre is transversely isotropic. Even if strongly polarizable side groups are present they are expected to be randomly oriented transverse to the fibre axis, so that no net anisotropy will be present in this plane. Thus interest focuses on the perfection of the orientation of main chain segments to the fibre axis. If the angle between the axis of the structural unit (e.g. the local direction of the molecular backbone axis within a polymer repeat unit) and the fibre axis is ϕ, then it can be shown that the birefringence is proportional to $(3\overline{\cos^2\phi} - 1)/2$ (known as Hermans' orientation function) where $\overline{\cos^2\phi}$ is the average value of $\cos^2\phi$ for all of the units within the volume under scrutiny. If there is perfect orientation $\overline{\cos^2\phi} = 1$ and the Hermans orientation function equals unity. For random orientation $\overline{\cos^2\phi} = 1/3$ so that the Hermans orientation function equals zero. If all of the units are perpendicular to the chosen axis then $\overline{\cos^2\phi} = 0$ and the Hermans orientation function equals $(-1/2)$. In the general case the birefringence is given by

$$\Delta n = \Delta n_0 (3\overline{\cos^2\phi} - 1)/2$$

where Δn_0 is the intrinsic birefringence, which is the value for perfect alignment. Thus if Δn_0 is known, measurement of the birefringence allows the Hermans orientation function to be calculated. This yields only the average of $\cos^2\phi$, which may correlate with some orientation-dependent properties, but for a fuller orientation assignment other techniques have to be used.

Sometimes the fabrication process imparts a biaxial orientation. In the general case the two in-plane orientations may be unequal, as will usually be the case with blown tubular film in which the machine direction orientation is caused by a combination of the flow during extrusion and the take off forces, whereas in the orthogonal direction orientation is caused by the hoop stresses that operate during inflation of the tube. These will generally not be in balance, though in some circumstances it is an advantage to have balanced orientation and the fabricator may seek operating conditions to achieve this. Another example of this kind of orientation will be produced when a sheet is drawn unequally in two orthogonal directions. If these directions are labelled y and z and the sheet normal direction is x, then the various birefringence values are given by

$$\Delta n_{zy} = \Delta n_0 \left\{ \frac{3\cos^2 \phi - 1}{2} + \frac{\sin^2 \phi \cos 2\phi}{2} \right\}$$

$$\Delta n_{zx} = \Delta n_0 \left\{ \frac{3\cos^2 \phi - 1}{2} - \frac{\sin^2 \phi \cos 2\phi}{2} \right\}$$

and

$$\Delta n_{xy} = \Delta n_0 \overline{\sin^2 \phi \cos 2\theta}$$

where θ is the angle made by the projection of the chain backbone part of the structural unit with the x axis and $\overline{\sin^2 \phi \cos 2\theta}$ is the average value of $\sin^2 \phi \cos 2\theta$ for all units. Thus if Δn_0 is known, both $\overline{\cos^2 \phi}$ and $\overline{\sin^2 \phi \cos 2\theta}$ can be obtained from measurements of birefringence taken in different directions.

(b) MOLECULAR ORIENTATION IN INJECTION MOULDINGS

When a polymer melt flows into a cold mould the molecules have a high degree of orientation. If the material is cooled rapidly a significant amount of this orientation may remain 'frozen-in'. The cooling rate near to the surface of the mould is likely to be adequate to preserve significant molecular orientation, whereas in the interior the cooling rate is slow, because of the low thermal conductivity of the polymer, and the molecules can return to a more random conformation. Thus it is expected that there will be considerable variation in birefringence through the thickness of a moulding. An example is shown in Fig. 11.8(a) which shows a moulded bar viewed between crossed polars.

In order to determine the birefringence as a function of depth within the moulding there are two possible approaches. One is carefully to cut away slices and measure their birefringence. The cutting is best achieved using a microtome (see section 11.4.2(b)) though it is possible that some orientation may be introduced during the process of cutting. The alternative also involves cutting away layers but measures instead the relative retardation of the remainder at each step. In the latter case material does not have to be removed using a microtome, but a high speed milling machine can be used instead, since it does not matter that the material removed is damaged by the cutting process. The relative retardation is plotted as a function of material remaining and a smooth curve is drawn through the data points. The gradient of this

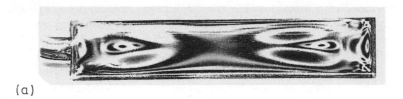

(a)

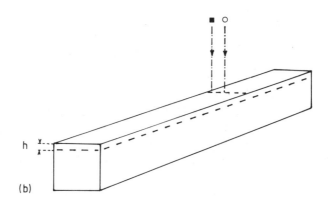

(b)

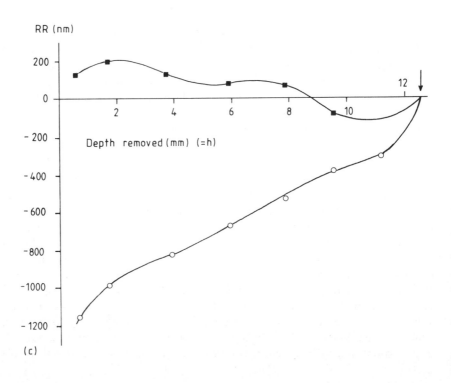

(c)

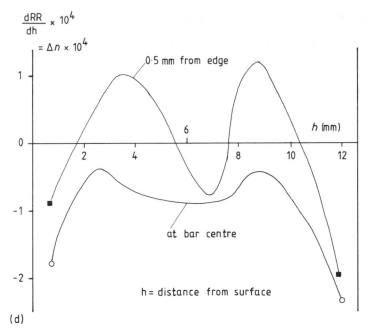

(d)

FIGURE 11.8 Birefringence in an injection moulding: (a) square-section bar, viewed between crossed polars (reversed contrast black and white copy of colour slide); (b) schematic of bar, showing locations for relative retardation measurements; (c) plot of relative retardation versus depth removed, measured at two locations: ○ through the central axis of the bar; ■ 0.5 mm from one edge. The arrow marks the bar thickness. (d) Through thickness birefringence at the two locations, derived from (c) (see Cuckson, I.M., Haworth, B., Sandilands, G.J. and White, J.R. (1981) *Intern. J. Polymeric. Mater.*, **9**, 21).

curve then gives the birefringence at the location at which it is measured (see Fig. 11.8(c, d)).

Although this procedure provides data that probably correlate well with the state of molecular orientation, the interpretation is not straightforward. During moulding, the gate freezes off before the material in the interior of the moulding has solidified, so preventing topping up from the injection system to counteract thermal shrinkage. There is a considerable temperature gradient across the moulding at this time and the skin is much nearer to room temperature than the core. There should therefore be differential shrinkage of the skin and core as the moulding cools. This sometimes leads to voiding or sinking, but if the moulding conditions are adjusted to avoid this happening there will be residual stresses set up instead, with tensile stresses in the interior, balanced by compressive stresses near the surface. These stresses produce stress birefringence and there are therefore at least two contributions to birefringence.

It should be noted that in the methods described above, material is cut away from the moulding and inevitably stresses will be at least partially relieved. This may be exploited to advantage if it can be arranged that cutting relieves all of the residual stresses within the removed portion, leaving only the orientation birefringence. Alternatively, the method which uses measurements of the relative retardation of the bar remainder could be improved by applying forces that returned it to its previous dimensions during the birefringence measurement; if this precaution is not taken the bar will both change length and bend. If, for example, the skin was in compression in the as-moulded state, when it is removed the bar will shrink and also bend so that the surface from which it was removed becomes concave. This relieves the imbalance of forces set up when the surface layer is removed, and results in modification of the residual stress distribution within the bar. Thus the stress birefringence changes at each layer removal. The advised procedure was not followed completely in producing the data in Fig. 11.8(c); the bar was straightened, but not returned to its original length for the relative retardation measurements. It can be shown that this deficiency is not important in this particular example because the dominant contribution to the birefringence is molecular orientation, and departure of the stress in the machined bar from the original value is fairly modest.

Even this is not the end of the story. It has been observed that the birefringence often increases in a moulding that has been ejected from the mould before cooling is complete. This seems to indicate that orientation birefringence in the interior is lost quickly (before the moulding is ejected) and

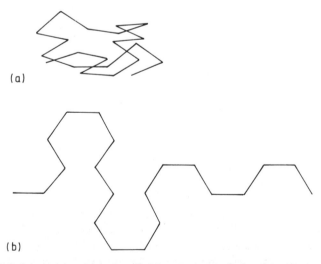

(a)

(b)

FIGURE 11.9 (a) an arrangement of 20 anisotropic links that show significant orientation but modest end-to-end separation; (b) 20 links showing little orientation but significant end-to-end separation.

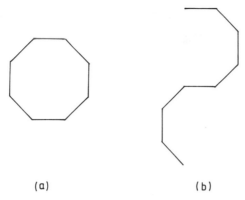

(a)   (b)

FIGURE 11.10 (a) 8 links arranged to form a closed octagon; (b) the same 8 links with the same orientation distribution, but displaced from the arrangement shown in (a), resulting in a large end-to-end separation.

that stress birefringence develops slowly while the residual stresses are becoming established.

Some experimental evidence points towards a second explanation in which it is hypothesized that the residual stresses influence the molecular segment conformation at the time of solidification and that localized orientation is produced. This idea is interesting because it emphasizes that orientation does not necessarily correlate with molecular extension (as might be revealed by annealing to produce shrinkage as the molecules restore random coiling) although the two will be expected to be closely linked in molecular flow in the melt state. For example, in Fig. 11.9, two arrangements of 20 freely linked chain elements are shown: (a) has little molecular extension (the end-to-end separation is small) but the links have a tendency to lie closely parallel to the horizontal, producing significant birefringence; (b) has significant molecular extension, but the birefringence is very small. The reason for this is clarified in Fig. 11.10 where eight units are shown forming an octagon in (a) (zero net birefringence and zero molecular extension) and reassembled in (b) in which the molecular extension is considerable but the net birefringence remains zero (the links have exactly the same orientation distribution as in (a)).

(c) CRYSTALLIZATION KINETICS

A simple apparatus for following crystallization kinetics of polymers is shown in Fig. 11.11. The specimen is in the form of a thin film of polymer, held between two microscope slides. This can be made by putting a glass slide on a hot plate and placing on it some polymer in the form of powder, or a small slice cut from the solid using a scalpel. A second glass slide is placed on top and pressure applied. The polymer melts and spreads under the applied pressure,

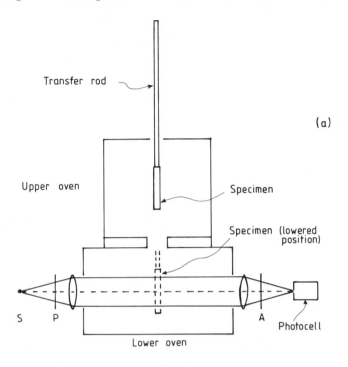

Transfer rod

(a)

Upper oven

Specimen

Specimen (lowered position)

S   P

A

Photocell

Lower oven

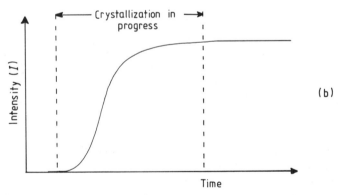

Crystallization in progress

Intensity (*I*)

Time

(b)

FIGURE 11.11 (a) Apparatus for following crystallization kinetics. S, source; P, polarizer; A, analyser; (b) Trace of transmitted light intensity (*I*) versus time during crystallization.

then is ready for use. The polymer is then remelted in the upper oven in the apparatus shown in Fig. 11.11(a) and the lower oven is heated to a lower temperature at which it is desired to study crystallization. The specimen is then transferred from the upper oven to the lower oven. The polarizer and analyser are set in the crossed position and no light reaches the photocell as long as the sample is in the (isotropic) melt state. When crystallization begins, the specimen becomes birefringent. The crystals themselves are anisotropic and birefringent, and form birefringence may also be present because the lamellar crystals have a different refractive index to the surrounding amorphous material. Thus, light will begin to reach the photocell, and the light intensity is taken to be proportional to the fractional crystallinity. The photocell signal can be recorded on a y/t pen recorder, as shown in Fig. 11.11(b). If a very fast response is required, the signal can be displayed on an oscilloscope and this method can be used to follow very rapid crystallizations.

11.4 Light microscopy

The light optical microscope is the most familiar tool for the study of the morphology of materials. It is very useful for a quick 'look see' but for best results proper alignment and the observation of many operational precautions are just as important as with the transmission electron microscope. There are many excellent texts on the subject and we wish only to emphasize the importance of following the well-tried procedures described therein in order to maximize performance. It is worth noting also that the ratio of the cost of the most expensive light microscope to the cost of the cheapest probably exceeds the equivalent ratio for transmission electron microscopes (and almost any other type of scientific instrument mentioned in this book) by at least an order of magnitude, and that there is a large variation in quality and capability. Thus the polymer microscopist should strive to get the best performance from the light microscope, but should also be mindful that there may be limitations that can only be overcome by moving to a better microscope. The resolution of the best microscopes is diffraction-limited, and is therefore of the order of the wavelength of light. If higher resolution is required the usual option is to go to an electron microscope. It should be remembered, however, that some light microscope contrast mechanisms do not have a direct equivalent in the electron microscope, and also that radiation damage occurs in the electron microscope and may limit the information obtained.

Just as important as the proper use of the microscope is specimen preparation. Many of the general microscope specimen preparation techniques are applicable to polymers and in addition some specialized procedures have been developed specifically for polymers and polymer composites. After dealing briefly with some of the image forming techniques that can be of value in polymer studies, the rest of this section will be arranged according to the specimen preparation procedures.

11.4.1 *Specialized image contrast techniques*

(a) TRANSMISSION MICROSCOPY

In transmission microscopy using white light the image obtained when the instrument is set up conventionally has contrast produced by differences in optical density (causing intensity variations) and/or differences in selected absorption of different wavelengths (giving rise to colour variations). In addition it is possible to obtain phase contrast, as described below.

(b) PHASE CONTRAST MICROSCOPY

In phase contrast microscopy light which is diffracted within the specimen and therefore travels in a different direction to that which is transmitted without deflection, is given a phase shift so that when the transmitted and diffracted components are recombined in the image some destructive interference takes place. Contrast will be obtained if the relative strengths of the diffracted and transmitted components vary from one location to another. Several phase contrast attachments are available, and the manufacturer's instructions for setting up and operation must be carefully followed for best results.

(c) THE INTERFERENCE MICROSCOPE

The interference microscope also relies on phase contrast. In this case the illuminating light passes through a beam splitter (e.g. a semireflecting plate), then the two components pass through the specimen along parallel adjacent paths. They are subsequently recombined in the image, and if they receive a differential phase delay on passing through the specimen interference occurs. Contrast is therefore especially strong at the boundary of an object, for example at the edge of a crystal grown within a thin polymer film, as a result of the very different optical paths taken by beams just within and just missing the crystal.

(d) DARK FIELD MICROSCOPY

With dark field microscopy a hollow cone of light is used and in the absence of an object no light enters the objective lens. When a specimen is in position light is scattered into the objective, providing contrast.

(e) POLARIZED LIGHT MICROSCOPY

This is probably the most common method for observing polymer specimens. Plane polarized light is used and an analyser is placed between the specimen stage and the eyepiece, and is rotated to the crossed position (zero light transmission) before inserting the specimen. The arrangement is then equivalent to that shown in Fig. 11.3 and light is transmitted into the image of those parts of the specimen that are birefringent, as described in section 11.3. The intensity at any particular position depends on the magnitude of the birefringence, the orientation of the birefringent components with respect to the plane of polarization of the illumination, and the specimen thickness, and it follows that the image contrast derives from variation in these from one location to another.

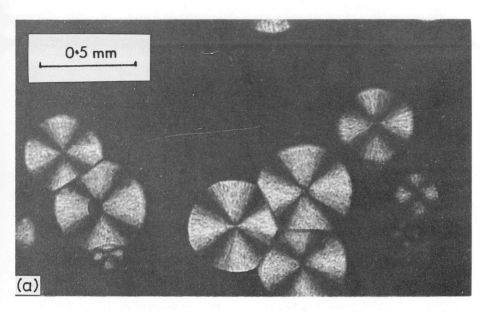

FIGURE 11.12 Spherulites in poly(butene-1): (a) early stage in their development; (b) the same area after impingement.

This method is especially good for studying spherulites because of the difference in birefringence between the crystal lamellae and the non crystalline regions, and because of the different orientations taken by the birefringent lamellae as they radiate out from the centre in all directions: this gives rise to a Maltese cross effect (Fig. 11.12). If the lamellae twist in phase with one another this gives rise to rings. Birefringence measurements can be made at chosen sites using a compensator inserted into the microscope tube.

(f) REFLECTION MICROSCOPY

Contrast enhancement can be achieved in reflection in ways analogous to those described above for transmission. Surface topography can be followed using the interference of the reflected light with a reference beam, derived from the same source and therefore coherent with it. The phase of the reflected light will differ according to path length, and contours corresponding to specimen heights for which the path length gives destructive interference with the reference beam are seen. These will be too close together to be of value except when dealing with very smooth surfaces.

Dark field microscopy can be used to observe coarser surface undulations. The illumination is arranged so that none of the light reaches the objective lens from regions on the specimen surface lying perpendicular to the microscope axis. Surface roughness causes reflection into the objective, producing contrast

11.4.2 *Polymer applications and specimen preparation for transmission microscopy*

For transmitted light microscopy it is generally necessary to prepare thin samples. This is often true even for transparent polymers because of the small depth of field of an optical microscope. A selection of techniques is presented below.

(a) MELT PRESSED FILMS

This method has been described in section 11.3.5(c) and involves melting the polymer and squeezing it between two glass slides to make it thin. This is commonly used to study crystallizing polymers, and on cooling, crystallization takes place. If cooling is rapid and follows immediately after squeezing the assembly, the structure normally varies from one region to another, because of the non-uniform cooling and differences in molecular flow (during squeezing) from one region to another. The specimen can now be transferred to the hot stage of a microscope, remelted, then cooled again at a slower, controlled, rate. If the melting temperature is high enough, memory of the molecular flow can be erased, and the subsequent crystallization should be fairly uniform. The

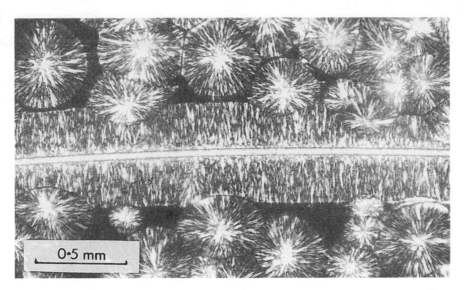

FIGURE 11.13 Transcrystalline growth in isotactic polypropylene in contact with a polyester fibre; crystallization temperature 135°C.

morphology is dictated by the inherent characteristics of the polymer (which determines whether or not it forms spherulites etc.) and on the nucleation density, which will be determined by the amount of undercooling (and hence on the cooling rate) and by the presence of any nucleating agent, whether added deliberately or present as impurity. Observation of the spherulites is normally most conveniently achieved between crossed polars and an example is given in Fig. 11.12. Alternatively, details can sometimes be shown under phase contrast conditions.

An extension of this technique enables examination of the effect of a foreign body on the crystallization behaviour by including a small amount along with the polymer. Of major interest is the kind of additive used in moulding resins, for example, a fibrous reinforcement. If a single fibre is used, its effect on nucleation can be examined most advantageously using a hot stage as before, and following crystallization after remelting. In a number of polymer-fibre combinations preferential nucleation on the fibre surface has been observed, leading to the so-called 'transcrystalline' morphology in which early impingement of crystal neighbours closely spaced along the fibre axis causes the growth direction to be radial and quite distinct from the spherulites growing in regions remote from the fibre (Fig. 11.13). A further extension of this method is to pull on the fibre just before crystallization, producing local shear in the viscous polymer melt adjacent to it, to study the effects that flow might have during a moulding operation.

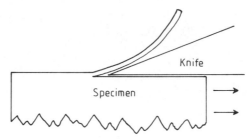

FIGURE 11.14 Microtome sectioning.

Thin films for the kind of studies outlined here could be cast alternatively from solution. This method is even suitable for producing very thin films suitable for transmission electron microscopy, and more details are given in section 9.7.2.

(b) THIN SECTIONS

The kinds of specimens described above are produced in the laboratory primarily for the purpose of basic scientific investigation and are not particularly useful for diagnostic purposes. If information is required about a real moulding the only course of action is to cut a thin section. This is best achieved with a microtome, which is an instrument used mainly for biological and biomedical preparations but also used widely in polymer laboratories. Most useful is the base sledge microtome in which the specimen is clamped firmly into a carrier (sledge) that can be advanced past a knife on a machine bed, (Fig. 11.14). A slice is taken from the complete section, which may be 10 mm × 10 mm or larger, then the carrier is retracted ready for the next section to be taken. It is normally arranged that the sample is raised automatically by a micrometer by a preset amount (e.g. 5–40 μm) at the end of the retraction stroke. As long as sectioning proceeds smoothly, and the sample does not deform in front of the knife making cutting uneven, the sections should be of consistent thickness, equal to the incremental advance. Ths sample should be propelled past the knife smoothly and although the motion is normally hand-driven, motor driven microtomes are commercially available. If the stroke is not smooth, or if the sample deforms during cutting, the section may contain 'chatter bands' lying parallel to the knife edge. If the cutting edge has imperfections these will leave marks on the sample parallel to the cutting direction.

It is sometimes necessary to support the sample in a mounting resin. For example, if a section of a thin-walled pipe is required, a resin must be used to prevent it from buckling when the knife makes contact. Epoxy resins are commonly used, and for polymeric samples it is normally advisable to select a cold curing resin to avoid heating the specimen to a temperature at which changes might occur. Epoxy mounting resins cut easily under conditions suitable for microtoming polymer samples and can usually be distinguished easily from the specimen in the microscope image.

The best sections are produced when the temperature is low enough to prevent ductile tearing, and should therefore be below the glass transition temperature. Cooling stages have been designed to facilitate this, and it is normally arranged that both the sample and the knife are cooled. Cooling is achieved either using the expansion of liquid carbon dioxide or using liquid nitrogen, and recently electrical devices based on the Peltier effect have been introduced and can often conveniently be used to convert existing equipment.

The sections are placed between a glass microscope slide and a cover slip for viewing and if a mounting liquid is used, care must be taken because some of the common mounting liquids attack certain polymers. If uncertain, a dummy specimen of the same material should be mounted in the desired liquid and carefully inspected for signs of swelling or changes in birefringence etc.

Some polymers are very difficult to cut using a microtome. Sometimes there is an easy direction (e.g. along the fibre axis) but difficulty is experienced transverse to this. Some success has been achieved by thinning a parallel-sided sample by grinding and polishing using the petrographic method. The initial section is cut using a diamond saw. It is sometimes an advantage to polish the surface of the sample before the second (parting) cut so that there is only one

FIGURE 11.15 Thin section viewed between crossed polars showing transcrystalline growth near the surface of a slowly cooled isotactic polypropylene compression moulded plaque.

saw cut to make smooth. The polished side of the sample is then glued on to a glass plate with an adhesive that does not contain components harmful to the sample. The sawn face is ground and polished using water or water plus detergent for lubrication and cooling, using first grinding papers then finally a polishing wheel. The procedure uses methods similar to those employed for producing polished surfaces for reflection microscopy, described below. With care, samples can be thinned to say 30 μm in this way. This method may be the best choice for reinforced polymers where the large difference in hardness between the reinforcing fibre or particle and the polymer causes tearing during microtome sectioning.

Sections often reveal the spherulitic structure in mouldings made from polymers. To do this a polarizing microscope is required and observations are conducted with the polarizer and analyser crossed as before. Generally there are at least two different morphologies present, with equiaxed spherulites in the centre and a second region near the surface that appears to contain either distorted spherulites of different size, or transcrystalline growth associated with nucleation at the mould cavity surface, or an ill-defined morphology. Sometimes three or more different morphologies are visible. An example of a section cut at room temperature from a compression moulded plaque made from polypropylene is shown in Fig. 11.15.

11.4.3 Polymer applications and specimen preparation for reflection microscopy

In its simplest form, reflection microscopy requires no special specimen preparation except, perhaps, the removal from a larger sample of a sample small enough to be placed on the specimen stage of the microscope. Used in this way the microscope can reveal details such as those discussed below.

(a) DETAILS REVEALED BY REFLECTION MICROSCOPY

(i) Texture

Texture may be revealed relating to crystallization (which may be influenced by the presence of a nucleating agent mixed into the polymer, or by nucleation promoted by the mould surface during the moulding operation) or relating to processing defects (e.g. sharkskin in an extrudate, where a characteristic pattern is formed when processing conditions are unfavourable).

(ii) Mould marks

Polymer moulding resins replicate the mould cavity surface with high fidelity and the machine marks on the mould can be recognized under the microscope. Coarse markings on the mould surface will limit the smoothness of the mould-ing and hence spoil the appearance, so that identification of this cause by microscopy is of value. On the other hand, the complete absence of mould marks is not an indication of a 'perfect' mould finish, but that the resin was not in contact with the mould surface at the point of solidification, either because of insufficient injection pressure, or because of shrinkage before solidification, resulting from insufficient hold-on pressure or premature freezing of the gate. Alternatively, surface degradation or the presence of a deposit of a foreign material might obscure mould marks, and might relate to the service conditions subsequent to moulding.

(iii) Filler particles, reinforcement fibres, pigments etc.

Even with an opaque polymer, some of these may be close enough to the surface to give contrast resulting from differences in colour or reflection amplitude. In addition, shrinkage of the polymer during solidification often causes filler particles and fibres to protrude from the surface.

(iv) Fracture surfaces

In the majority of these examples contrast results from surface topography. If the amplitude of the topographic contours is small, dark field or phase contrast conditions should be employed, giving strong enhancement of contrast. If the amplitude is large, as is often the case with fracture surfaces then a scanning electron microscope (SEM) should be used if possible because of its greatly superior depth of field and resolution, although it requires some specimen preparation. Examples of the application of the SEM to reflection microscopy of polymer subjects are given in Chapter 10.

Not all reflection microscopy of polymers is conducted on as-received specimens, and the techniques that may be used (often in combination) to reveal certain structural details are presented below.

(b) POLISHING

If it is required to examine the internal structure of a material it is customary to section the sample at the chosen location then to polish the exposed surface, as in the preparation of metallographic specimens of metals and metal alloys. With polymers certain precautions have to be taken, and the procedure departs somewhat from that developed for metals.

The polishing operation involves the removal of material from the surface by rubbing it repeatedly against an abrasive surface. If the cross section of the sample is small there is a tendency for it to tilt during this operation, causing the edges to be rounded, and in most cases it must be mounted in a resin so that the combination of resin plus the centrally mounted specimen presents a larger surface area to the abrasive surface. When the subject is a polymer the mounting resin selected must be cold curing because even a modest temperature rise (say 40°) may modify the polymer structure.

The polishing process is normally conducted in two stages. Firstly, the rough saw cut is removed using emery paper, held flat against a supporting base. The specimen may be hand held and is rubbed repeatedly across the surface. The surface temperature should not be allowed to rise significantly and this is ensured by flowing water across the emery paper surface. This acts partly as a lubricant, reducing the amount of heat produced (and reducing also the polishing rate, which is therefore necessarily low) and partly as a coolant to take away the heat generated, the thermal conductivity of the polymer and mounting resin being inadequate. Emery papers with progressively finer grit size are used, and at each stage the sample is turned through 90° to ensure that the polishing grooves left by the abrasive particles of the previous coarseness are successfully polished away.

Secondly, a finer polish is applied using a polishing wheel with a synthetic cloth carrying abrasive particles. Alumina particles have been used successfully, and are available in various sizes, the smallest being 0.05 μm diameter ('γ-alumina') for a very fine polish. Once again the coarsest size is used first and progressively finer polishes (each on a fresh cloth) used to abrade away the grooves left by the previous one. A mixture of water and a detergent (e.g. 'Teepol') is used for lubrication and cooling. Many materials used for metal polishing are unsuitable for use with polymers, for example diamond abrasive paste contains organic components that may attack some polymers, and the alcohol lubricant/coolants may act as swelling or crazing agents. Polishing of polymers is an extremely slow process and an alternative method of structural/morphological study will normally be chosen if available.

Although polishing is more difficult when two or more components of different hardness are present in the section, this technique can be used to reveal details of filler or fibre distribution in polymer composites.

(c) ETCHING

If the section contains only one phase there will be practically no contrast after polishing. In the case of semicrystalline polymers contrast can be enhanced by using an organic solvent. This process is accelerated by using a slightly elevated temperature (for example, if polyethylene is submerged in xylene at 65°C, significant etching occurs within 3 min). The etchant normally attacks the non-crystalline regions preferentially, leaving the spherulites showing as humps on the surface. Etching may also be employed to reveal the phase distribution of phase-separating block copolymers and polymer blends, choosing an etchant that attacks one of the components but not the other. More discussion of etching is found in the chapter on SEM where the objective is normally to reveal details on a finer scale.

(d) COATING AND SHADOWING

Coating the surface of a specimen with a thin reflective film (e.g. silver or aluminium) can enhance the contrast in reflection. Coating is normally conducted by evaporation in a vacuum chamber or by sputtering at low pressure, (see section 10.3.1(a)). Sometimes topographic detail may be enhanced by 'shadowing' in which the evaporation source is positioned to one side of the sample instead of directly above it so that the evaporant arrives at an oblique angle, and deposits preferentially on those surfaces tilted locally toward the source; evaporant cannot reach those areas of the surface that are in the shadow of high points. This procedure is more often associated with transmission electron microscopy, and is discussed more fully in Chapter 9.

More micrographs illustrating the application of light microscopy to polymer subjects can be found in [1].

11.5 Small angle light scattering

When light passes through a medium the oscillating electric vector induces the electrons to vibrate in phase, and they emit 'scattered' electromagnetic radiation of the same frequency. In a homogeneous material the electrons (and hence the sources of this radiation) are uniformly distributed and the scattered contributions tend to cancel because they are not in phase. This is what happens in a transparent material, so that the forward propagating wave passes through but light is not scattered to the sides. If the electron density is not uniform then complete destructive interference will not take place and finite intensity is observed in directions inclined to that of the incident beams. This is the reason why most semicrystalline polymers (e.g. polyethylene, polypropylene) are translucent or opaque: light is scattered as a result of the difference in density (and hence electron density) between the crystalline and non-crystalline regions. One semicrystalline polymer, poly(-4-methyl pentene-1), is almost as transparent as polystyrene or poly(methyl methacrylate). This is because the molecules are arranged into a helical conformation in the crystals and are not packed particularly closely together, so that the crystal has a very low density, almost identical to that in the non-crystalline regions.

Small angle light scattering (SALS) can be used to provide morphological information about polymers. Light scattering theory is much more complicated than the diffraction theory leading to the Laue conditions or Bragg's law in X-ray or electron diffraction and the results are presented here without derivation. The arrangement of apparatus suitable for SALS is shown in Fig. 11.16. It has a certain resemblance to Fig 11.3, but here the polarizer and analyser are either vertical or horizontal (not at $+45°$ and $-45°$ respectively to the vertical as in Fig. 11.3). Also the most effective light source is a laser (often making the polarizer unnecessary). If the specimen has a 'unique' axis (e.g. an extension axis) this is placed vertically. The intensity distribution, projected on to a screen (or photographic film) has been calculated for a number of model specimen morphologies and for combinations of polarizer

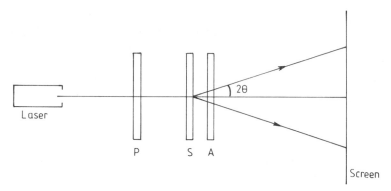

FIGURE 11.16 Small angle light scattering arrangement.

horizontal/analyser vertical ('H_v') and polarizer vertical/analyser vertical ('V_v'). By computing the scattering from spheres, it is predicted that the H_v pattern for spherulites will have the appearance of a four leaf clover (Fig. 11.17) with maxima at $45°$ to the vertical. If the location of the maxima within each lobe is obtained at a scattering angle $2\theta_m$, then the average radius

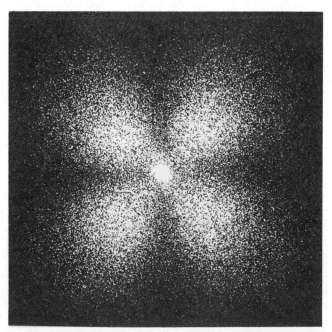

FIGURE 11.17 H_v pattern obtained with linear polyethylene prepared by the method described in section 11.4.2(a).

of the spherulites is given by

$$\bar{R} = \frac{4.1\lambda}{4\pi \sin \theta_m} \qquad (11.5)$$

where λ is the wavelength of the light. Thus the larger the spherulite size the smaller is the scattering angle, and the resemblance of this formula to Bragg's law is clear. The significance of $\bar{R}$ must be regarded with caution, however, especially if there is a variation of spherulite size because larger spherulites contribute more to the summation that leads to equation (11.5) than smaller ones, and the result given is not strictly the arithmetic mean radius.

If a different morphology is known (or suspected) to be present, the scattering pattern must again be computed before any interpretation can be attempted. For example the case of anisotropic rods has been examined, to account for the scattering observed with polymers believed to crystallize into units of this shape, including polymers in the early stages of spherulite formation.

The scattering pattern is normally recorded on a photographic plate or film placed at the screen position in Fig. 11.16. If changes in morphology are to be followed, for example during crystallization or during deformation, then a series of photographs can be recorded at appropriate intervals. Alternatively, some investigators have scanned the pattern using a photometric detector, recording the signal continuously. If a four leaf clover H_v pattern is expected the detector can be set to travel repeatedly backwards and forwards along the line at 45° to the vertical, though insufficient information will be recorded to enable a check to be made that the pattern is truly the expected one.

Reference

1. Hemsley, D.A. (1984) *The Light Microscopy of Synthetic Polymers*, Oxford University Press/Royal Microscopical Society, London.

Further reading

1. Longhurst, R.S. (1973) *Geometrical and Physical Optics*, 3rd edn, Longman, London.
2. Bradbury, S. (1984) *An Introduction to the Optical Microscope*, Oxford University Press/Royal Microscopical Society, London.

Advanced

3. Read, B.E., Duncan J.C. and Meyer, D.E. (1984) Birefringence techniques for the assessment of orientation, *Polymer Testing* **4**, 143.
4. Sawyer, L.C. and Grubb, D.T. (1987) *Applied Polymer Microscopy*, Chapman and Hall, London.

Exercises

11.1 Birefringence studies. Try to obtain two pieces of polaroid. Ideally, use two large sheets and a viewing box, but old sunglass spectacles may be a suitable source and you can use light reflected from a white surface for illumination. Find the crossed position and now view a selection of transparent plastic articles placed between them. Note the effect of rotating the object about the viewing direction. Suitable objects include transparent rulers, transparent disposable catering cups, or boxes for sweets, cocktail sticks, sticking plasters etc. When viewed in white light between crossed polars fringe patterns will be seen if the object is birefringent. Look out for regions in which the contours are very close together, and see if you can find a feature such as the witness mark of the gate to the mould nearby. Check the effect of applying forces to the mouldings. Do the same thing with film. Try to find some film that necks and draws when pulled–this can often be done with the securing film at the top of six-pack canned drinks. Try to inspect some engineering components or 'quality' domestic articles which have much more exacting requirements than the single trip disposables.

11.2 Light microscopy. If you have access to a light microscope try examining some moulded polymer surfaces in reflected light using as many different imaging techniques as are available (dark field, interference contrast etc.). Examine also some fracture surfaces. Include an example of a brittle polymer such as polystyrene that gives a shiny mirror-like fracture surface, and a polymer composite, for example a short fibre reinforced polymer.

If you have the use of a transmission microscope try looking at some polymer films between crossed polars. Even with a suitable polymer (e.g. polyethylene) it is unlikely that spherulites will be visible in a commercially produced film because orientation effects are usually dominant, but remelting, as described in section 11.4.2(a) can be achieved without sophisticated equipment and may provide a suitable specimen.

11.3 Small angle light scattering. If you have access to a laser (make sure you observe the laser safety code) try setting up the arrangement shown in Fig. 11.21, using a piece of white paper as a screen. The best subject is a thin film of a spherulitic polymer, and again a remelting procedure may be necessary, though there is no need to place the film between glass slides. The magnification of the pattern is directly proportional to the specimen screen separation, but the intensity falls rapidly as this is increased. Examine the effect of deforming the film.

If the pattern shown in Fig. 11.17 was taken with a He–Ne laser of wavelength $0.633\,\mu m$, using a specimen to film separation of 400 mm, estimate the average spherulite diameter.

12

Thermal analysis

12.1 Introduction

Thermal analysis embraces all methods in which measurements are made of a property that changes as the temperature changes. The equipment used ideally consists of a measurement cell in which the specimen is placed, the measuring apparatus, and the means to change the temperature of the specimen within the measurement cell in a controlled manner, preferably according to a pre-set programme. A constant rate of change of temperature is normally preferred, and it is desirable to make the property measurements continually. Some methods, such as differential thermal analysis (DTA) and differential scanning calorimetry (DSC) are specifically designed to measure the thermal properties of the material under examination (enthalpy or entropy changes etc.), whereas others simply monitor a property that changes with temperature and may produce data that are used directly without reference to the underlying thermal physics. An example of such a method is dynamic mechanical thermal analysis (DMTA) in which the dynamic mechanical properties of a specimen (storage modulus, loss tangent) are measured continually while the temperature is changed; the data may be used directly for design purposes and in many applications will never be processed any further. By making measurements at different frequencies the method can be used for the estimation of activation energies or compared with the predictions of unified time–temperature interdependence relationships such as the Williams–Landel–Ferry (WLF) equation.

Further reference to such methods will be made only briefly here, for a full discussion belongs more fittingly in a treatise on polymer mechanical property measurement and analysis. Thus the first topics, the closely related DTA and DSC methods, receive the most attention, followed by thermomechanical analysis (TMA). Thermogravimetric analysis (TGA) as the name implies monitors weight losses as the temperature is increased. Weight losses occur as the result of driving off volatiles and, at higher temperatures, degradation products. Thus this technique can be used both to examine the state of the material (i.e. the presence and quantity of volatiles, often water) and to investigate the process of degradation (identifying temperatures at which significant changes occur and, by controlling the atmosphere and/or by analysing the products evolved, obtaining information about the chemistry of the degradation process). Other methods that will be touched upon briefly are DMTA and dielectric thermal analysis (DETA).

The temperature of the specimen is normally monitored using a thermo-couple located close to it, and there will often be a small difference between the true specimen temperature and the measured value. This can be minimized using appropriate calibration experiments, possibly conducting a dummy run with a thermocouple embedded in a specimen placed in the normal position. In techniques such as TGA, DMTA this prevents the measurement of weight loss, dynamic mechanical properties etc. of the polymer and is done simply to obtain a correction for the temperature readings recorded during the normal operation of the equipment.

In all thermal techniques the heating or cooling rate is of great importance and must always be taken into account when assessing the results. If the rate is too high there will be thermal lag. The specimen temperature may fall behind the programme 'target' temperature. Furthermore, a temperature gradient will always be present in the specimen, especially in a poor thermal conductor such as a polymer, and this will be greatest at the highest heating (or cooling) rates. If, on the other hand, the heating rate is too small, the specimen may suffer physical ageing during the test, thus changing its response in the later stages. For example, a rapidly quenched glassy polymer has quite different properties to one aged at a temperature just below T_g, and this can be revealed by various techniques including DTA, DSC and DMTA. Ageing during the test may cause the characteristic differences between a quenched sample and an aged one to disappear. Most tests are conducted at a compromise rate (often in the range $1-10$ K/min) at which some element of both effects is probably present. Attention should be given to specimen size, small sections reducing thermal lag effects. A method developed by Richardson [1] for quantitative analysis of DSC data seeks to avoid the effects of changes taking place during the test run by monitoring the responses over a wide temperature interval on either side of the transition, establishing 'reference' states that enable a more accurate assessment of the data obtained near to the transition to be attempted (see section 12.2.2(a)).

12.2 Differential thermal analysis and differential scanning calorimetry

DTA and DSC are conveniently dealt with together because there are strong similarities in both equipment and application. The essential features of the apparatus used for DTA are shown in Fig. 12.1. The sample is placed in cell S, located in a block that can be heated (or cooled) at a programmed rate. A reference sample in an identical cell is located close to the sample cell in the uniform temperature block; its temperature is T_r. The reference sample chosen

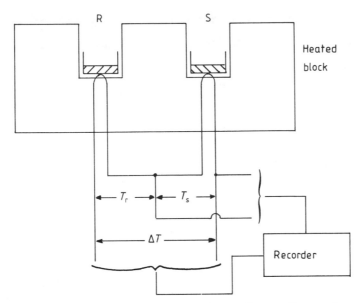

FIGURE 12.1 Schematic of a DTA apparatus: R, r, reference; S, s, specimen.

should be free from transitions within the range of temperature under investigation. Its purpose is to provide a direct comparator for temperature measurement for the sample, assisting the minimization of inaccuracies due to thermal lag in the equipment. When the sample passes through a transitional state its temperature (T_s) departs from that of its surroundings. For example, if the programme is set for heating, at an endothermic transition, such as crystal melting, T_s falls below the programme temperature and the reference temperature, and $\Delta T(=T_s - T_r)$ is negative. The size of ΔT depends on the thermal properties of the equipment, particularly the thermal capacity of the cell, as well as on the sample, for which the thermal conductivity is important as well as the change in enthalpy associated with the transition. For this reason it is difficult to extract quantitative measurements of the thermal properties of the sample using DTA, though the temperatures at which transitions occur can be located fairly accurately.

Careful consideration must be given to the handling of the data, for it is easy to use an erroneous criterion for locating a specific temperature of interest, e.g. the glass transition temperature (see section 12.2.2(a)). More direct measurement of thermal properties is possible using DSC and it is generally preferred for quantitative analysis. In this method the sample and the reference are provided with independent heaters (Fig. 12.2). Background heating of the block may be provided separately to give approximate adherence to the programmed rate, but the individual heaters are used to maintain the sample

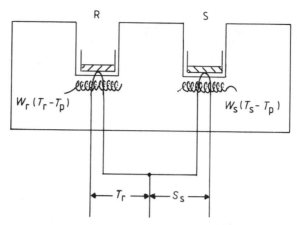

FIGURE 12.2 Schematic of a DSC apparatus: R, r, reference; S, s, specimen.

and reference cells at the programme temperature $(T_p(t))$. The temperature of each cell is measured continuously and compared with the instantaneous value of $T_p(t)$. It is arranged that the power delivered to the sample and reference cells via the individual heaters is a function of the departure from the programme temperature, i.e. $W_s(T_s - T_p)$ and $W_r(T_r - T_p)$ respectively. The differential power requirement, $\{W_s(T_s - T_p) - W_r(T_r - T_p)\}$, is the quantity plotted and can be presented as a function of T_p, T_r or T_s. With this arrangement T_p, T_r and T_s should be very close together even near a transition, and therefore much closer than T_s and T_r in the DTA method whenever thermal changes are taking place.

12.2.1 *Quantitative analysis in DTA and DSC*

Consider the heat flow in the analysis cell. The sample temperature is T_s, assumed uniform, while the block temperature is T_p. The block temperature may be slightly different to the set (target) value of T_p, but we shall refrain from introducing yet another subscript; the (measured) block temperature is what should be substituted for T_p here if there is any departure. Let the total heat capacity of the sample plus the cell be C_s, and the thermal resistance to heat flow between the cell and the block be R (Fig. 12.3). When heat flows into the sample from the surroundings (at a rate dQ_s/dt) the energy balance gives

$$\frac{dQ_s}{dt} = C_s \frac{dT_s}{dt} - \frac{dH_s}{dt} \tag{12.1}$$

where H_s is the enthalpy of the sample. The rate of heat flow can alternatively be given as

$$\frac{dQ_s}{dt} = \frac{(T_p - T_s)}{R} \tag{12.2}$$

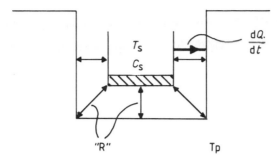

FIGURE 12.3 Heat transfer in a DSC cell: R represents the thermal resistance to heat flow between the cell and the block.

Hence

$$\frac{dH_s}{dt} = C_s \frac{dT_s}{dt} - \frac{(T_p - T_s)}{R}$$ (12.3)

If a suitably inert reference sample has been chosen, the equivalent expression for the reference cell can be written as

$$\frac{dH_r}{dt} = 0 = C_r \frac{dT_r}{dt} - \frac{(T_p - T_r)}{R}$$ (12.4)

where subscript r stands for reference cell, and R is assumed equal for the (identical) sample and reference cells.

$$R \frac{dH_s}{dt} = (T_s - T_r) + R(C_s - C_r)\frac{dT_r}{dt} + RC_s \frac{d(T_s - T_r)}{dt}$$

i.e.

$$R \frac{dH}{dt} = \Delta T + R(C_s - C_r)\frac{dT_r}{dt} + RC_s \frac{d\Delta T}{dt}$$ (12.5)

where $\Delta T = (T_s - T_r)$, and the subscript s is dropped from H since it is redundant when changes in enthalpy occur only in the sample. Remembering that the DTA method produces a plot of ΔT versus T (or t) then, in principle, equation (12.5) can be used for quantitative analysis. In practice it is neither convenient nor accurate to do this, for it requires a knowledge of R, the thermal resistance, which depends on several things including the conductivity of both sample and reference. These values not only alter with temperature, but show marked changes on either side of a transition.

For DSC we can take equation (12.1) and the equivalent equation for the reference cell and find

$$\frac{d(Q_s - Q_r)}{dt} = C_s \frac{dT_s}{dt} - C_r \frac{dT_r}{dt} - \frac{dH}{dt} = \frac{dQ}{dt}$$

where Q is the difference in heat supplied to the two cells i.e.

$$\frac{dH}{dt} = \frac{-dQ}{dt} + (C_s - C_r)\frac{dT_r}{dt} + C_s\frac{d(T_s - T_r)}{dt} \tag{12.6}$$

Now from equation (12.2) we have

$$\frac{dQ_s}{dt} = \frac{(T_p - T_s)}{R} \quad \text{and} \quad \frac{dQ_r}{d} = \frac{(T_p - T_r)}{R}$$

so that

$$\frac{d(Q_s - Q_r)}{dt} = \frac{dQ}{dt} = \frac{T_r - T_s}{R} \tag{12.7}$$

Substitution of $(T_r - T_s)$ from equation (12.7) into equation (12.6) gives

$$\frac{dH}{dt} = \frac{-dQ}{dt} + (C_s - C_r)\frac{dT_r}{dt} - RC_s\frac{d^2Q}{dt^2} \tag{12.8}$$

If R is made sufficiently small then the final term in equation (12.8) can be made negligible; this can be done without affecting the sensitivity of the method, whereas inspection of equation (12.5) shows that with DTA the sensitivity depends on R (i.e. $\Delta T \propto R$). Using DSC, if ΔQ is the difference in heat capacity between the sample and reference cell then the measured heat flow (Q_1) when both pans are empty will be

$$Q_1 = K\Delta Q \tag{12.9}$$

where K is a constant for the apparatus. If the same measurement procedure is now used with the sample in position the difference in heat capacity between the two cells becomes $(\Delta Q + m_s C_{p,s})$ where m_s is the mass of the sample and $C_{p,s}$ is the specific heat capacity of the sample, and the corresponding measurement is

$$Q_2 = K(\Delta Q + m_s C_{p,s}) \tag{12.10}$$

If now the sample is replaced by a calibrant (c) (e.g. alumina) and the procedure is repeated the measurement becomes

$$Q_3 = K(\Delta Q + m_c C_{p,c}) \tag{12.11}$$

From equations (12.9), (12.10) and (12.11) it follows that

$$C_{p,s} = \frac{(Q_2 - Q_1)\,m_c C_{p,c}}{(Q_3 - Q_1)\ \ m_s} \tag{12.12}$$

Thus $C_{p,s}$ versus temperature curves can be obtained. These can be integrated to give enthalpy changes (see section 12.2.2(a)).

12.2.2 Applications of DTA and DSC to polymer studies

(a) MEASUREMENT OF T_g

The glass transition temperature is a very important characteristic of a polymer. Approximate values can be obtained from a large range of methods including measurements as a function of temperature of density, of linear expansion (TMA), of mechanical properties (e.g. DMTA), of IR absorption and of NMR. DTA and DSC can be added to this list, and T_g values can be read directly from the ΔT versus T, or $C_{p,s}$ versus T traces much in the same way that T_g is obtained with the other methods. T_g is a non-equilibrium property and is kinetic in nature, and the measurements made are sensitive to the heating rate and the method used. Thus the value of T_g obtained differs from one technique to another, over a range of several degrees, but this often does not matter. For example, if the value of T_g is required merely to test the suitability of a polymer for operating within a certain range of service temperatures the exact value is not critical since the safety margin allowed would generally be much larger than the range of measurements by different techniques. On the other hand, the T_g for a particular polymer changes with processing conditions, molecular weight and other detailed compositional changes etc. Such changes in T_g are small (of the order of a degree or so) yet are of great interest as they may reveal details about the structure or thermomechanical history of the polymer. They can be detected with careful experimentation using a number of the methods cited above, but to measure them accurately requires a method that is independent of the measuring conditions. Such a method has been proposed by Richardson and Savill and can be used to study subtle changes in T_g and relate them to detailed differences between samples, as described below.

It is found that on either side of a transition the specific heat content is linear with temperature, i.e. $C_{p,s} = a + bT$. Thus for the glassy region

$$H_g(T) = \int C_{p,s}\, dT = aT + 0.5bT^2 + c \qquad (12.13)$$

and for the liquid (or rubbery) region

$$H_l(T) = \int C_{p,s}\, dT = AT + 0.5BT^2 + C \qquad (12.14)$$

a, b, A and B can be determined from the DSC data, then if equations (12.13) and (12.14) are evaluated and subtracted, a value for the integration constant $(c - C)$ can be found. If the two lines represented by equations (12.13) and (12.14) are extrapolated towards one another they meet at $T = T_g$, and setting $H_g(T_g) = H_l(T_g)$ allows these equations to be solved to find T_g. In this way T_g is determined from data obtained well away from the transition temperature and is thus unlikely to be influenced by experimental conditions such as heating rate and is independent of the path taken from the glassy to the liquid state.

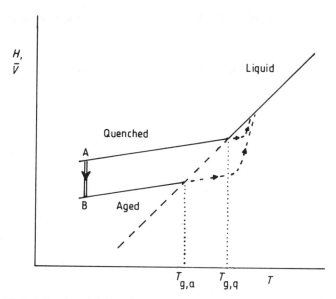

FIGURE 12.4 Schematic of the relationship between enthalpy, H, and temperature for a glassy polymer. On quenching from the liquid state the indicated glass transition temperature is $T_{g.q}$. After ageing the glass transition temperature becomes $T_{g.a}$, but if a fast heating rate is used this will not be immediately apparent because of the overshoot. Specific volume, $\bar{V}$, follows a similar pattern.

To illustrate how T_g can be affected by the processing conditions, consider Fig. 12.4 which shows schematically the relationship between enthalpy (H) and temperature for both a quenched polymer and an aged (or slowly cooled) sample of the same polymer. Specific volume, $\bar{V} (= 1/\text{density})$, shows a similar relationship with temperature. When the polymer is quenched from the melt state it follows the liquid line until it passes T_g, then follows the line marked 'quenched'. If when cooling ceases (e.g. position A on Fig. 12.4) the material is allowed to reside at a temperature below T_g for an extended period, both H and $\bar{V}$ fall slowly. This process is known as physical ageing, and the rate is too slow to be observed if the temperature is too low, but is normally significant at temperatures in the range $(T_g - 50°)$ to T_g.

For the purpose of the present discussion let us assume that the sample is reheated at a constant rate as soon as the quenching is complete. It then follows the 'quenched' line in reverse, and the enthalpy and specific volume both increase linearly until the molecules have sufficient freedom to move in a way characteristic of the liquid state. When this occurs it follows the liquid line. The exact path from the 'quenched' line to the liquid line depends on the heating rate for there is some overshoot followed by a catching up process as the temperature passes through T_g (see the dashed lines with arrowheads on Fig. 12.4). The process of ageing (often called annealing if it takes place above room temperature) allows the sample to densify so that the specific volume and enthalpy are both closer to the thermodynamic equilibrium line, which is the liquid line extrapolated through T_g into the region $T < T_g$. Thus, after ageing, the state of the material may be described by point B, and on heating it follows the 'aged' line, parallel to the 'quenched' line. Overshoot and catching up can be expected to occur again. The accepted values for T_g read from the data presented in the manner shown in Fig. 12.4 are thus $T_{g,q}$ for the quenched sample and $T_{g,a}$ for the annealed sample, and it is seen that the glass transition temperature for the quenched sample is above that for the annealed one.

(b) ANALYSIS OF MATERIAL MODIFIED THROUGH FABRICATION AND AGEING

The remarks at the end of the last section have already indicated that modifications to the structure of a glassy polymer that take place on ageing modify the thermal behaviour. During ageing the material moves closer to its thermodynamic equilibrium state. The extent of the initial departure from equilibrium depends on the thermomechanical history suffered during fabrication. Therefore there is normally some memory of the fabrication conditions that influences the thermal properties. With thermoplastics the most common shaping operations (injection moulding, extrusion, blow moulding, fibre and film production, and thermoforming) all involve molecular orientation in the melt or heat softened state, followed by rapid cooling. Rapid cooling through T_g causes the material to be well removed from equilibrium, and in all but the thinnest articles the rate of cooling differs considerably at different depths, causing the state of the material to be very different from one location to another. In addition there are considerable differences in frozen-in orientation at different depths. As a result there may be significant differences in DSC thermograms from material extracted from different positions in the same moulding, but as yet no quantitative assessment of these differences has been developed. Applications are normally confined to comparisons within a closed set of samples made from the same material but treated with different thermomechanical histories (including fabrication).

An example is given in Fig. 12.5 which shows DSC thermograms obtained for injection moulded polystyrene cut respectively from the skin and core of identical bars. Shown for comparison in both cases is a repeated run on the same specimen after cooling down in the DSC equipment. The rerun traces are considerably displaced from those obtained from the as-moulded material near to T_g and for about 30° before T_g, indicating that there is a marked effect attributable to processing. In the case of the skin sample the T_g peak is shifted relative to that on the reference (rerun) trace. With both skin and core samples a pre-T_g peak was present in the as-moulded state but disappeared in the rerun. The pre-T_g peak is much more prominent with the skin sample. The rerun curves for skin and core are almost exactly superimposable.

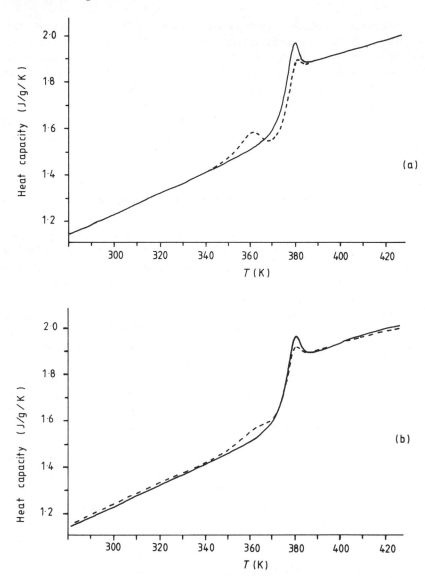

FIGURE 12.5 DSC thermograms for polystyrene samples cut from an injection moulded bar: (a) skin; (b) core. The re-run (reference) curves are shown as solid lines, and the initial runs as broken lines (courtesy of M.J. Richardson and C.S. Hindle).

(c) GLASS TRANSITION TEMPERATURE OF COPOLYMERS AND POLYMER BLENDS

The glass transition temperature of a random copolymer containing two monomer species ('1' and '2') can be expected to lie between the glass transition temperatures of the corresponding homopolymers ($T_{g,1}$ and $T_{g,2}$). If

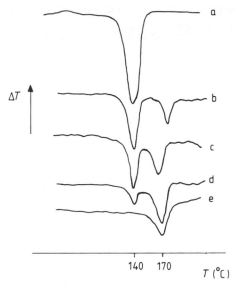

FIGURE 12.6 DTA traces obtained from blends of linear polyethylene, PE, and isotactic polypropylene, PP, with the following compositions: (a) PE homopolymer; (b) 70PE: 30PP; (c) 30PE: 70PP; (d) 10PE: 90PP; (e) PP homopolymer (see Robson, P. Sandilands, G.J. and White, J.R. (1981) *J. Appl. Polym. Sci.* **26**, 3515).

the weight fractions of the two components are w_1 and w_2 then the glass transition temperature of the copolymer $(T_{g,c})$ is normally given by an expression of the form

$$T_{g,c} = \frac{a_1 w_1 T_{g,1} + a_2 w_2 T_{g,2}}{a_1 w_1 + a_2 w_2} \qquad (12.15)$$

where a_1 and a_2 depend on the monomer type, and would both equal unity if the simple rule of mixtures applied.

If the two components phase separate, as is often the case with both graft copolymers and block copolymers, then the T_g of both phases can be identified, lying close to the values for the individual homopolymers. A similar observation is made with many polymer blends, where intermixing at the molecular level rarely occurs. This is true even with a blend of linear polyethylene and polypropylene, despite the similarity in molecular structure (Fig. 12.6).

(d) CRYSTAL MELTING BEHAVIOUR

The crystal melting temperature is an important characteristic of a polymer. It is of relevance when deciding the service temperature or processing temperature for the material, and may reveal something about the structure of the polymer. For example, with some polymers, polyethylene included, the melting temperature increases with increasing molecular weight and crystallinity. The melting temperature may not be discrete if a range of crystal sizes is present, and the breadth of the endothermic peak in the C_p versus temperature trace may indicate the extent of size variation.

If the polymer had been cooled rapidly when the sample was formed then on reheating in the thermal analysis equipment it may show 'premelting crystallization' where at an elevated temperature (still below the melt temperature) the kinetics of crystallization are sufficiently speeded up to allow some (parts) of the molecules that did not crystallize during the initial solidification to do so. This is another example where it is especially valuable to conduct an initial experiment with a constant heating rate, followed by controlled cooling within the thermal analysis cell, then by a repeat run using the same heating conditions. The second run can act as a reference, and reveals clearly those features in the first run for which an explanation must be sought that takes account of the previous thermomechanical history.

Some polymers are polymorphic and the sample may contain two or more crystal types. For those polymers that can form alternative crystal structures, the likelihood of this happening is usually greatest when the material is rapidly cooled and/or when different regions have different thermomechanical histories, as in the skin and the core of an injection moulding. In this case a number of melting signals will be obtained, and once again a reference thermogram from the same sample, obtained on reheating after controlled cooling in the analysis equipment, will assist interpretation.

The enthalpy of fusion, ΔH_f can be measured using the C_p versus T curve from a calibrated DSC instrument by integrating under the curve over the melting range. If the curve is fairly sharp, with the peak temperature at T_m, then the entropy of fusion can be estimated as $\Delta S_f = \Delta H_f / T_m$. The melting behaviour of a polymer crystal is sensitive to the lamellar thickness because of the significance of the surface free energy in a thin crystal, though it is not yet possible to measure lamellar thickness distributions in this way.

(e) CRYSTALLINITY

If the enthalpy of fusion of a semicrystalline polymer sample is H_f then the fractional crystallinity is given by $f_c = H_f / H_{f,c}$, where $H_{f,c}$ is the enthalpy of fusion of a completely crystalline sample of the same polymer. This quantity can rarely be determined directly because of the difficulty of obtaining a totally crystalline sample. For polyethylene the value obtained from crystalline dotriacontane ($C_{32}H_{66}$) has been used, since the paraffin crystal structure is identical to the core of a polyethylene crystal so it is indistinguishable from a polyethylene crystal with the fold surfaces sliced off. Apart from the problem of obtaining $\Delta H_{f,c}$, there are other difficulties. The heat of fusion is temperature dependent, and melting can occur over a wide temperature range. Quantitative analysis may sometimes be more straightforward with C_p data, noting that the measured value can be written as

$$C_p = C_{p,c}f_c + C_{p,a}(1 - f_c) \qquad (12.16)$$

where the subscripts c and a stand for crystalline and amorphous respectively. $C_{p,a}$ can be obtained by making measurements on the liquid at temperatures above T_m and extrapolating back into the melting range. $C_{p,c}$ is a function of temperature, and can be determined from a series of experiments on samples of different known crystallinities (sometimes almost as difficult to find as a completely crystalline sample!).

12.3 Thermomechanical analysis

Thermomechanical analysis (TMA) is the name given to a method capable of measuring the thermal expansion coefficient of materials in its most refined

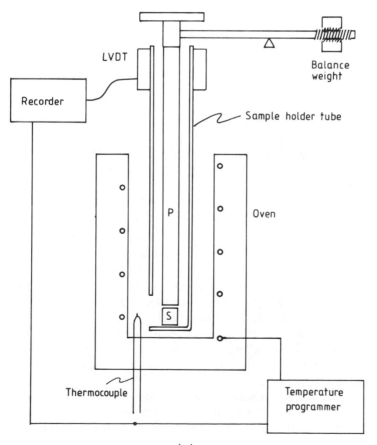

(a)

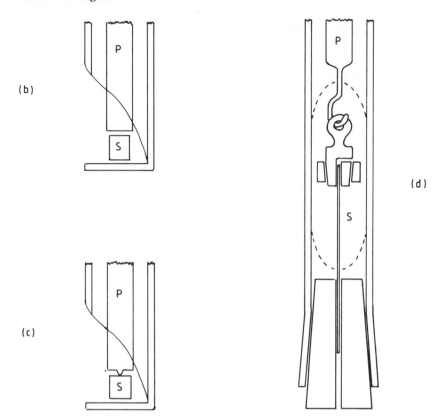

FIGURE 12.7 (a) Schematic of TMA: S, sample; P, probe; LVDT, linear variable displacement transducer. (b) Detail of flat-ended probe. (c) Detail of "pip" probe. (d) Detail of the fibre/film attachment. The broken line indicates the sample tube cut away for access to the specimen.

form, but which is sometimes used simply to determine the location of the glass transition temperature. Figure 12.7(a) shows the general layout of the apparatus. The probe (P) is made from quartz, chosen because it has a low thermal expansion coefficient that is constant within the range of temperatures employed for polymer studies. In the simplest form of the apparatus, the bottom end of the probe is flat and rests on the sample , located on the platform at the bottom of the sample holder tube, also made from quartz. The upper end of the probe is attached to the core of a linear variable differential transformer (LVDT) which accurately monitors position, the signal normally being displayed on a chart recorder. Above this is a counterbalancing arrangement that can be adjusted so that the vertical (gravitational) force on the probe is any chosen value. A convenient way of doing this is to counterbalance the weight of the probe so that it just floats above the surface of the sample, then to add a weight (e.g. 500 mg or 1 g) to the platform at the top. This is sufficient to ensure that the probe remains in contact with the sample and small enough to allow the elastic compression strain to be neglected.

The lower part of the apparatus where the sample is located is surrounded by a constant temperature chamber. This may consist of a cylindrical furnace element within a vacuum flask. If liquid nitrogen is used as a coolant this arrangement is capable of providing temperature scans from $-100°C$ to $+200°C$ and higher. A temperature programmer controls the furnace, giving a preset rate of change, usually between $1°C/min$ and $10°C/min$. The temperature is normally displayed on the same chart recorder as the LVDT signal. If an $X-Y$ recorder is used a direct plot of displacement versus temperature can be obtained, but in practice it is more common to use a two pen $Y-t$ recorder. If a linear increase in temperature is indeed obtained the LVDT signal easily converts to displacement versus temperature. If a linear increase in temperature is not obtained, individual displacement versus temperature readings must be read off and replotted for analysis. This displacement is equal to the difference in expansion between the sample and an equal length of (quartz) tube. Thus, the linear expansion coefficient of the sample equals the slope of the expansion measurement (expressed as a strain) versus the temperature plus the linear expansion coefficient of quartz.

The linear thermal expansion coefficient of an isotropic material is equal to one third of the volumetric expansion coefficient ($= d\bar{V}/dT$ in Fig. 12.4), and the remarks made in section 12.3 about determining T_g are equally valid here. A simple modification to the TMA that can be used to give a sensitive if approximate measure of T_g is shown in Fig. 12.7(c). The flat ended probe is replaced by one with a pip at the bottom. The pip concentrates the load applied to the probe and when the temperature passes through T_g the probe starts to penetrate the softened surface. Thus the direction of motion of the probe reverses and is easily detected on the TMA trace. This probe is not recommended for thermal expansion coefficient measurements, even below T_g.

The size of sample used in the TMA is fairly small (e.g. a cube of 3 to 4 mm edge length is suitable) and the expansion behaviour of samples can be examined in different directions. For example if a sample of these dimensions is extracted from an injection moulded article the expansion coefficient can be measured parallel to the flow direction, transverse to it and through the wall thickness. Significant anisotropy can often be revealed in this way. Such studies are also of interest in fibre reinforced polymers, both aligned fibre composites and those filled with short fibres, especially if the fabrication process causes preferred orientation (again injection moulding is an example).

Sometimes the thermal expansion coefficient differs markedly from one location to another on a scale even finer than the 3 mm level. For example, an injection moulding normally has a 'skin' that sets very rapidly early on in the moulding process that is very different in many respects to the 'core' material that cools and solidifies much more slowly. The skin thickness is often between 0.2 and 0.4 mm and although suitable samples of this thickness can be

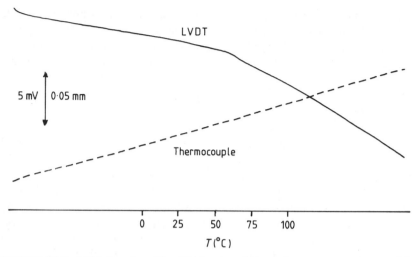

FIGURE 12.8. TMA for glass-fibre filled nylon 6, 6 cut from an injection moulding and measured in the thickness direction. LVDT (solid line) shows the probe displacement, with the sensitivity shown by the scale bar representing 0.05 mm. The broken line shows the thermocouple output, with the sensitivity shown by the scale bar representing 5 mV (courtesy of M.W.A. Paterson).

removed (e.g. by high speed milling) they bend too easily to permit measurements to be made using the arrangement shown in Fig. 12.7(b). An alternative method is to keep the sample under a small amount of tension and a suitable modification to the equipment is shown in Fig. 12.7(d). This can also be used to test (polymer) fibres.

It should be noted that although TMA equipment has high sensitivity, accurate measurements of thermal expansion coefficients may sometimes be hampered by end effects.

A typical result from a TMA run is shown in Fig. 12.8. One of the traces follows the millivoltage output from the thermocouple, plotted in the y-direction with the chart paper running at a constant speed. These values have been converted to temperature and plotted at the corresponding positions on the x-axis. A small horizontal shift was made to take account of the relative positions of the two pens, which are displaced to permit them to cross over so that the whole paper width can be used for both. The results shown are for a thin strip of short glass fibre reinforced nylon-6, 6 cut from an injection moulded bar and tested in the thickness direction with a flat ended probe. It shows the glass transition, which is not as prominent as in a non-crystalline polymer, near to 60°C. The thermal expansion coefficients estimated from this trace are approximately $10^{-4}/°C$ below T_g and $2.3 \times 10^{-4}/°C$ above T_g. Much smaller values were obtained in the bar axis direction (approximately $3.6 \times 10^{-5}/°C$ and $3.7 \times 10^{-5}/°C$) because of the constraints imposed by the fibres (33% by weight), which were aligned preferentially in this direction.

Unfilled nylon-6, 6 moulded and tested in a similar way showed much less anisotropy.

12.4 Thermogravimetric analysis

With thermogravimetric analysis (TGA) the mass of the sample is recorded continuously while the temperature is increased at a constant rate. Weight losses occur when volatiles absorbed by the polymer are driven off, and at higher temperatures when degradation of the polymer occurs with the formation of volatile products. The design of the equipment is most exacting, not only because the weight losses to be measured are very small, demanding a precision weighing mechanism, but also because of the need to avoid convective forces arising within the heating chamber and because of the changes in the density of the gaseous environment. It is important to ensure that volatiles do not condense on the weighing apparatus. It is also necessary to control the atmosphere when this has an influence on the process of degradation.

Instruments can be obtained that make TGA and DTA measurements simultaneously, giving more comprehensive data in a single test. Another useful combination is TGA plus mass spectroscopy. In this case a mass analysis of the products given off by the sample is made, assisting identification of the species evolved at various temperatures, whether volatiles or degradation products.

A TGA run made on a nylon-6, 6 sample cut from an injection moulding grade granule that had been standing in laboratory air for several months before testing is shown in Fig. 12.9. The spikes are superimposed at five minute intervals. It is evident that the predominant feature is weight loss and that this begins to become important above 120°C. The weight loss is attributed to water, and the total amount driven off by the end of the run, terminated after

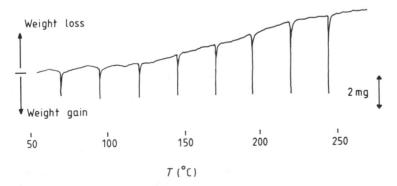

FIGURE 12.9 TGA trace for nylon 6, 6 grade (courtesy of M.W.A. Paterson).

crystal melting, was approximately 1.1% of the original weight of the sample, a value typical of that obtained by other techniques for the water content in similar samples.

12.5 Dynamic mechanical thermal analysis

In dynamic mechanical analysis (DMA) the sample is deformed cyclically, usually under forced vibration conditions. By monitoring the stress–strain

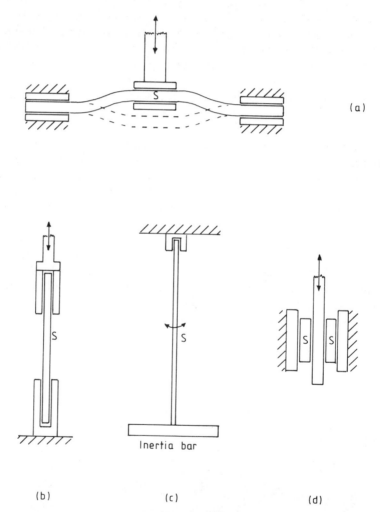

FIGURE 12.10 Loading options for dynamic mechanical testing: (a) double cantilever (NB single cantilever is also used); (b) reversed uniaxial tension; (c) torsion; (d) shear. In the case of shear it is essential that the specimens S are identical.

relationship while changing temperature, information can be obtained about the relaxation behaviour of the test piece material. The method then becomes dynamic mechanical thermal analysis (DMTA), though the abbreviation DMA is often used even when the temperature variation is an important feature of the test. Many modes of vibration are possible, but the most popular are reversed bending (e.g. the double cantilever, Fig. 12.10(a)), axial tension (usually from zero load to maximum, Fig. 12.10(b)), torsion (Fig. 12.10(c)) and shear (Fig. 12.10(d)).

The simplest of all of the dynamic test arrangements is the torsion pendulum, in which the sample in the form of a wire or a thin narrow strip hangs vertically and supports an inertia bar or disc which is rotated about the sample axis and is then allowed to execute free vibrations and undergoes damped oscillations. Although the sample can be surrounded by a temperature jacket and the experiment repeated for a series of temperatures to provide 'thermal analysis', it does not lend itself very readily to automated thermal analysis, and for this a forced vibration system is preferred. Although there is no fundamental reason why torsion should not be chosen, it is more popular to use arrangements based on those shown schematically in Fig. 12.10(a), (b) and (d). The vibration chosen is usually sinusoidal. If the amplitude is sufficiently small a sinusoidal load gives rise to a sinusoidal deformation, and for the purpose of interpretation it does not matter whether the test is run under load control or deformation control. There are often practical reasons for choosing one rather than the other. In either case the strain lags behind the stress and this is described by a phase angle (δ).

12.5.1 Dynamic mechanical parameters

The stress (σ) can therefore be described by the equation

$$(\sigma - \sigma_m) = \sigma_0 \sin \omega t$$

where σ_0 is the stress amplitude, ω is the cyclic frequency and t is time. σ_m is the mean stress, and for reversed loading in which the tensile and compressive amplitudes are of equal magnitude it is zero. For zero to tensile testing, $\sigma_m = \sigma_0$, and the corresponding strain is

$$(\varepsilon - \varepsilon_m) = \varepsilon_0 \sin (\omega t - \delta)$$

where ε_0 is the strain amplitude, and ε_m is the mean strain, having special cases as described for stress.

The dynamic mechanical properties of the material are described in terms of a complex dynamic modulus:

$$E^* = (E'^2 + E''^2)^{1/2} = E' + iE'' \tag{12.17}$$

where E' is called the storage modulus and is a measure of the recoverable strain energy in the deformed body and E'' is called the loss modulus and is related to the hysteresial energy dissipation. The phase angle (δ) is given by

$$\tan \delta = E''/E' \tag{12.18}$$

Thus the stiffness and damping properties of the material can be described by any two of the quantities E', E'' and $\tan \delta$ (the 'loss tangent').

12.5.2 Test details

For a forced vibration test it is necessary to impose a chosen load or deformation programme and to measure load and deformation continuously. There are several commercial instruments that do this and although it is prudent to question the accuracy of the load and deformation measurements it will rarely be found that they are ultimately responsible for limiting the absolute accuracy of the derived material parameters. It is much more likely that the absolute accuracy of the technique will depend on other factors. The nature of the deformation depends on both the material properties and the sample shape, and is very sensitive to the sample clamping arrangement. For example, the nature of the deformation in the double cantilever vibration shown in Fig. 12.10(a) is very different from that in three point bending, which it resembles in some ways. The relationship between the load applied and the deflection of the centre of the beam is quite different for these two cases, and can be accounted for if the beam truly deforms in the manner specified. In practice, it is not possible to ensure that the deformation is purely one of these because of problems associated with deformation within the clamps and slipping in the clamps. Errors associated with clamping also occur in the tensile, torsion and shear alternatives. Clamping errors can sometimes be reduced by testing samples of different lengths and extrapolating to infinite length.

Further problems arise when the temperature is changed. Returning to the double cantilever arrangement, thermal expansion will cause the bar to slacken between the grips, and may even lead to a toggle effect as the bar passes through the zero displacement position. A similar problem may arise in deformation-controlled tensile-mode testing in which buckling may occur in a thermally expanded bar. If tensile-mode testing is conducted under load control, this problem does not arise. In bending, one way to reduce the problem of thermal expansion is to apply tension to the bar when mounting it in the clamps so that on raising the temperature the expansion no more than compensates for the initial elastic strain. This remedy is usually applied to remove the more serious problems, such as the toggle effect, but may introduce secondary (often unknown) effects which are rarely allowed for, so that there may be an error in the absolute stiffness values derived from the measurements.

Thus, unless a careful assessment is made of the mechanics of the deformation, the absolute values will be of doubtful accuracy, and it is not expected that absolute modulus values can be routinely obtained using DMTA. With an appropriate choice of geometry for the test, and with care taken to reproduce sample size and shape, and the clamping procedure, reproducible results can be obtained that permit comparisons to be made between different materials. In many of the commercial instruments it is possible to test very small samples only a few millimetres long and a fraction of a millimetre thick. This is an advantage when significant property variations are present; for example samples can be extracted from an injection moulding from the skin and from the core for comparison. The anisotropy of such samples can be examined, for measurements can be made transverse to the flow direction as well as parallel to it using samples cut in the appropriate ways.

Although the modulus values may not be very accurate, relaxation temperatures can be determined with good accuracy using DMTA. By conducting tests at different frequencies a time–temperature transformation can be performed. Data for this may be gathered in a single run by using a microprocessor controlled DMTA.* The vibration frequency can be programmed to change automatically at the same time as the temperature is ramping, and results for several different frequencies can be recorded and displayed. The advantage of this, apart from the obvious one of time saving, is that all data are obtained from the same sample and any clamping error will be common to all sets of data at each frequency.

12.5.3 Interpretation

The remarks of the previous section indicate that the modulus measurements obtained using DMTA instruments often contain serious errors. If precautions are taken, however, the values may be used for comparison purposes within a set of tests conducted under strictly controlled conditions and an example is given in the following section. It is quite common, however, to use DMTA simply to identify important relaxation temperatures without any attempt to acquire accurate modulus data. Thus in the testing of a glassy polymer the measured value of E' shows a steep fall as the temperature is raised through T_g and, for the heating rate and vibration frequency used, the value of T_g is reproducible and fairly accurate. At secondary transitions there is usually a recognizable change in the slope of the E' versus temperature plot, but the temperature of the transition is usually most clearly identified by the position of the peak in the $\tan \delta$ plot.

The glass transition temperature is, of course, associated with conformational changes of the molecule main chain, whereas the secondary transitions involve conformational changes that do not require main chain

---

*As supplied, for example, by Polymer Laboratories Ltd, Loughborough, UK.

movement, for example the rotation of a pendant ester group about the C—C bond that connects it to the main chain. The preferred direction of such motions may be biased by stress just as happens with main chain relaxations, and under reversed loading conditions a particular group of atoms may change conformation repeatedly back and forth across an energy barrier, hence taking energy from the driving system. This effect is at a maximum at the so called transition temperature, and diminishes if the temperature is changed (in either direction) or, alternatively, if the frequency of the driving system is changed (without a compensating change in temperature). It is not possible to tell from the mechanical tests what conformational changes give rise to the various relaxations that are observed. This must be deduced with the help of information from other relaxation experiments or, perhaps, by comparing results obtained from similar polymers with systematic differences in structure that lead to corresponding changes in relaxation behaviour. Interpretation at this level, that is the identification of the relaxation mechanisms, is a rather specialized activity and will not be explored further here; discussion can be found in [2].

12.5.4 *Applications of DMTA*

The publicity of the manufacturers of DMTA equipment gives many diverse examples of applications of the technique to polymers. Damping properties can be compared and a polymer having a high $\tan \delta$ value over the frequency/temperature range of interest shows strong damping. It is especially important to know this characteristic for materials to be used for tyres, for high damping can lead to failure. High temperatures may build up on rolling along the road, and if the damping is high at the elevated temperature still more heat may be dissipated, leading to even higher temperatures. The effect of the presence of reinforcing fillers can be conveniently investigated using DMTA.

The curing of thermoset resins can be examined, the modulus increasing with cure. The $\tan \delta$ value usually rises at first then falls rapidly as the crosslink density becomes higher and progressively inhibits conformation changes. For investigations of this kind a properly designed experimental programme is essential to take account of the time–temperature effects in the chemical reaction.

Copolymer characteristics can often be explored. If the different components phase separate, separate T_g values associated with the separate phases can be distinguished. In a terpolymer such as acrylonitrile-butadiene-styrene, several transitions can be found. If samples with different thermomechanical histories are examined, it is found that they give rise to different loss peaks. Thus if a sample is tested over a wide temperature range then cooled and retested it will often produce a different result the second time. Such information may be of direct value in warning that changes may occur in service, if exposure to

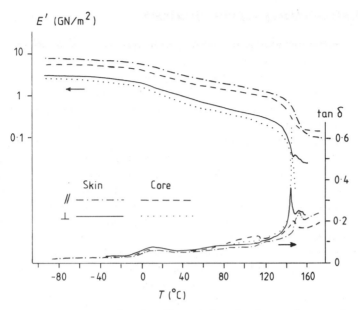

FIGURE 12.11 DMTA traces for glass-fibre reinforced polypropylene specimens cut from the skin and core of an injection moulded bar and tested parallel and transverse to the bar axis.

similar temperatures is possible, and this may lead to property changes. Combined with microscopic examination, the DMTA results may help to indicate what kind of molecular reorganization is taking place during annealing. The compatibility of blends and the dependence of their properties on thermomechanical history can be examined in a similar way.

An example of the use of DMTA to investigate the properties of injection moulded polypropylene reinforced with short glass fibres (20% by weight) is shown in Fig. 12.11. Results are presented for samples cut from the skin and the core of end gated bar mouldings, and for both the flow direction and transverse to it. Even when the remarks about the absolute reliability of the E' values are heeded, the stiffness is seen to be considerably greater in the bar axis direction at all temperatures; this effect is attributed to fibre alignment. (Note the E' axis is logarithmic.) The T_g for this material should be near to 0°C, and is indicated by a change in slope in the E' plots and by a peak in tan δ. The E' curve falls steeply as the temperature is increased through the crystal melting temperature and tan δ peaks in this range. Much larger effects occur in this range with unreinforced polymer. A much more prominent T_g effect is observed with non-crystalline polymers. The E' trace falls steeply so the temperature rises through T_g and a large tan δ peak is obtained at this position.

12.6 Dielectric thermal analysis

When a polar polymer is placed in an electric field the permanent electric dipoles along the molecule attempt to align parallel to the field direction. Their realignment is retarded by the neighbouring parts of the molecule and by neighbouring segments of other molecules. Thus if the temperature is below T_g realignment is very difficult and takes place very slowly or not at all. In essence, the dependence of realignment on temperature will show the same behaviour as mechanical relaxation when the driving force is an externally applied stress rather than one that derives from the interaction of charge with an electric field. If an alternating field is applied the dipoles will be encouraged to move in opposite directions on alternate half cycles. If the temperature is too low or the frequency of the field too high it is impossible for them to follow. If the temperature is high enough, however, the dipoles can realign repeatedly, and this causes energy to be withdrawn from the driving field. The polarization lags behind the energizing electric field and the angle, δ and the corresponding tan δ values have a similar significance to that described for DMTA (section 12.5). Also analogous to the mechanical case, the properties of the material are described in terms of a complex dielectric constant (ε^*), that contains a storage component (ε') and a loss component (ε''):

$$\varepsilon^* = \varepsilon' - i\varepsilon'' \tag{12.19}$$

The negative sign in equation (12.19) follows from the relationship between the dielectric constant and the applied field (E). The electric displacement (**D**) which includes the contribution from the polarization (**P**) is given by

$$\mathbf{D} = \mathbf{E} + 4\pi\mathbf{P} = \varepsilon^*\mathbf{E} \tag{12.20}$$

whereas in the mechanical case (recall section 12.5.1)

$$\sigma = E^*\varepsilon \tag{12.21}$$

Having noted that ε and E have entirely different meanings in the mechanical and dielectric cases, it is evident that in equation (12.20) it is **E** that provides the driving force, whereas in equation (12.21) it is σ that fulfils a similar function. Thus ε lags behind σ (mechanical case) and D lags behind E (dielectric) so that ε^* has a significance equivalent to $1/E^*$.

N.B.

$$\frac{1}{E^*} = \frac{1}{E' + iE''} = \frac{E' - iE''}{E'^2 + E''^2}$$

The value of tan δ can be expected to show a similar temperature dependence as the mechanical loss tangent, since both depend on the ease of molecular relaxations, though the intensity of the relaxation depends on the strength of the coupling to the electric field, that is, on the dipole moment of the relaxing group.

By using a range of different methods to measure the dielectric properties an enormous frequency range can be covered (from 10^{-4} Hz to 10^{10} Hz). The majority of the data found in the literature have been obtained using apparatus designed and assembled in the laboratories in which the measurements were made. This demands considerable specialist knowledge, and a brief discussion of the principles would serve little purpose here. Dielectric test apparatus specifically designed for polymer analysis can now be purchased, however, and the Polymer Laboratories* 'DETA' equipment produces tan δ values over a range of frequencies from 20 Hz to 100 kHz. This complements very well the DMTA technique, which extends the frequency range to much lower values but provides a considerable overlap within which the expected agreement between the two techniques can be checked. The Polymer Laboratories DETA tests disc-shaped samples of 33 mm diameter and 0.1 to 2 mm thick, and permits the temperature to be scanned automatically through a wide range. Thus the DETA provides rapid assessment of the relaxation spectrum of polar polymers.

Finally, it should be recognized that commercial plastics contain many additives, and if these are polar they may provide coupling to the electric field that can give a significant response even with a non-polar polymer. The opportunities for realignment of the polar impurities are governed by the movements of the neighbouring segments of the polymer in which it is embedded, so that the tan δ peaks correspond to relaxations of the polymer.

12.7 Summary

In surveying the methods of thermal analysis discussed above, it is clear that they have a wide range of application. Data obtained using TMA and DMTA can be used directly in engineering design, for which the stiffness variation and thermal expansion must be taken into account whenever a component is subjected to a wide range of temperatures in service. In both TMA and DMTA the equipment used routinely for thermal analysis does not provide the most accurate data, but they are adequate for all but the most exacting requirements. TGA can be used to identify the temperature at which chemical degradation of the material commences, and is also valuable in detecting and measuring absorbed volatiles such as water. The glass transition temperature can be determined using DSC, DTA, DMTA, DETA, and, with less precision, TMA. DMTA and DETA are the most useful methods for examining secondary relaxations and for measuring the damping properties.

*Polymer Laboratories Ltd, Loughborough, UK.

References

1. Richardson, M.J., and co-workers (1984) *Polymer Testing* **4**, 101.
2. McCrum, N.G., Read B.E. and Williams, G. (1967) *Anelastic and dielectric effects in polymeric solids*, Wiley, London and New York.

Further reading

1. Hay, J.N. (1982) in *Analysis of polymer systems* (eds L.S. Bark and N.S. Allen), Applied Science Publishers, London, Chapter 6.
2. Read, B.E. and Dean, G.D. (1978) *The determination of dynamic properties of polymers and composites*, A Hilger, Bristol.

13

Other techniques

In this chapter we deal with techniques that we consider do not merit a chapter of their own. The reasons for this may be quite diverse (see Chapter 1) and the subjects included here do not establish any kind of theme.

13.1 Density

The density of a polymer sample can indicate its crystallinity or, in the case of a glassy polymer, its thermomechanical history. The most popular method for measuring the density of a polymer is by flotation. The idea is simply to find a liquid in which the polymer just floats, then to measure the density of the liquid. Normally the measuring liquid is a mixture of two liquids of different densities in various proportions to give a continuous range of density. Polymers have densities in a range for which several suitable liquid combinations can be found. The density of the mixture at which flotation occurs can be measured in a number of ways, including measuring the refractive index, which has a unique relationship, usually linear, with the composition and hence the density, but the most popular method described below involves making a density column.

The density column is made in a tall graduated glass tube by mixing the two liquids in a controlled manner during filling. Although the various components that are required to build a density column are fairly easily obtained separately it is recommended that a commercially produced apparatus is used since this is unlikely to be much more expensive unless 'false economies' are made. If the manufacturer's instructions are carefully followed during filling the result is a liquid column which is denser at the bottom than at the top, with a linear gradient over most of the depth. The most convenient way to calibrate the density in the column is to use marker floats of known density. A calibration graph is made of density versus the height at which the floats settle (read from the graduated scale). If the graph is not linear the column should be rejected. If a sample is placed in the column it settles at the position at which the liquid is of equal density, and the density can be read from the calibration graph.

Although the essential parts of the measurement are thus very easily understood, there are several practical tips that must be heeded.

1. The liquids must be out-gassed before preparing the column; if air bubbles form and attach to marker floats or samples serious errors occur;
2. Marker floats and samples should be wetted with the denser of the liquids before introduction into the column, again to reduce the chance of trapping an air bubble;
3. The column should be surrounded by a constant temperature jacket.

In selecting liquids for the column it is essential to avoid those which would swell the polymer or interact with it in any other way. Aqueous potassium iodide solution is suitable for many polymers and offers a range of densities from $1000 \, \text{kg/m}^3$ to $1600 \, \text{kg/m}^3$. For a particular density range two solutions with different concentrations giving densities at either end of the range are chosen. If the likely density of the sample is known fairly closely beforehand a column with a small range (and high accuracy) can be made using two solutions of fairly close concentrations. If the column is kept free from vibration it can last for several weeks, though it is essential to check the height of the marker floats regularly, and to up-date the calibration graph if necessary. It is even possible to sweep out the samples and floats and to reuse the column. This must be done when there are so many samples in the column that there is a danger of them fouling one another. To do this it is required to have a wire basket that fits snugly within the column in place before the column is made, then to raise it very slowly under the control of a geared-down motor, to which it is attached by a fine thread. The old samples are discarded and the basket is lowered again slowly. The operation can be performed without disturbing the density gradient.

13.1.1 *Polymer applications*

If the polymer contains additives this may have an influence on the density, especially in the case of fillers. It is possible to obtain a rough guide to the filler content from a density measurement if the filler density is known. In the following applications it is assumed that the samples do not contain additives that would alter the results.

(a) CRYSTALLINITY MEASUREMENT

If it is assumed that a semicrystalline polymer consists of two phases, crystal and amorphous, each with its own density, then the density of the mixture is given by

$$\rho = f_c \rho_c + (1 - f_c) \rho_a \tag{13.1}$$

where f_c is the fractional crystallinity, ρ_c is the crystal density, which can be calculated accurately from the crystal unit cell dimensions (from X-ray diffraction) knowing the atomic constituents of the unit cell, and ρ_a is the density of the amorphous phase. It is not possible to measure ρ_a directly, and indeed it is likely to vary according to the thermomechanical history of the material, but in many instances it is sufficiently different from ρ_c that a value deduced indirectly is sufficiently accurate to put into equation (13.1) and obtain meaningful results. It must be remembered, however, that there will almost certainly be some non-crystalline material that has some order and will contribute to the result in a different (and incalculable) way to the 'genuinely' amorphous material. This 'third phase' may have a different influence on measurements of f_c made by X-ray diffraction (also with reference to a two phase model).

(b) DENSITY IN NON-CRYSTALLINE POLYMERS

Large variations in density can be found in non-crystalline polymers as a consequence of differences in their thermomechanical history. The skin and core of an injection moulding contains material in very different states of organization. This is a result of a combination of causes:

1. Flow-induced orientation is frozen-in in those regions near the surface of the moulding that cool most rapidly, but orientation relaxation takes place in the slowly cooled interior;
2. Differential cooling rates cause density variations quite separate from the effect described in 1. (see section 12.2.2);
3. The material near the surface solidifies under high pressure applied from the injection system, whereas the material in the interior does not solidify until after the gate admitting material into the mould cavity has frozen, decoupling it from the injection system so that solidification takes place at a much lower pressure.

Unless special precautions are taken in preparing a polymer sample it will normally be in a non-equilibrium state. It will thus tend to change density on holding it at any temperature. If the temperature is too low the rate of change may be too small to be detected. Many polymers show ageing at room temperature, however, and density may be seen to change within the column if observed over an extended period. Some of these points are illustrated in Fig. 13.1 which shows measurements made on samples machined from the skin and the core of injection moulded polystyrene bars. The skin is denser than the core, and material from both regions are seen to densify at a similar rate at room temperature. The different symbols represent samples cut from bars cooled differently after removal from the mould. Samples placed in the column after a period of ageing outside the column produced results falling on the appropriate line showing that the density variations are not caused by interaction of the sample with the column medium.

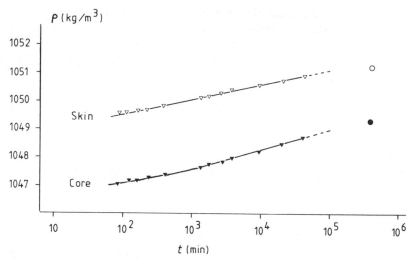

FIGURE 13.1 Densification of polystyrene cut from the skin and the core of a newly moulded bar and immersed in a density column. Readings were taken at the times shown. The densities of specimens measured after ageing for 10 months at room temperature are also shown ($\bigcirc$, $\bullet$) (see A.V. Iacopi and J.R. White (1987), *J. Appl. Polym. Sci.*, **33**, 577).

13.2 Electron spectroscopy for chemical application (ESCA)

This technique is sometimes known as X-ray photoelectron spectroscopy (XPS) and is based on the observation that electrons are emitted by atoms under X-ray bombardment. If the target is a solid the energy of the emitted electrons is measured and this enables the binding energy to be determined. The exciting radiation is a monoenergetic beam of soft X-rays, for example the $K_{\alpha 1,2}$ radiation of Al or Mg, whereas in a related technique, ultraviolet photoelectron spectroscopy, UV radiation is used, though this provides information only about the lower energy valence bands. Since the electrons are emitted with low energy they are easily recaptured, and only those produced very close to the surface have a significant chance of escape. Only these can contribute to the signal and the information comes from a very small depth, of the order of nanometres.

The atomic processes that lie at the heart of the technique can be fairly complicated. They involve the core orbitals of the atoms of the target and are essentially localized on the atoms. The energies measured are characteristic of the elements, but are sensitive to the electronic environment of the atom, and can provide information about its binding state. There are three alternative processes. In the simplest, photoionization, a secondary electron is ejected from the core, as shown in Fig. 13.2. In the other two processes the perturbation caused by the loss of an electron from the core stimulates the valency electrons

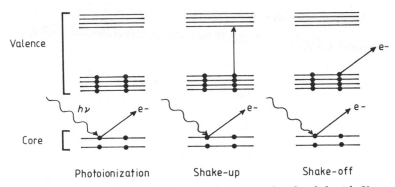

Valence

Core

Photoionization   Shake-up   Shake-off

FIGURE 13.2 Energy level scheme for a target atom bombarded with X-rays, $h\nu$, showing the processes (photoionization, shake-up, shake-off) by which characteristic photoelectrons are produced.

to reorganize. This may take the form of further ionization, with the emission of an electron from the valence band ('shake-off') or as the promotion of a valence electron to a higher unoccupied level ('shake-up') (Fig. 13.2). These rearrangements require energy, and the photoelectron emitted from the core in the shake-up process has less kinetic energy than that produced in unmodified photoionization, and the core electron emitted in the shake-off process has still smaller kinetic energy.

Energy analysis of the secondary electrons is achieved by passing them through a magnetic field analyser or, more usually, an electrostatic field analyser, that causes them to travel on curved paths, with the curvature determined by the velocity (Fig. 13.3). The path of the electrons from the sample into and through the analyser and on into the detector must be unimpeded by gas atoms, which readily scatter the slow moving low energy electrons, and the instrument must be operated under high vacuum conditions, preferably at 10^{-8} torr or better. This is considerably greater than the vacuum required for most electron microscopy, and means that rapid sample exchange is much less easy to arrange, even with an airlock system that

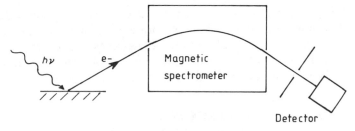

FIGURE 13.3 Schematic layout of an ESCA machine.

permits prepumping of the chamber through which the sample is inserted. Samples are preferably in thin film form, either solution cast or melt pressed. If information is required about a polymer surface on a sample that has been subjected to a specific treatment (e.g. outdoor weathering) it can be extracted by cutting or machining. Great care must be taken in handling and preparing ESCA samples for it must be remembered that the information comes from the surface and that contamination must be completely avoided.

The energy of the X-ray photon becomes divided as follows

$$hv = E_{BE} + KE + \phi + E_R$$

where E_{BE} is the binding energy, KE is the energy of the electron ejected (which is measured by the spectrometer), ϕ is the work function of the spectrometer and E_R is the recoil energy which is negligible except when the target is a hydrogen atom.

13.2.1 Practical details

Most polymers are insulators and become charged during ESCA observation, causing a shift in the energy scale. For the most exact data some form of referencing back to the Fermi level must be made. This can be done by depositing a thin coating of a suitable reference material that will provide a signal corresponding to a precisely known binding energy within the range of interest but not overlapping a sample signal. Examples are C_{1s} from hydrocarbons at 285 eV and the $4f_{1/2}$ levels of gold at 84 eV. Sometimes the correction is only a few eV and can be ignored.

If contamination is present on the surface uneven charging can occur, altering the signals from the sample in addition to introducing unwanted signals from the contaminant itself. Even if the sample can be prepared in clean conditions and can be introduced into the ESCA machine free from contamination, contamination builds up slowly even in the high vacuum provided in ESCA machines, and it is recommended that the core level binding energies should be measured as soon as observations on the sample commence. This may take 30 min, during which time the contamination build up should be negligible in a typical machine, and the results can then be compared to those obtained later after more prolonged observation.

Thus the strategy will generally be: first to take a low resolution wide range scan to determine which elements are present; secondly to measure the relative intensities of the signals present, allowing estimation of the relative abundance of the elements present (taking account of their different sensitivities); and finally to measure the absolute binding energies.

13.2.2 Interpretation of ESCA data

The chance of a particular atomic reaction taking place when the target atom is bombarded by an X-ray is measured in terms of the 'cross-section' that it presents to the incoming photon. The cross-section depends on the wavelength of the incoming X-ray and theoretical predictions have been made of the relationship between the photoionization cross-section and photon energy for ejection of electrons from different orbitals. For example, Clarke [1] shows how the cross-sections for the 1s, 2s and 2p levels in carbon vary with photon energy, and orbitals can be distinguished by studying changes in peak intensities using X-rays with different energies.

Theoretical predictions can be used as a guide only and are not sufficiently accurate to give unambiguous interpretation of peak shifts caused by different bonding states, for example. Thus, as in other spectroscopic techniques, the method adopted has been to build up a data bank of ESCA spectra using first well-characterized simple samples, extending to more complex structures as experience expands. The data obtained is of three kinds: the absolute binding energy (determined by the element and its bound state), the shift (giving details of the binding state) and relative peak intensity.

In hydrocarbon polymers there is just one carbon core level, and shifts between carbons in sp^3, sp^2 and sp bonding states are extremely small. Thus the C_{1s} levels with polyethylene, polypropylene and polystyrene are almost identical, but polystyrene has an additional shake-up peak for the $\pi^* \leftarrow \pi$ transition in the aromatic carbons. Shake-up satellites can indicate the presence of unsaturated structures within the polymer main chain or in side groups.

Another common polymer that has a C_{1s} spectrum that gives a useful illustration of peak shifts is poly(ethylene terephthalate):

A prominent peak occurs at about 285 eV and corresponds to the aromatic carbons, another peak is at about 287 eV and is attributed to the methylene carbons, and a third is located between 289 and 290 eV and comes from the carbons in the carbonyl groups.

Another piece of information that can be sought by ESCA is the variation in composition as a function of depth near to the surface of a sample. This can be done by comparing the intensities of different peaks and is known as 'depth profiling'. The escape depth of electrons produced with different kinetic energies is significantly different and when this is taken into account it is often possible to deduce whether or not the abundance of a particular element in a particular state varies according to the distance below the surface. Another way of obtaining this type of information is to study the variation in signal strength with take-off angle since this varies the escape path length for a particular depth.

13.2.3 *Application of ESCA to polymer studies*

Many applications are cited in the literature, but some of these have been executed as much to test the limits and versatility of the technique as to provide unique and necessary information. It is probably too early to judge what applications will turn out to be of most importance, and although we have tried to choose examples of general interest, our list is not comprehensive and may rapidly become dated.

(a) OXIDATION OF POLYMERS

Oxidation is a common contributor to failure in polymers. It can occur during processing at elevated temperature. In service oxidation is again associated with elevated temperatures, but often involves in addition other aspects of the environment. For example, in the case of a rolling car tyre 'mechanico-oxidation' is stress aided, whereas many polymers degrade outdoors by a process of photo-oxidation promoted by the UV component in sunlight. These reactions are expected to take place primarily near the surface, and it is here that the damage causes the most serious reduction in mechanical properties. Hence ESCA is an appropriate technique to study the development of oxidation.

It is normal practice to examine both the O_{1s} and C_{1s} spectra. The O_{1s} spectrum contains a peak corresponding to a binding energy of about 533 eV which often has a shoulder at about 535 eV. It is found that the shoulder disappears if the sample is first placed in a desiccator over P_2O_5, and is thus attributed to absorbed water. The 533 eV peak is attributed to C=O and may be broadened since carbonyl groups may exist in several different environments. C=O shows as a small shoulder on the high binding energy side of the main C_{1s} peak in polypropylene.

(b) SURFACE FLUORINATION

Surface fluorination can be used to improve the chemical resistance of polyethylene and is less expensive than making an article from a fluorocarbon with similar chemical resistance. ESCA can be used to monitor the reaction and determine the depth of fluorination. The C_{1s} and F_{1s} intensities reveal the composition as a function of depth, and the F_{1s} and F_{2s} peaks can be used to determine the thickness of the fluorine layer. Binding to fluorine induces a large shift in core levels, including C_{1s}, assisting recognition in studies with these materials.

(C) COPOLYMER COMPOSITION

If each of the different units in a copolymer has at least one peak associated with it exclusively, then a reasonably accurate compositional analysis can be performed. The most favoured sequences of the different units can also be determined, for the nearest neighbours of each main chain carbon determine its core level so that each different unit sequence leads to a different value.

In a particular investigation of a copolymer of polystyrene and poly(dimethyl siloxane) (PDMS), cast from solution, the surface in some samples was found by ESCA to be pure PDMS. From a consideration of escape depths it was deduced that the lower limit of the thickness of the PDMS at the surface was 4 nm. In some samples prepared in different ways (e.g. cast from a different solvent) some polystyrene was detected near to the surface, but the major component was still PDMS, indicating once again preferential segregation of this component to the surface. This effect is quite separate from the normally experienced localized phase separation in block copolymers.

13.3 Neutron diffraction

To conduct neutron diffraction measurements a neutron beam source is required. This demands the provision of a nuclear reactor, thus restricting the opportunity for this kind of analysis. For the high flux beams demanded by other than the most basic work, specially designed reactors are required and these are available only at national or international centres. Few readers will have cause to use such equipment, and it is therefore not appropriate to present a lengthy discussion. Furthermore, neutron diffraction is applied to topics such as the study of the diffusional motion of rubbers and testing the theory of rubber elasticity that lie outside our remit. The technique is capable of providing quite unique information, however, and cannot be neglected altogether.

The neutron has a mass of 1 amu, the same as the proton, or hydrogen nucleus, and it has no charge. Neutrons are produced by fission in nuclear reactors and, after collisions within the moderator, have a characteristic distribution of velocities and are called thermal neutrons. Recalling the principle of particle wave duality first mentioned in Chapter 9, the distribution of velocities can be converted into an equivalent distribution of wavelengths, and this has a peak at a wavelength < 0.2 nm. For many diffraction applications, especially some polymer studies, longer wavelengths are required (0.5 to 1 nm) and this can be achieved by cooling using liquid nitrogen.

Large momentum transfers can occur in collisions of neutrons with nuclei within the sample, because of the similarity in size of the neutron and the nuclear targets. The scattering cross section for hydrogen is extremely large, much larger even than that for deuterium, a fact that is used to advantage in many neutron diffraction studies.

One of the most challenging applications of neutron diffraction is in the examination of the state of a molecule in the bulk. Light scattering can be used to determine the radius of gyration* of a molecule in dilute solution, where individual molecules are separated. Such a prospect seems untenable in the bulk because of the intimate entanglement of identical molecules. Neutron diffraction offers a solution, however, for if a small fraction (say $\leqslant 1\%$) is deuterated, then the 'tagged' molecules are likely to be separated from one another within the bulk. Most polymers contain a significant number of hydrogen atoms and can be treated this way. Significant diffraction effects are observed when the sample is bombarded by neutrons since the absence of strong scattering centres at a particular location (i.e. the sites where deuterium has replaced hydrogen) is just as potent as the presence of highly scattering centres within a uniform background of low scattering power (complementarity principle). The latter case could, of course, be achieved by deuterating 99% of the molecules and leaving 1% highly scattering protonated molecules. Small angle neutron scattering can be used to determine the radius of gyration of the isolated tagged molecules. The interpretation of the results must be conducted with great care, however, for there is evidence that in some cases the molecular dispersion of the deuterated species is not perfect so that the conformational state of the molecules that provide the scattering effects is not necessarily the same as if they were not deuterated, and segregation may occur, causing serious modification to the diffraction.

Neutron diffraction studies have shown that in the amorphous phase, coiling of the molecules is similar to that predicted and observed to be present in dilute solution, with the radius of gyration proportional to $\bar{M}_w^{1/2}$ where $\bar{M}_w$ is the weight average molecular weight. Melt grown crystals are found to have a similar radius of gyration as that in the melt, and this places some restrictions on the models for crystallization that can be accepted. Annealing does not alter the radius of gyration even if the crystallinity changes. The radius of gyration obtained with sedimented mats of solution-grown crystals from the same polymer is different, and the rate of crystallization appears to have an important influence on molecular dimensions.

Another application of neutron diffraction is in the identification and study of torsional vibrations. Although these produce a signal that can be observed using IR spectroscopy, identification is not always easy because a number of vibrations lie in the same energy range. In the case of neutron diffraction the intensity of the interaction is proportional to the square of the displacement of the atom, which is much larger for torsional vibrations than for those involving bond stretching or bond angle distortion. Thus the signal for a torsional vibration is relatively much larger for neutron diffraction and this can help to identify the source of the various relaxations obtained from a polymer.

Finally it should be noted that voiding can have a significant effect on

---

*The radius of gyration is a measure of the compactness of the molecule, and is equal to the root mean square distance of the atoms from the centre of mass.

neutron diffraction and its effect is difficult to subtract. This can be a problem especially with polymer samples that have been subjected to large deformations.

Thus neutron diffraction has a special place in polymer analysis. Despite the experimental difficulties and the need for quite sophisticated data handling procedures, the method is limited currently primarily by the uncertainties in the interpretation of the data rather than measurement accuracy. These problems will doubtless be solved and neutron diffraction can be expected to continue to provide unique information about polymers.

Reference

1. Clark, D.T. (1976) in *Structural Studies of Macromolecules by Spectroscopic Methods* (ed. K.J. Ivin), Wiley, London.

Further reading

1. Clark, D.T. (1982) *Pure. Appl. Chem.* **54**, 415.
2. Sadler, D.M. (1984) in *Structure of Crystalline Polymers* (ed. I.H. Hall) Elsevier Appl. Sci., London, Chapter 4.
3. Allen, G. Higgins, J.S. and Wright, C.J. (1976) in *Structural Studies of Macromolecules by Spectroscopic Methods* (ed. K.J. Ivin) J. Wiley, London, Chapters 1, 2, 3.

Exercise

13.1 Calculate the density of an orthorhombic polyethylene crystal using $a = 0.742$ nm, $b = 0.494$ nm and $c = 0.254$ nm. (There are four CH_2 units per unit cell.) If the density of amorphous polyethylene is 855 kg/m^3, calculate the fractional crystallinity of (i) a linear polyethylene with a measured density of 975 kg/m^3 and (ii) a branched polyethylene with a measured density of 930 kg/m^3. Construct a graph of density versus crystallinity for polyethylene.

Why is it not possible to measure the crystallinity of poly(-4-methyl pentene-1) using density measurements?

14

Multiple characterization applications

This chapter is intended as a review, showing how the various techniques described in previous chapters can be combined together to investigate particular problems, providing powerful complementary information. The examples chosen are not meant to be comprehensive, but illustrate how characterization studies can be of value in the development of a deeper and broader understanding of polymeric materials on the one hand, or can provide invaluable information that may be used to remedy problems of a practical nature on the other.

14.1 The structure of a spherulite

Much of the understanding of polymer crystals came from studies of solution grown single crystals, with the TEM as the principal tool for characterization. The crystal orientation was determined using electron diffraction, and an approximate measurement of the crystal thickness was obtained from the shadow length in shadowed samples. It was quickly realized that the thickness is much less than the molecular length and this led to the idea of chain-folded crystals [1-3]. Although this discovery did not prove that such structures exist in bulk crystallized material, it did provide crucial insight into polymer crystal habits and is of great historical importance.

Many polymers were known to be semicrystalline long before the concept of spherulitic morphology evolved. X-ray diffraction showed crystals to be present, but the high level of scattering into regions far removed from the crystal peaks shows non-crystalline material to be present too. Density measurements confirm that the material must contain a mixture of crystalline and non-crystalline regions. The fringed micelle model [1, 2] would appear to be adequate to account for these experimental observations, and it required more detailed microscopical examination to deduce the true arrangement of the crystals.

Polarized light microscope study of thin melt crystallized samples shows the crystalline regions to have radial symmetry, and this is difficult to reconcile with the fringed micelle model. The direction of birefringence within the spherulite can be used to determine the molecular chain orientation and in many common polymers, including polyethylene, it is found to be arranged tangentially. This leads to the model of a spherulite as consisting of a radial array of crystalline lamellae, with the molecule axis lying in tangential directions.

Next the connection is made between these lamellae and solution grown crystals, for it is found that the lamellar thickness in bulk crystallized samples lies in the same range as in thin solution-grown crystals. This can be confirmed by X-ray line broadening and SAXD measurements. TEM of stained solution-cast samples and studies of etched and stained samples by both SEM and TEM (of replicas) have provided convincing verification of the proposed model for spherulite structure. These methods have also revealed that the lamellae often twist and that under certain crystallization conditions the twist period is very regular, with all lamellae twisting in phase. This gives rise to rings in the image in the polarized light microscope. In addition, the work by Bassett and co-workers [4], using specimens stained by the chlorosulphonation technique and viewed in the CTEM, has demonstrated clearly the presence, in some cases, of a bimodal lamellar thickness distribution, leading to still further refinement of the model for the structure. This last feature might be inferred from DSC studies, but the TEM provides the most positive structural details.

Thus it can be seen that several techniques have contributed to the development of our current picture of a spherulite. Many of the studies mentioned above were based on specially prepared laboratory specimens, however, and together helped to develop a general model for the structure of a spherulite. It is a quite different matter to deduce the structure of a particular sample of a melt crystallized polymer. For this task a suitable starting point would be to prepare thin microtome sections for polarized light microscopy. If the polymer cuts statisfactorily this will reveal the size of the spherulites. An estimate of the average spherulite size over a larger sample could be made using small angle light scattering from thin sections. SAXD could then be used to measure the long period and wide angle X-ray diffraction to obtain the lamellar thickness from line broadening measurements. For detailed study staining followed by sectioning for TEM observation may be the best option. The chlorosulphonation technique for polyethylene may be suitable for some other polymers, but other stains may need to be developed. Similarly, the permanganate etching procedure, followed by replication for TEM study, or direct inspection in the SEM, is not suitable for all spherulitic polymers and other etchants may need to be tested.

It is disappointing that very little success has been achieved with direct diffraction contrast imaging of thin sections of spherulites in the TEM. It would be expected that this technique should be capable of providing the most detailed and unambiguous information, but experimental problems seem to be difficult to surmount. If this could be achieved successfully it would show whether a lamella consists of a small number of long yet twisted and defective crystals, or whether it contains many smaller crystals.

Finally, if a more complete characterization of the spherulites than simply a morphological assessment is required, then DSC or DTA can be used to measure T_g and the crystal melting temperature. TMA, IR and NMR are other techniques by which T_g may be found. Light microscopy could be used to investigate whether nucleation had taken place on a foreign body within the material. An idea of the lamellar packing could be obtained by estimating fractional crystallinity from density or X-ray diffraction measurements.

14.2 Skin–core characterization in moulded polymers

Injection moulded polymers contain a skin–core morphology. The skin forms when the polymer melt contacts the cold mould wall, freezing very rapidly, whereas the core cools much more slowly. Polymer molecules become oriented during flow, and the skin normally sets too quickly to permit much recovery, whereas the core cools sufficiently slowly to permit substantial (sometimes complete) recovery and is often nearly isotropic. Skin and core have quite different morphology and properties and the overall properties of the moulding depend on these properties and on the division of the material between the two regions. For example, in a simple tensile test on polypropylene the oriented skin often fails and separates from the core which continues to cold draw.

Again it is sensible to start a characterization study by examining a microtome section in the light microscope. The skin/core boundary is normally clearly visible, with best contrast usually in polarized light. With semicrystalline polymers the core often contains equiaxed spherulites, whereas the crystals in the skin have preferred orientation and a different morphology. Sometimes two layers can be distinguished within the skin, and sometimes more than three layers may be visible between the surface and the centre of the moulding. In the case of non-crystalline polymers contrast at the skin/core boundary is a consequence of a very large change in molecular orientation. Birefringence measurements can be made at this stage to investigate the state of molecular orientation as a function of location and should be conducted on sections cut both transverse and longitudinal to the flow direction.

The skin is typically of the order of 0.3 mm thick and parts of it can be removed using a microtome for study in isolation from the core. Alternatively high speed milling using a single point tool with a fly cutting action can be used to remove material from a flat moulded surface, leaving at the end just the skin on the lower face mounted on the machine bed. Samples of the core can be made by removing material from both sides. Samples containing only skin or only core can then be used for a range of measurements as follows:

1. Density measurements in a density column: in some non-crystalline polymers the skin is found to be significantly denser than the core. There is even more scope for variation in semicrystalline polymers.

2. DSC measurements: small yet significant differences in T_g are found. Furthermore, enthalpy measurements reveal differences that seem to correlate with the different pressures prevailing at the different parts of the moulding cycle during which the material in the respective regions solidified.
3. DMTA measurements: significant differences in Young's modulus are observed between skin and core, and measurements parallel and perpendicular to flow reveal anisotropy in the skin. The difference in T_g between skin and core is confirmed.
4. TMA measurements: differences in orientation lead to differences in linear thermal expansion coefficient; the skin again shows anisotropy, and a significant difference exists between skin and core. This will cause residual stresses to develop when the temperature of the moulding changes in service, and may lead to distortion or warping.
5. Wide angle X-ray diffraction: this can be used to reveal differences in crystal orientation, and with those polymers that are polymorphic may show that different crystal structures develop in the two different regions.
6. FTIR: measurement of the dichroic ratio can be used as an indication of preferred orientation.

In addition, the methods discussed in section 14.1 could be used here in the same way if the subject is a semicrystalline polymer in order to provide a more detailed analysis of the morphology. A more comprehensive discussion of the strategies recommended for morphological investigations is given in [5].

If the moulding material contains a filler it is normally not possible to produce satisfactory thin sections for transmitted light microscopy. Thicker sections may be used for microradiography in which the fibre orientation distribution in fibre-filled samples is revealed by shadows cast when X-rays are transmitted through them. Alternatively cross sections can be prepared for reflection light microscopy by sectioning and polishing using metallographic procedures.

14.3 Weathering of polymers

Most polymers degrade when used outdoors. The major hazard is generally UV radiation, though some polymers are more vulnerable to this than others. Other causes of problems are extremes of temperature (and temperature cycling), high humidity (and large variations in humidity), erosion by rainfall or wind-borne particles (sand etc.), and pollutants. Weathered polymers have been the subject of numerous studies, the motives for which have been to identify the mechanism(s) of degradation, so that appropriate remedies can be devised, and to compare the weatherability of different materials. The remedies often involve adding a suitable UV absorber, and comparisons are as likely to be made between different formulations based on the same polymer as they are to be made between different polymers.

Even before a polymer article breaks in outdoor service it often shows clear signs of change, and inspection in the light microscope may reveal tiny flaws or even fissures developing within the surface. These can be examined in much more detail in the SEM and can reveal the depth to which serious degradation has penetrated. Samples can be extracted at different depths (e.g. using methods mentioned in section 14.2). Molecular scission takes place in those polymers most prone to UV degradation and this is clearly shown by a reduction in the molecular weight. The T_g may also fall significantly at the surface. IR spectra may reveal the presence of functional groups that form during the degradation process. The ATR technique is most appropriate for this application. Most common is the carbonyl group that forms on many polymers (e.g. polyethylene) as a result of chemical reactions that follow the initial hydrogen abstraction caused by the UV irradiation. Alternatively, ESCA can be used to characterize the chemical state of the surface and to follow the changes that take place on exposure to the weather. Hydroxyl groups are also often produced by degradation reactions and can be observed by these techniques.

Some polymers, such as polystyrene, develop a conjugated polyene structure as a consequence of weathering, and this can be detected using UV spectroscopy. Polymers containing UV stabilizers have strong UV spectral responses coming from the additives and these must be recognized before deductions can be made about the polymer.

14.4 Particular studies

The previous sections have dealt with fairly general areas of interest in which different studies by different workers often with quite different objectives have helped to develop an overall understanding of a polymer system. In this section we consider some particular examples in which many different techniques were used in a single coordinated study.

We begin by giving examples of investigations conducted by a single group of workers, using several techniques to provide complementary information, and finish with a study to which several groups contributed in a coordinated manner, bringing together a range of expertise and specialized equipment beyond the resources of a single laboratory.

When any new polymer is synthesized or when an existing polymer is modified chemically it is essential to characterize it in order to explain fully the properties of the material. Detailed characterization invariably necessitates use of a number of different analytical techniques. The examples given are intended to illustrate how various techniques can be used to advantage. Note that sometimes contrary indications may arise from different techniques, emphasizing the importance of conducting a comprehensive characterization, and not relying on too few measurements.

14.4.1 Poly(bismethylene hydroquinone) [6]

There is considerable interest in polymers possessing particular electrical and/or chemical properties yet sufficiently tractable to permit processing. To this end a linear, soluble poly(bismethylene hydroquinone) has been reported. It is an example of a ladder polymer. Characterization of the polymer involved measurements of intrinsic viscosities to give a measure of molecular weight. It is worthy of note that attempts to use GPC were less successful due to apparent polymer association in the solvent.

FTIR spectroscopy on the product obtained at different stages in the reaction indicated that the reaction proceeded via intermediates containing reactive methylol ($-CH_2OH$) groups. The characteristic methylol absorption at $1040 \, \text{cm}^{-1}$ decreased during the course of the reaction. This was confirmed by ^1H and ^{13}C NMR spectroscopy in solution and in the solid state using magic angle spinning and cross polarization techniques.

FTIR also showed the existence of quinone structures in the polymer (also indicated by colour changes during polymerization). This was not confirmed in the ^{13}C NMR which indicated only a small amount of quinone present. The discrepancy can be reconciled if it is presumed that any quinones present couple with hydroquinones to give semiquinone structures. Electron exchange would then be expected between adjacent semiquinones. This exchange would relax the carbonyl carbons involved in the interchange so rapidly that they would not be detected in the solid state NMR. An alternative explanation could be that the strong IR absorption arises due to a large extinction coefficient for the quinones.

The morphology of the polymer in the solid state was investigated using polarized light microscopy and X-ray diffraction. Light microscopy of samples crystallized by solvent evaporation indicate the existence of spherulitic structure although larger crystalline structures are believed to be present. X-ray diffraction patterns show a high degree of crystallinity with little amorphous content. Detailed interpretation of the X-ray patterns was not made but the relatively sharp rings observed should permit detailed structure determination.

14.4.2 Polyelectrolytes [7]

Another example concerns the synthesis of potential polymeric electrolytes by chain extension of poly(ethylene glycol)s by oxymethylene groups, and the subsequent characterization of the products. Use was made of GPC both for molecular weight determination and to separate individual fractions for subsequent characterization of the products.

The synthesized polymers comprised a range of molecular weights. Partial separation of individual fractions could be achieved by fractional precipitation using mixed solvents. The resulting fractions were then investigated using GPC in tetrahydrofuran. High molecular weight fractions with molecular weights up to 10^6 were obtained. Low molecular weight fractions consisted of linear chains and ring structures with degrees of condensation from 1 to 15. ^{1}H and ^{13}C NMR gave resonance absorptions characteristic of the ethylene glycol chains and oxymethylene units. Differences were observed between the high and low molecular weight fractions and were consistent with the molecular weight distributions. The linear chains gave resonance absorptions due to chain end groups which were absent in spectra of ring structures. The higher molecular weight polymers showed different morphologies depending on the nature of the ethylene glycols used in the synthesis.

14.4.3 Isotactic polypropylene [8]

Investigations of the crystal structure and chain packing in different forms of isotactic polypropylene afford another example of the use of a combination of experimental techniques. Methods employed include X-ray diffraction, DSC and ^{13}C NMR.

Isotactic polypropylene exists in three different crystal forms:

1. thermodynamically stable α-form
2. metastable β-form obtained by controlled crystallization
3. smectic form which is produced by grinding at low temperatures

The well defined X-ray diffraction patterns from the α- and β-forms permit determination of the crystal structures. The α-form is monoclinic with right- and left-handed 3_1 helices arranged in pairs. The β-form also consists of 3_1 helical chains but with hexagonal packing. The smectic form exhibits only two diffuse reflections indicative of a lower degree of order.

The predominantly β samples also contain contributions from the more stable α-form and the relative amounts of each can be estimated from the intensities of the peaks in the diffractograms.

DSC measurements show the expected endothermic melting peaks for the α- and β-forms whereas the smectic samples exhibit an exothermic peak due to a solid state transformation to the α structure. The degree of crystallinity of each form can be estimated from the DSC traces. ^{13}C NMR of the solid polymers using MAS and cross polarization give additional detailed information on chain conformations in the different crystal forms. Thus for the α-form the doublet splitting of the methyl and methylene carbon resonances is evidence for the existence of two non-equivalent sites produced by pairing of helices of opposite handedness. The different values of chemical shift are presumed to be due to differences in the ease of rotation of the methyl groups. The single absorption peaks observed for the β-form indicate that the right and left helices pack together in groups such that they correspond to one of the non-equivalent sites in the α-form.

The ^{13}C shifts of the smectic material are similar to the β-form and might indicate equivalent packing.

The spin relaxation times (T_1) for the β-form and for the smectic form were also found to be very similar which appears to be consistent with the above. It should be noted that the T_1 values for the amorphous regions in each type were similar.

Although the similarity of the X-ray diffraction and ^{13}C NMR for the β-form and smectic form would tend to indicate similar helical chain packing, there is one significant problem. On heating the smectic form a solid state transformation to the α-form occurs and presumably can only involve small-scale chain movement. There is still some uncertainty about the details of the helical chain structure of the smectic form. One explanation is that the chains are similar to the α-form but slightly displaced so that the restrictions on the rotation of the methyl groups are reduced.

14.4.4 Poly(p-phenylene benzobisthiazole)

Poly(p-phenylene benzobisthiazole) is a rigid rod molecule that is soluble in strong acids, forming lyotropic liquid crystal solutions that can be processed to form fibres or films:

The fibres have very high stiffness and strength, values above $300\,GN/m^2$ and $3\,GN/m^2$ respectively having been reported. When divided by density to give specific stiffness and strength values the mechanical properties are extremely attractive, and the material can be used at relatively high temperatures (about $500^\circ C$ is probably possible for extended periods). Thus it was quickly seized upon as a potential aerospace material, and studied intensively in a joint programme in which a number of laboratories participated, with contributions from universities, industrial fibre producers and the US Air Force. The number of papers appearing in the literature and the number of authors involved are too great to list here, and the volume of work is too vast to attempt even a brief summary or to review the relative importance of the many techniques used to study this material. It is, however, of value to note the techniques that have been used, and we choose to present them more or less in the order that they appear in this book. Even though the list is long we doubt whether it is comprehensive. The remarks made refer to experimental materials in their early stages of development and do not necessarily relate to latest levels of refinement.

Molecular weight estimates were made by making intrinsic viscosity measurements. Caution is necessary when using rigid rod molecules because of their special flow properties. UV–visible studies were made on dilute solutions of low molecular weight model compounds that resembled the polymer, using chloroform as solvent (the polymer dissolves only in concentrated acids). These studies, together with Raman spectroscopy indicated the level of electronic delocalization between the phenyl and heterocyclic rings: this depends on the twist angle between the phenyl and heterocyclic rings, and is maximum when the rings are coplanar and is completely interrupted when they are mutually perpendicular.

FTIR was used with the sample oriented in different directions to investigate molecular orientation in fibres and films. It was also used to check whether any acid solvent remained after processing into fibre or film, and whether any absorbed water was present. The movement of the ring structures when the molecules were stressed was followed using a miniature tensile test rig on a FTIR spectrometer.

X-ray diffraction and electron diffraction were used to investigate molecular packing in fibres and film. Possible crystal structures were proposed, but the diffraction information was not quite complete enough to permit certain identification. This is because the regions within which full three dimensional order was established were not very extensive transverse to the molecule axis. This was indicated by X-ray line broadening. Fragments of the fibres thin enough for electron transmission showed fibrillation to be present, limiting the maximum extent of lateral order. The electron beam resistance was sufficiently good to permit lattice images to be recorded on the best samples. At least one example was found in which fringes could be seen perpendicular to the fibre axis, indicating that the rigid rod molecules were not only parallel and packed regularly (giving fringes parallel to the molecule axis) but that they were in register along the axial direction too, though the region within which this could be positively identified was quite small (tens of nanometers only). At lower levels of resolution, coherently diffracting domains of similar size were visible in dark field; this does not prove the presence of fully developed three dimensional order, for regular lateral packing would provide sufficiently strong diffraction to enable imaging by this method. SEM images confirmed the presence of fibrillation in peeled fibres, and were used to examine the surface structure of fibres both as-produced and after various treatments, including mechanical testing to fracture. Polarized light microscopy revealed voids in fibres, and reflected light microscopy of films showed blisters that were present after heat treatment: these were thought to be voids caused by water evolved from the acid from which the polymer was coagulated. Small angle X-ray diffraction was used to analyse microvoids, and showed that many voids were present at the 10 nm level. Scattering from voids was found to obliterate information about molecular conformation when solid samples were studied by small angle neutron scattering.

Although we have not seen reports of TGA studies, these might have been valuable, for weight losses (about 3%) were reported to occur on high temperature heat treatment of fibres. This can probably be attributed to loss of water or acid from the coagulation process, but TGA would appear to be a valuable aid in studying the eventual degradation of this material at high temperature.

14.5 Conclusions

Sections 14.1, 14.2 and 14.3 give examples of the ways in which several characterization techniques can help the development of an understanding of a polymer system. Sometimes this understanding develops over a period of many years and is in some ways an inductive process, with information provided by one technique helping to give insight into the meaning of results obtained by another so that the interpretation is continually being updated and refined. In a similar way, dedicated studies of the kind presented in section 14.4 require a broad-fronted approach. Thus, although the examples have not yet led to unequivocal characterization of the polymers they do demonstrate the necessity of employing a number of different techniques. As we have seen there are situations where relying on data from single experimental techniques can lead to possible erroneous interpretations. It is the accumulation of information which permits more reliable assignments to be made.

References

1. Powell, P.C. (1983) *Engineering with Polymers*. Chapman and Hall, London.
2. Billmeyer, F.W. Jr, (1984) *Textbook of Polymer Science*, 3rd Edn, Wiley-Interscience, New York.
3. Young, R.J. (1981) *Introduction to Polymers*. Chapman and Hall, London.
4. Bassett, D.C. (1981) *Principles of Polymer Morphology*. Cambridge University Press, Cambridge.
5. Sawyer, L.C. and Grubb, D.T. (1987) *Polymer Microscopy*. Chapman and Hall, London.
6. Bi, X-T and Litt, M.H. (1987) *Polymer* **28**, 2346.
7. Craven, J.R., Nicholas, C.V., Webster, R. *et al.* (1987) *Br. Polym.* 5, **19**, 509.
8. Gomez, M.A., Tanaka, H. and Tonelli, A.E. (1987) *Polymer* **28**, 2227.

APPENDIX

Solutions to exercises

Chapter 2

2.1 1 mole AZBN contains $2 \times 6.03 \times 10^{23}$ initiator radicals. 1 count corresponds to $(2 \times 6.03 \times 10^{23})/10^7$ radicals, and so 105 counts corresponds to $105 \times 12.06 \times 10^{16}$ radicals. From the mode of termination each polymer molecule contains 1.5 initiator radicals. Therefore the number of polymer molecules in $3.5\,g = (105 \times 12.06 \times 10^{16})/1.5\,\bar{M}_n$, the mass of polymer containing Avogadro's number of molecules is then

$$\frac{3.5 \times 6.03 \times 10^{23} \times 1.5}{105 \times 12.05 \times 10^{16}} = 2.5 \times 10^5$$

2.2 Plot π (in cm toluene)$/c$ versus c. Intercept $(\pi/c)_{c \to 0} = 1.000$.

$$\bar{M}_n = \frac{RT}{(\pi/c)_{c \to 0}} = \frac{8.314 \times 378}{78.5 \times 10^{-3}} = 4 \times 10^4$$

2.3 (i) Calculate $(t - t_0)/t_0 c = \eta_{sp}/c$ for each concentration;
(ii) Plot η_{sp}/c versus c. Intercept $(\eta_{sp}/c)_{c \to 0} = [\eta]$
From Mark Houwink equation: $[\eta] = K\bar{M}_v^\alpha$; $\bar{M}_v = [\eta]^{1/\alpha}/K = 4.48 \times 10^5$

2.4 (i) Calculate $K = (2\pi^2 n^2/\lambda^4 N_0)(dn/dc)^2 = 2.46 \times 10^{-7}$;
(ii) Calculate Kc/R_θ for each value in table;
(iii) Plot

$$\frac{Kc}{R_\theta} \quad \text{versus} \quad \sin^2\left(\frac{\theta}{2}\right) + k'c$$

(where k' is a scaling factor; let $k' = 100$ to give reasonable spread) for each angle and concentration.
(iv) To extrapolate to $c = 0$, $\theta = 0$ note that where $c = 0$, $\sin^2(\theta/2) + k'c \to \sin^2(\theta/2)$ for each angle and when $\theta = 0$, $\sin^2(\theta/2) + k'c \to c$

for each concentration. Extrapolation gives $Kc/R_\theta = 4 \times 10^{-6} = 1/\bar{M}_w$; $\bar{M}_w = 2.5 \times 10^5$

Chapter 6

6.1 $\dfrac{N_u}{N_L} = 1 - \dfrac{g_N \beta_N B_0}{kT}$

At 1.5 T, $N_u/N_L = 1 - 6.8 \times 10^{-6}$; at 12 T, $N_u/N_L = 1 - 81.73 \times 10^{-6}$

6.2 $v = \dfrac{g_N \beta_N B_0}{h}$

(i) for ^{1}H, $v = 97.91$ MHz; (ii) for ^{2}H, $v = 15.023$ MHz; (iii) for ^{19}F, $v = 92.18$ MHz; (iv) for ^{31}P, $v = 39.97$ MHz.

6.3 $B_0 = \dfrac{hv}{g_N \beta_N}$

(i) for ^{1}H, $B_0 = 1.5$ T;
(ii) (a) for ^{19}F, $B_0 = 1.59$ T; (b) for ^{13}C, $B_0 = 5.966$ T.

6.4 The spectrum consists of two pairs of equally spaced lines centred on 297 and 400.2 Hz, with spacings of 0.7 Hz. Hence
(i) (a) chemical shifts 297 and 400 Hz; (b) hyperfine coupling constant 0.7 Hz, and
(ii) 492.6, 494.0, 666.3, 667.7 MHz.

6.5 (i) Isomer A contains 4 different types of protons, whereas isomer B contains only 2 different types of proton. The spectrum is assigned to isomer B.
(ii) The peak intensities indicate that the chemical shifts are as follows:

$$-CH_3 - 64.8 \text{ Hz}; \ -CH_2 - 87.6 \text{ Hz}.$$

The lack of hyperfine interaction is due to the rapid rotation of the CH_3 groups making them equivalent and to the fact that the protons are separated by two sigma bonds.
Additional hyperfine interaction would be expected at low temperature when the rotation of the methyl groups is hindered.

Chapter 7

7.1 $\dfrac{N_\beta}{N_\alpha} = 1 + \dfrac{g\beta H}{kT}$

	N_β/N_α
0.3 T, 100 K	1.00403
0.3 T, 300 K	1.00134
1.0 T, 100 K	1.0134
1.0 T, 300 K	1.00478

Increased H and low T gives higher population in lower energy state and favours resonance absorption.

7.2 (i) 4 equally spaced lines – $1:3:3:1$
(ii) 7 equally spaced lines – $1:3:6:7:6:3:1$
(iii) 8 equally spaced lines – $1:3:5:7:7:5:3:1$
(iv) 7 equally spaced lines – $1:2:3:4:3:2:1$

7.3 Seven equally spaced triplets with intensity ratios $1:6:15:20:15:6:1$ due to interaction with six equivalent methyl protons. The triplets lines are of equal intensity due to interaction with the nitrogen nucleus.

7.4 (i) The sextet spectrum arises when two of the protons become equivalent. The pattern implies that the equivalent protons have the lowest value of coupling constant, i.e. $a_{H_1} = a_{H_3} = 1.8 \times 10^{-3}\,T$; $a_{H_2} = 2.9 \times 10^{-3}\,T$

(ii) The polymer radical contains one α proton and two β protons. The β protons are essentially isotropic whereas the α proton exhibits marked anisotropy. The differences in the β proton coupling constants are due to their conformational angles with respect to the unpaired electron orbital.

(iii) From the equation (7.32) $a_\beta^H = B_1 + B_2 \cos^2 \theta$, $\theta_2 = 45°$; $\theta_3 = 55°$.

7.5 In general $\dfrac{1}{T_2} = \dfrac{1}{2T_1} + \dfrac{1}{T_2^1}$

For pure spin–lattice relaxation the line shapes are approximately Lorentzian and the peak to peak separation is $2/T_2\sqrt{3}$, $(10^{-5} T \equiv 2.8 \times 10^5\,Hz)$

$$T_1 = \frac{1}{\sqrt{3} \times 2.8 \times 10^5} = 2.06 \times 10^{-6}\,s$$

For pure spin–spin relaxation the line shape is Gaussian with peak to peak separation of $2/T_2$; $T_2 = 2/2.8 \times 10^5 = 0.71 \times 10^{-5}\,s$.

Chapter 8

8.1 (i) From Fig. 8.16, $2\theta_{110} = 21.6°$, $2\theta_{200} = 24°$. Hence from Bragg's law $(\lambda = 0.1542\,nm)$, $d_{110} = 0.411\,nm$, $d_{200} = 0.371\,nm$. Thus, $a = 2d_{200} = 0.742\,nm$. For an orthorhombic crystal, $d_{110} = ab/\sqrt{(a^2 + b^2)}$, Hence $b = ad_{110}/\sqrt{(a^2 - d_{110}^2)} = 0.494\,nm$.

(ii) Breadths at half peak intensity (B), of the peaks in the intensity versus 2θ curve are $B_{110} = 0.8°$ and $B_{200} = 1.0°$. Using the Scherrer equation (8.10) with $K = 0.9$, and remembering to convert the breadths measurements into radians, the corresponding crystal diameters are $L_{110} = 10\,\text{nm}$ and $L_{200} = 8\,\text{nm}$. The discrepancy may be a consequence of unequal growth, with the crystal diameter in $\langle 110 \rangle$ directions exceeding that in the [100] direction.

(iii) The ratio of the weight of the portion of the curve above the broken line in Fig. 8.16 to the weight of the total curve above the background line is taken to equal the fractional crystallinity, and equals approximately 0.56.

8.2 Unstrained peak position is given by $2d_{002} \sin \theta_{002} = 0.1542$; $d_{002} = 0.127\,\text{nm}$, and $\theta_{002} = 37.379°$.

When a stress of $1\,\text{GN/m}^2$ is applied, the strain in the chain axis direction is $1/285$, and $2 \times 0.127\,(1 + 1/285) \sin \theta'_{002} = 0.1542$. Solving gives $\theta'_{002} = 37.226°$. Hence the change in the Bragg angle is $0.153°$ and the shift in the peak is $0.306°$ in terms of the deviation angle (2θ).

8.3 Four CH_2 units are required to represent the repeat unit. Two are associated with each corner of the unit cell plan in Fig. 8.4. Each corner is shared by four similar cells, so one pair of CH_2 units must be included in the representative set. The molecule passing through the body centre is not shared by any of the lateral neighbours of the chosen cell and this

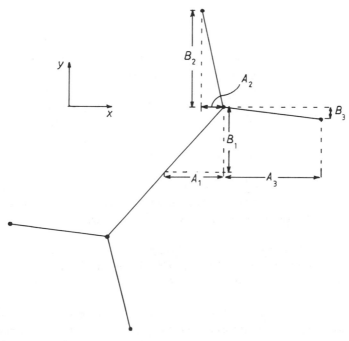

FIGURE A8.1

provides the second pair of CH_2 units. If the origin of the coordinate system is chosen to be at the centre line of the corner molecule at the bottom left hand corner, then, remembering the stagger of the carbon atoms along the chain axis, the placement of the carbon atoms that can be used as a representative set can be described by:

p_x	p_y	p_z
A_1	B_1	0.75
$1 - A_1$	$1 - B_1$	0.25
$0.5 + A_1$	$0.5 - B_1$	0.00
$0.5 - A_1$	$0.5 + B_1$	0.5

where A_1 and B_1 are defined in Fig. A8.1.

By substituting these values into $\sum f_n \exp(-2\pi i(p_x h + p_y k + p_z l))$, and setting $l = 0$, it can be shown that the contribution from carbon atoms is

$$f_c \exp[-\pi i(h + k)][\cos\{\pi(h + k) - 2\pi(A_1 h + B_1 k)\}$$
$$+ \cos 2\pi(A_1 h - B_1 k)]$$

where f_c is the scattering factor for carbon atoms. This expression is zero for all combinations of h and k for which $(h + k)$ is odd. The hydrogen scattering obeys the same rule. This can be confirmed using the same process with the eight hydrogen atoms corresponding to the four carbons chosen above. The coordinates can be written as:

p_x	p_y	p_z
$A_1 - A_2$	$B_1 + B_2$	0.75
$A_1 + A_3$	$B_1 - B_3$	0.75
$1 - (A_1 - A_2)$	$1 - (B_1 + B_2)$	0.25
$1 - (A_1 + A_3)$	$1 - (B_1 - B_3)$	0.25
$0.5 - (A_1 - A_2)$	$0.5 + (B_1 + B_2)$	0.00
$0.5 - (A_1 + A_3)$	$0.5 - (B_1 - B_3)$	0.00
$0.5 + (A_1 - A_2)$	$0.5 - (B_1 + B_2)$	0.5
$0.5 + (A_1 + A_3)$	$0.5 + (B_1 - B_3)$	0.5

Substituting into $\sum f_n \exp(-2\pi i(p_x h + p_y k + p_z l))$ with $l = 0$ leads to an expression for the contribution from hydrogen atoms that is zero when $(h + k)$ is odd, and non-zero when $(h + k)$ is even.

8.4 For small angles, Bragg's law becomes $2d\theta = \lambda$. Using this equation to predict the position of the scattering maximum for a periodic structure with $d = 15\,\text{nm}$ gives $2\theta = 0.59°$.

8.5 Taking Fig. 8.12(c) first, this pattern indicates a highly oriented structure. The equatorial reflections, which index as $hk0$, reveal the degree of perfection of the orientation of the [001] axis to the draw

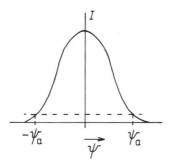

FIGURE A8.2

direction. The angular spread apparently differs from one reflection to another. This is because of the difference in brightness, and the high intensity reflections, which are grossly overexposed, contain sufficient intensity at positions significantly displaced from the equator to be observed. The angle subtended at the centre by the equatorial arcs varies from approximately $4°$ to approximately $15°$, according to the reflection chosen. A longer exposure would increase both limits and a shorter one would reduce it. For a more quantitative assessment a series of diffraction patterns should be taken at different exposures, and measurements made only on those reflections recorded within the range of linear response of the photographic emulsion (i.e. optical density proportional to exposure: this may require separate determination). Working from the negatives, the optical density versus azimuthal angle (ψ) should be determined and a cut-off intensity chosen (Fig. A8.2). The ratio of the area of the curve within $-\psi_a < \psi < \psi_a$ to the total area will represent the fraction of material orientated with the [001] axis within ψ_a of the fibre axis.

For a sample possessing fibre orientation, as shown in Fig. 8.12(c), the form of orientation information presented in Fig. A8.2 is probably the most convenient for relating to properties. With Fig. 8.12(b) the arcing shows significant orientation to be present, but the analysis is far less straightforward than with Fig. 8.12(c). This is because some of the arcs are split, some reflections form complete rings, and the variation in the angles subtended by different arced reflections at the centre does not appear to be a consequence simply of the intensity/exposure effect discussed above. To assess the texture in this sample properly a full pole figure determination should be made.

Chapter 9

9.1 Twin patterns for both {310} and {110} twinning are given in a paper by J.R. White (1974) *J. Mater. Sci.* **9**, 1860.

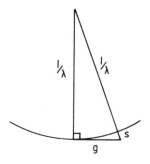

FIGURE A9.1

9.3 First calculate $d_{110} = ab/\sqrt{(a^2 + b^2)} = 0.411$ nm (see question 8.1)
Also $d_{200} = a/2 = 0.371$ nm

From the Ewald sphere construction for the symmetric position (Fig. A9.1).

$$(1/\lambda + s)^2 = 1/\lambda^2 + g^2$$

(2θ is very small and s can be measured along $\mathbf{k}'$ without introducing serious error.)

Thus $\quad 2s/\lambda + s^2 = g^2 = \dfrac{1}{d^2} \quad$ and $\quad s = \dfrac{\lambda}{2d^2} \quad$ (neglecting s^2)

Hence $s_{110} = 0.011$ nm $\quad$ and $s_{200} = 0.0134$ nm^{-1}

As $s \to 0$, $\qquad \dfrac{\sin^2 \pi st}{(\pi s)^2} \to t^2, \propto I_{max}$

$$\frac{I_{110}}{(I_{110})_{max}} = \frac{\sin^2 \pi s_{110}t}{(\pi s_{110}t)^2} = 0.91$$

Tilt required (for the tip of the reciprocal lattice spike, distance $1/t$ from the centre, to just touch Ewald sphere) is $\alpha = (1/t - s_{110})/g$

$$= \left(\frac{d_{110}}{t} - d_{110}s_{110} \right) \text{ radians} = 1.32°$$

9.4 Defocus shift $= \Delta f \cdot 2\theta \simeq 3.1\,\mu$m (in object space: multiply by the magnification for the shift in the image).

Chapter 11

11.3 The lobe maxima are located at approximately 9 mm from the centre, giving $2\theta_m = 9/400$. Using equation (11.5), $\bar{R} = 18.4\,\mu$m, giving an average spherulite diameter approximately $37\,\mu$m.

Chapter 13

13.1 Taking the atomic mass constant to be 1.66×10^{-27} kg gives a total mass for the unit cell of $(4 \times 12 + 8) \times 1.66 \times 10^{-27}$ kg, and dividing this by the volume, *abc*, gives density 998 kg/m^3. Hence the rule of mixtures gives

$\rho = 998 f_c + 855(1 - f_c)$

giving (i) $f_c = 0.84$; (ii) $f_c = 0.52$

The graph is a straight line: $\rho = 143 f_c + 855$.

The crystal and amorphous densities of poly(-4-methyl pentene-1) are approximately equal, and density is not sensitive to changes in crystallinity (see section 11.5).

Index